Lecture Notes in Economics and Mathematical Systems

Managing Editors: M. Beckmann and W. Krelle

234

B. Curtis Eaves

A Course in Triangulations for Solving Equations with Deformations

Springer-Verlag
Berlin Heidelberg New York Tokyo 1984

Author

Prof. B. Curtis Eaves
Department of Operations Research
School of Engineering, Stanford University
Stanford, California 94305, USA

ISBN 3-540-13876-5 Springer-Verlag Berlin Heidelberg New York Tokyo
ISBN 0-387-13876-5 Springer-Verlag New York Heidelberg Berlin Tokyo

Printed in Germany

Printing and binding: Beltz Offsetdruck, Hemsbach/Bergstr.
2142/3140-543210

TABLE OF CONTENTS

1. INTRODUCTION

The basic version of an important method for solving equations is described in the following homotopy principle.

Homotopy Principle: To solve a given system of equations, the system is first deformed to one which is trivial and has a unique solution. Beginning with the solution to the trivial problem a route of solutions is followed as the system is deformed, perhaps with retrogressions, back to the given system. The route terminates with a solution to the given problem. □

A primary driving force for development and application of this principle has been the solution of equations corresponding to economic equilibrium models. The deformations, that is, homotopies, for the method have been either PL (piecewise linear) or differentiable. For the PL approach subdivisions and triangulations provide the understructure on which to build the PL deformation. The path of solutions followed, as described in the homotopy principle, proceeds through a sequence of adjacent cells or simplexes of the subdivision or triangulation, see Figure 1.1. Our interest is in a class of triangulations used for this purpose. This class, called variable rate refining triangulations, and denoted $\mathbf{S}$ and $\mathbf{S}_+$, are coarse near the trivial system or starting point and are fine near the given system and, furthermore, the rate of refinement is variable and can be selected by the user, as the computation proceeds, see Figure 1.1 again.

Variable rate refining triangulations also play an essential roll in computing a path of solutions for a path of problems wherein one wants to refine at some rate and then settle upon a fixed rate, or even, encoarse the triangulation.

Herein is a careful development of a sequence of subdivisions and triangulations leading to a class of variable rate refining triangulations. Our progression begins with the basic triangulation and carefully and gradually builds to obtain the class.

Although the principle focus is triangulations, the study of these triangulations is greatly enhanced by the use of certain subdivisions. Consequently we find ourselves interested in both triangulations and subdivisions. Indeed, as the notion of a subdivision includes that of a triangulation, we often cast our statements in the language of subdivisions even though our intended application is to triangulations.

Our progression to the variable rate refining triangulations, $\mathbf{S}$ and $\mathbf{S}_+$, can be viewed as four major stages.

a) Preliminaries (Sections 1-4).

b) Freudenthal Triangulation $\mathbf{F}$ (Sections 6-12).

c) Subdivision $\mathbf{P}$ and triangulations $\mathbf{V}$, and $\mathbf{V}[r,p]$ (Sections 13-16).

d) Stacking copies of $\mathbf{V}[r,p]$'s to construct $\mathbf{S}$ and $\mathbf{S}_+$ (Sections 17-18).

In the preliminaries, the motivation, goals, organization, mathematical background, definitions of subdivisions, and elementary properties of subdivisions are discussed. From this point on, the manuscript is self contained and considerable effort has been extended to make the arguments complete and clear. In the second stage is an extensive study of the Freudenthal triangulations **F**. This triangulation is the single most important nontrivial subdivision used in the solution of equations with PL homotopies, indeed, it has played a role in virtually every triangulation for such purposes, however, to the point at hand, a full understanding of it represents almost half the effort toward an understanding of our class of variable rate refining triangulations. The third stage involves the construction of subdivisions **P**, **V**, and **V**[r,p]. Using **P** and the Freudenthal triangulation **F**, the triangulation **V** is constructed, and by restricting, squeezing and shearing **V** the triangulations **V**[r,p], which contain the final local structure, are derived. In the last stage the triangulations **V**[r,p] are stacked in an orderly fashion, using the Freudenthal triangulation **F**, to form the variable rate refining triangulations **S** and $\mathbf{S}_+$; a portion of $\mathbf{S}_+$ for one dimension is shown in Figure 1.1.

Along with each (triangulation and) subdivision we encounter in our progression, we shall develop representation and replacement rules, namely,

a) a representation set

b) a representation rule

c) a facet rule, and

d) a replacement rule.

It is these instruments which enable one to move about in the subdivision in order to follow a path, or in other words, these devices enable one to generate locally portions of the (triangulations or) subdivisions as they are needed.

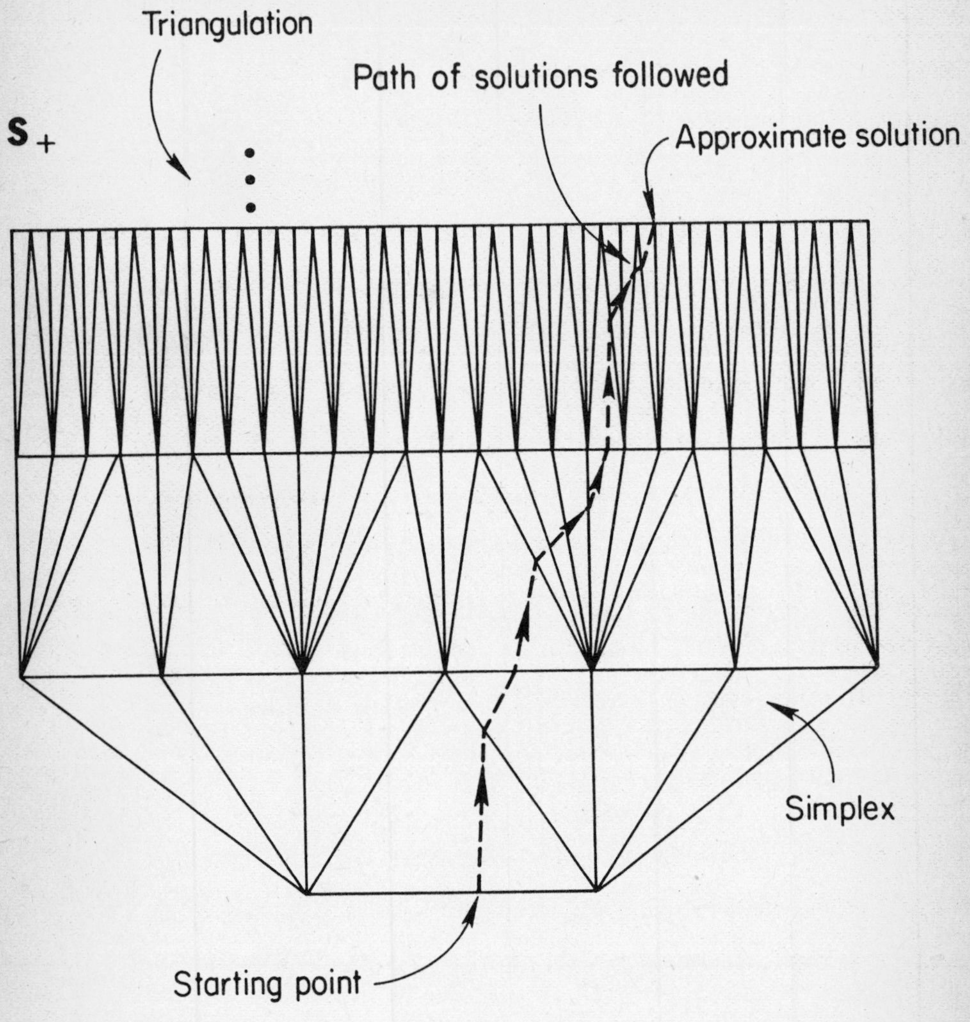
Triangulation
Path of solutions followed
S_+
Approximate solution
Simplex
Starting point

Figure 1.1

Particular subdivisions will be superceded and forgotten, and perhaps, such a fate awaits some of those we discuss. However, the ideas we have employed are fundamental and have, and will continue to serve as blue prints for subdivisions for solving equations with PL homotopies. Of course, although the triangulations $\mathbf{S}$ and $\mathbf{S}_+$ have performed quite well, especially for smooth functions, there is always the hope that better ones will be found; the author believes that such an improvement will have a structure much like $\mathbf{S}$ or $\mathbf{S}_+$.

Because they are not required for our progression some important topics have been omitted. Namely, certain triangulations, certain subdivisions for special structure, and measures for comparing the quality of triangulations. If at some point this manuscript is extended these items will be the first to be included.

Although most of the material herein is known to researchers in this field, many results are new, but more importantly, this material has not previously been assembled, organized, and given a uniform view.

1.1 Bibliographical Notes

The use of PL homotopies for global computation of solutions of systems of equations was introduced in Eaves [1971a,1972]. This suggestion was couched in a vast background of PL and differential topology for example, including Poincaré [1886], Sperner [1928], and Davidenko [1953], and more recently Hirsch [1963], Lemke and Howson [1964], Lemke [1965], Scarf [1967,a1967b], Cohen [1967], Kuhn [1968], and Eaves [1970,1971]. Recent general treatment of solution of equations by homotopies can be found in any one of Eaves and Scarf [1976], Eaves [1976], Todd [1971a],

Allgower and Georg [1980], and Garcia and Zangwill [1981]. Other important recent references include Scarf [1973], Merrill [1972], Kellog, Li, and York [1976], Eaves and Saigal [1972], and Smale [1976], and Hirsch and Smale [1974]. The computation of solutions of general economic equilibrium models on a theoretically sound basis began with Scarf [1973].

Triangulations and subdivisions are nearly as old a notion as mathematics itself. Triangulations are probably most familiar to the general mathematical community through tiling problems and the simplicial approximation theorem. The simplicial approximation theorem underlies our interest in refining triangulations.

The Freudenthal triangulation was introduced in [1942] and was brought to the solution of equations in Kuhn [1968] and Hansen [1968]. Refining triangulations were introduced in Eaves [1972] and Eaves and Saigal [1972] and improved upon in Todd [1974]. The subdivision **P** and triangulation **V** are a product of van der Laan and Talman [1979] and Todd [1978b]. Shamir [1979] and van der Laan and Talman [1980b] introduced the variable rate refining triangulation **S**. Bárány [1979], Kojima and Yamamoto [1982a] and Broadie and Eaves [1983] have introduced variable rate refining triangulations which generalize the refining triangulation of Todd [1974]; how these latest variable rate refining triangulations, which are not discussed herein, relate computationally to **S** or $\mathbf{S}_+$ is not yet known. Engles [1978] used triangulations which could refine, encoarse, or remain constant in order to compute paths of solutions to paths of systems. Other important triangulations and subdivisions, also not discussed herein, can be found in Todd [1978a,1978c], Wright [1981], van der Laan and Talman [1981] and Kojima [1978]. Nevertheless, these triangulations and subdivisions are built upon the Freudenthal triangulation and have structures very close to those discussed here.

The matter of measuring the quality of triangulations has not been covered; see Saigal [1977], Todd [1976b], van der Laan and Talman [1980a], Eaves and Yorke [1982], and Eaves [1982]. Nevertheless, such reasoning underlies the existence of refining triangulations.

Our triangulations are geometric, however, an abstract treatment is available; consider Kuhn [1967], Gould and Tolle [1974], and recently Freund [1980].

This manuscript was written in the academic year of 1979-80 when the author was on a sabbatical. The bibliographical notes at the end of each section have been updated to include pertinent developments in the interim. □

1.2 Acknowledgments

The author would like to express his appreciation to the many doctoral students in Operations Research at Stanford University who have contributed in one fashion or another to this manuscript, to the Guggenheim Foundation, National Science Foundation, Department of the Army, and Stanford University for their support during the period this manuscript was written, and to Gail Stein and Audrey Stevenin for their dedicated word processing and preparation of the manuscript. □

2. MATHEMATICAL BACKGROUND AND NOTATION

It is assumed that the reader is acquainted with the elementary notions of set theory, linear algebra, linear inequalities, matrix theory, and topology. Here we state our basic notation and discuss cells, that is, closed polyhedral convex sets, their faces, unimodularity, and an existence theorem from linear complementary theory.

Given a finite set α we let $\#\alpha$ denote the number of elements in it. Given two sets α and β we let $\alpha\backslash\beta$ denote the set of those elements in α but not β.

Let $\mathbb{Z}$ be the set of integers $i = 0, \pm 1, \pm 2, \ldots,$ $\mathbb{Q}$ be the set of rationals, $\mathbb{R}$ the reals, and $\mathbb{G}$ any ordered field contained in $\mathbb{R}$, for example, the rationals $\mathbb{Q}$ or reals $\mathbb{R}$. Unless stated otherwise, our notation and results are with respect to the field $\mathbb{G}$; for example $[0,+\infty)$ denotes the set $\{x \in \mathbb{G} : x \geqq 0\}$. Let $|a|$ indicate the absolute value of the scalar $a \in \mathbb{G}$. Let $\nu \triangleq \{1, \ldots, n\}$ and $\mu \triangleq \{1, \ldots, n+1\}$. Let $\mathbb{G}^{m\times n}$ and $\mathbb{Z}^{m\times n}$ denote the set of $m \times n$ $\mathbb{G}$ and $\mathbb{Z}$ matrices, respectively. We write $\mathbb{G}^n$ for $\mathbb{G}^{n\times 1}$. We say x in $\mathbb{G}^n$ is integral, if x is in $\mathbb{Z}^n$. For x in $\mathbb{G}^n$ let

$$\|x\|_1 = \sum_{\nu} |x_i|$$

$$\|x\|_\infty = \max_{\nu} |x_i|$$

$$\|x\|_2 = \left(\sum_{\nu} x_i^2\right)^{1/2}$$

be the usual norms and let $\|\;\|$ be, consistently, any one of these. A neighborhood of x in $\mathbb{G}^n$ is defined to be any subset of $\mathbb{G}^n$ which contains the set of y in $\mathbb{C}^n$ with $\|x-y\|_\infty < \varepsilon$ for some positive scalar ε.

For a in $\mathbb{C}^{m\times n}$ we let a_i, a^j, and a_i^j denote the ith row, j^{th} column, and the ij^{th} element of a. For $\alpha \subseteq \{1, \ldots, m\}$ and $\beta \subseteq \{1, \ldots, n\}$ a_α^β denotes the submatrix of a indexed by $\alpha \times \beta$.

For a and b in $\mathbb{C}^n$, $a \leqq b$ denotes $a_i \leqq b_i$ for i in $\nu \triangleq \{1, \ldots, n\}$. Further $a \leq b$ denotes $a \leqq b$ and $a \neq b$. Finally, $a < b$ denotes $a_i < b_i$ for i in ν. We say a is lexico positive if $a \neq 0$ and the first nonzero element a_i for $i = 1, \ldots, n$ is positive. We say a is lexico greater than b if $a-b$ is lexico positive, etc.

Given a nonempty set σ in $\mathbb{C}^n$ we denote the convex, affine, tangential, conic, and linear hulls $\mathrm{cvx}(\sigma)$, $\mathrm{aff}(\sigma)$, $\mathrm{tng}(\sigma)$, $\mathrm{cone}(\sigma)$, and $\mathrm{lin}(\sigma)$ of σ to be the collection of all

$$\sum_{i=1}^{k} \lambda_i x^i$$

with $\lambda_i \in \mathbb{C}$, $x^i \in \sigma$, $k = 1, 2, \ldots$ and

$$\alpha = \sum_{i=1}^{k} \lambda_i$$

where (a) $\lambda_i \geqq 0$ and $\alpha = 1$, (b) $\alpha = 1$, (c) $\alpha = 0$, (d) $\lambda_i \geqq 0$ and $\alpha \in \mathbb{C}$, and (e) $\alpha \in \mathbb{C}$, respectively. For $\sigma = \phi$, the empty set, we define $\mathrm{cvx}\ \phi = \phi$, $\mathrm{aff}\ \phi = \phi$, $\mathrm{tng}\ \phi = \{0\}$, $\mathrm{cone}\ \phi = \{0\}$, and $\mathrm{lin}\ \phi = \{0\}$. A set is defined to be convex, affine, tangential, conic, or linear, if it is empty, or if $\mathrm{cvx}\ \sigma = \sigma$, $\mathrm{aff}\ \sigma = \sigma$, $\mathrm{tng}\ \sigma = \sigma$, $\mathrm{cone}\ \sigma = \sigma$, or $\mathrm{lin}\ \sigma = \sigma$, respectively.

For a subset σ of $\mathbb{C}^n$, the relative interior $ri(\sigma)$ of σ and relative boundary $\partial(\sigma)$ of σ are defined to be the interior of σ and boundary of σ, respectively, where $aff(\sigma)$ is viewed as the space. For σ non-empty the dimension $\dim(\sigma)$ is defined to be the number of linearly independent vectors in $tng(\sigma)$; the dimension of a singleton is zero, and the dimension of the empty set is defined as -1. It can be shown that a finite number of convex sets of dimension k or less cannot cover a convex set of dimension $k+1$ or more.

A subset ρ of σ is defined to be an extreme set in σ if $\sum_{i=1}^{k} \lambda_i x^i \in \rho$, $\sum_{i=1}^{k} \lambda_i = 1$, $\lambda_i \geq 0$, and $x^i \in \sigma$, for $i = 1, \ldots, k$ implies $x_i \in \rho$ for $\lambda_i > 0$. We regard the empty set ϕ as an extreme set of σ. If an extreme set contains only one point, the point is called an extreme point. Clearly, the union of extreme sets is extreme, and σ is an extreme set in σ. An extreme convex subset ρ of σ is defined to be a face of σ. If τ is a face of ρ and ρ is a face of σ, then clearly, τ is a face of σ. As the empty set is convex and extreme in σ, it is a face of σ. If τ is a convex subset of σ, and ρ is a face of σ, then $\tau \cap \rho$ is a face of τ. By an m-face we mean a face of dimension m.

By a closed halfspace in $\mathbb{C}^n$ is meant a set of form $\eta = \{x \in \mathbb{C}^n : ax \leqq b\}$ where $a \in \mathbb{C}^{1\times n}$, $b \in \mathbb{C}$, ax is the inner product of a and x, and $a \neq 0$. We call $\partial\eta = \{x \in \mathbb{C}^n \; ax = b\}$ a hyperplane and denote the closed halfspace $\{x \in \mathbb{C}^n : ax \geqq b\}$ by η^-. If σ is a subset of the closed halfspace η we say that the hyperplane $\partial\eta$ supports σ.

The intersection of a finite number $k = 0, 1, 2, \ldots$ of closed half-spaces in $\mathbb{C}^n$ is defined to be a cell of $\mathbb{C}^n$, in particular, $\mathbb{C}^n$

and ϕ are cells. An m-cell is a cell of dimension m. A face of a cell is a cell. An (m-1)-face of an m-cell is called a facet of the m-cell. The empty set has no facets. By a proper face ρ of σ we mean a face with $\phi \neq \rho \neq \sigma$. A proper face of a cell σ is an intersection of facets of σ, and any intersection of facets of σ is a face of σ. A proper face of σ can be expressed in the form $\rho = \sigma \cap \eta$ where η is a supporting hyperplane of σ; and if ρ is of this form, it is a face of σ. The relative interiors of all faces of a cell partition the cell. An extreme point of a cell is sometimes called a vertex of the cell.

The convex, affine, tangential, conic, and linear hull of a finite number of points in $\mathbb{C}^n$ is a cell. In fact, σ is a cell if and only if $\sigma = \tau+\rho$ where τ is the convex hull of finitely many points and ρ is the conic hull of finitely many points. An m-cell σ is defined to be an m-simplex or simplex if σ is the convex hull of m+1 points $v^1, \ldots, v^{m+1}$. Let σ be the convex hull of $v^1, \ldots, v^{m+1}$, if the v^i are vertices of σ, or equivalently, if the points $(v^1,1), \ldots, (v^{m+1},1)$ are linearly independent, then σ is an m-simplex. An m-simplex has m+1 facets, and each facet is obtained by taking the convex hull of m of the vertices.

We shall from time to time be concerned with simplicial cones. Let $w = (w^1, \ldots, w^k)$ be an $n \times k$ matrix and observe that $\dim\{wx : x \geqq 0\} \leqq k$. If $k = 0$ we take $\{wx : x \geqq 0\}$ to be the singleton $\{0\}$. An m-cell of form $\sigma = \{wx : x \geqq 0\}$ is defined to be a simplicial cone if w has m columns; in this case any positive scale of a column of w is called a generator of σ. If the columns of w are linearly independent then $\{wx : x \geqq 0\}$ is a simplicial cone. A face of a simplicial cone ρ

is simplicial, and indeed, if $\rho = \{wx : x \geqq 0\}$ and the columns of w are linearly independent, then τ is a face of ρ if and only if τ is empty or has the form $\{w^\alpha x_\alpha : x_\alpha \geqq 0\}$ for a subset α of $\{1,\ldots,k\}$; for facets of ρ we have $\alpha = \{1, \ldots, k\}\backslash t$ for $t = 1, \ldots, k$.

We say that two cells σ and τ meet in a common face, if $\sigma \cap \tau$ is a face of σ and of τ. We say that a hyperplane ρ separates σ and τ if there is a closed halfspace η with $\sigma \subseteq \eta$, $\tau \subseteq \eta^-$, and $\partial\eta = \rho$. Two cells σ and τ meet in a common face, if and only if there is a separating hyperplane ρ with $\sigma \cap \partial\eta = \tau \cap \partial\eta = \sigma \cap \tau$. We say that two distinct n-cells are adjacent if they meet in a common facet.

If σ is an n-simplex with vertices $v^1, \ldots, v^{n+1}$ and x is in $\mathrm{aff}(\sigma)$, then we can express $x = \sum_i \lambda_i v^i$ where $\sum_i \lambda_i = 1$. The $\lambda = (\lambda_1, \ldots, \lambda_{n+1})$ are called the barycentric coordinates of x and are unique. Further x is in σ, if and only if $\lambda \geqq 0$, and x is in $\mathrm{ri}\,\sigma$ if and only if $\lambda > 0$.

A matrix A in $\mathbb{C}^{m\times n}$ is said to be unimodular if every square submatrix of A has a determinant of 0, -1, or +1; observe that each A_i^j is -1, 0, or +1 for such A.

The following theorem states a condition in which the vertices of a cell are integral.

2.1 Lemma: If a is integral and A is unimodular, each vertex of the cell $\sigma = \{x \geqq 0 : Ax = a\}$ is integral.

Proof: If x is a vertex of σ, then for some $\alpha \subseteq \nu$ and $\beta = \nu\backslash\alpha$ we have $x_\alpha > 0$, $x_\beta = 0$ and $A^\alpha x_\alpha = a$, where the columns of A^α are linearly independent. Select a subset δ of $1, \ldots, m$ where A_δ^α is square and nonsingular. Then $A_\delta^\alpha x_\alpha = a_\delta$, and from Cramer's rule it follows that x is integral. □

In the next lemma let D_i be a signed permutation matrix, that is, each row and column has exactly one nonzero element which is a -1 or $+1$. Let I be the identity matrix.

2.2 Lemma: The following are equivalent

(a) A is unimodular.

(b) A^T is unimodular.

(c) (A,I) is unimodular.

(d) D_1AD_2 is unimodular. □

The next lemma concerns matrices whose rows are of form

$$(0, \ldots, 0, 1, \ldots, 1, 0, \ldots, 0), (1, \ldots, 1, 0, \ldots, 0),$$
$$(0, \ldots, 0, 1, \ldots, 1), (1, \ldots, 1) \quad \text{or} \quad (0, \ldots, 0) .$$

2.3 Lemma: Let A be a matrix of zeros and ones such that the ones in each row, if any, occur consecutively, then A is unimodular.

Proof: Any $k \times k$ submatrix B of A will have the same properties. For $k = 1$ it is obvious that $\det B = 0$ or 1 and let us assume the result holds for $(k-1) \times (k-1)$ matrix. Now rearrange the rows of the $k \times k$ matrix B in a lexico decreasing fashion to obtain C, for example:

$$\begin{bmatrix} 1 & 1 & 1 & 0 \\ 0 & 1 & 1 & 1 \\ 0 & 1 & 1 & 0 \\ 0 & 0 & 0 & 1 \end{bmatrix} .$$

If the matrix has zero ones the result is clear. Let us assume that the matrix has ℓ ones and the assertion is true for fewer than ℓ ones. If $C_1^1 = 0$, then $C^1 = 0$ and, clearly, det $C = 0$. If $C_1^1 = 1$ and $C_2^1 = 0$ the result follows by induction on k by computing det C by expanding by cofactors on the first column. If $C_2^1 = 1$, subtract the second row from the first to obtain a new $k \times k$ matrix in the form of the lemma, but with fewer than ℓ ones. The result follows by induction on ℓ. □

The following lemma indicates our usage of unimodularity. Let σ be an n-simplex and τ a closed convex set in $\mathbb{C}^m$. Notice that $\sigma \cap \tau$ is a face of σ if and only if each extreme point of $\sigma \cap \tau$ is also an extreme point of σ. Let v^i for i in μ be the vertices of σ and let ρ be the collection of all $\lambda = (\lambda_1,\ldots,\lambda_{n+1}) \geqq 0$ such that

$$\sum v^i \lambda_i \in \tau ,$$

$$\sum \lambda_i = 1 .$$

That is, ρ is the collection of barycentric coordinates with respect to σ of points in τ.

2.4 Lemma: The set $\sigma \cap \tau$ is a face of σ, if and only if the extreme points of ρ are integral.

Proof: Suppose $x \in \sigma \cap \tau$, then $x = (v^1, \ldots, v^{n+1})\lambda$, $e\lambda = 1$, and $\lambda \geqq 0$. Select $\alpha \subseteq \mu$ so that $\lambda_\alpha > 0$ and $\lambda_{\mu\backslash\alpha} = 0$. Since the extreme points of ρ are integral there is an x^i in $\sigma \cap \tau$ for i in α with $\lambda_i = 1$, that is, $x^i = v^i$ is in $\sigma \cap \tau$. Thus, each point of $\sigma \cap \tau$ is a convex combination of vertices of σ also in τ. As τ is convex, we have that $\sigma \cap \tau$ is the convex hull of a set of vertices of σ, which is to say, $\sigma \cap \tau$ is a face of σ. On the other hand, if $\sigma \cap \tau$ is a face of σ, letting $\alpha = \{i : v^i \in \tau\}$ then ρ is the collection $\{\lambda : e\lambda = 1, \lambda_\alpha \geqq 0, \lambda_\beta = 0\}$ which clearly has the integral extreme point property. □

As $0 \leqq \lambda_i \leqq 1$ above, by integral we mean each component is 0 or 1. In our application of this line of argument we shall also need the following result regarding projections. For a set ρ in $\mathbb{C}^n$ and $\alpha \subseteq \nu$ let $\rho_\alpha = \{x_\alpha : x \in \rho\}$ be the projection of ρ to the α-coordinates.

2.5 Lemma: If ρ is bounded and has integral extreme points, then ρ_α inherits these properties. □

Our final remarks in this section are concerned with the linear complementarity problem, that is, with an LCP. Let I be the $n \times n$ identity matrix, A any $n \times n$ matrix, and a an $n \times 1$ vector. The system of equations

$$Ix + Ay = a$$

$$x \geqq 0, \quad y \geqq 0, \quad xy = 0$$

where xy is an inner product is called an LCP. Observe that the constraints imply $x_i = 0$ or $y_i = 0$ for $i = 1, \ldots, n$. For the next theorem let $d \geqq 0$ be a vector in $\mathbb{G}^n$ where the rows of (d,a) are lexico positive. The next theorem is nontrivial and is a fundamental result in linear complementarity theory.

2.6 Theorem: The LCP has a solution, if the following system has no solution.

(a) $Ix + Ay - dz = a$

(b) $I\bar{x} + A\bar{y} - d\bar{z} = 0$

(c) $(x,y,z,\bar{x},\bar{y},\bar{z}) \geqq 0$

(d) $(x+\bar{x})(y+\bar{y}) = 0$

(e) $(\bar{x},\bar{y},\bar{z}) \neq 0$

(f) $(y,\bar{y}) \neq 0$

(g) $z > 0$. □

Theorem 2.6 will be used only once, namely, in Section 13 to prove that **P** is a subdivision.

2.7 Bibliographical Notes: For a study of cells see Grünbaum [1967]. For a study of unimodularity see Veinott and Dantzig [1968] or Hoffman and Kruskal [1956]. For the LCP theorem see Lemke [1965], Eaves [1971], or Garcia [1973]. The LCP is introduced in Cottle [1964]. □

3. SUBDIVISIONS AND TRIANGULATIONS

In this section we recount pertinent facts regarding general subdivisions and briefly discuss two familiar subdivisions of $\mathbb{C}^n$.

Let $\mathbf{M}$ be a nonempty collection of sets in $\mathbb{C}^n$. Let $\mathbf{M}^{-2} \triangleq \phi$ and for $k = -1, 0, 1, \ldots, n$ let $\mathbf{M}^k$ be the collection of all sets of $\mathbf{M}$ of dimension k. Clearly $\mathbf{M} = \cup_{k=-2}^{n} \mathbf{M}^k$. By the dimension of a nonempty collection $\mathbf{M}$, denoted $\dim \mathbf{M}$, we mean the largest k with $\mathbf{M}^k \neq \phi$.

Given the collection $\mathbf{M}$ we call the union of all sets in $\mathbf{M}$, denoted $M = \cup \mathbf{M}$, the carrier of $\mathbf{M}$. We say the collection $\mathbf{M}$ is locally finite, if for each point x in the carrier M there is a neighborhood about x which meets only finitely many sets of $\mathbf{M}$.

Let $\mathbf{M}$ be a nonempty countable collection of cells in $\mathbb{C}^n$. For $m = -1, 0, 1, 2, \ldots$ we define an m-subdivision.

Definitions: $\mathbf{M}$ is an m-subdivision if

(a) each face of each cell of $\mathbf{M}$ is in $\mathbf{M}$

(b) each cell of $\mathbf{M}$ is a face of a cell of $\mathbf{M}^m$

(c) any two cells of $\mathbf{M}^m$ meet in a common face

(d) any cell of $\mathbf{M}^{m-1}$ lies in at most two cells of $\mathbf{M}^m$. □

Figure 3.1 illustrates a 2-subdivision. Note $\mathbf{M}$ is a (-1)-subdivision if and only if $\mathbf{M} = \mathbf{M}^{-1} = \{\phi\}$.

A 0-subdivision $\mathbf{M}$ has the form $\mathbf{M} = \{\phi, \{x\}\}$ or $\mathbf{M} = \{\phi, \{x\}, \{y\}\}$ where x and y are points, for otherwise, the (-1)-cell ϕ would be contained in too many 0-cells. We call elements of $\mathbf{M}^0$ the vertices of

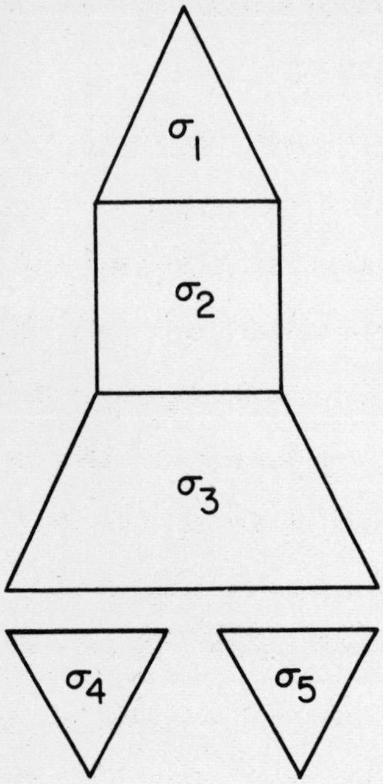

$$\mathbf{M}^2 = \{\sigma_1, \sigma_2, \sigma_3, \sigma_4, \sigma_5\}$$

Figure 3.1

the subdivision; such a vertex is a singleton, but we shall also use the word vertex to represent the element of the singleton. The carrier $M = \cup \mathbf{M}$ of a subdivision is referred to as a manifold, and we say that M is subdivided or m-subdivided by $\mathbf{M}$. If each cell of the subdivision $\mathbf{M}$ is a simplex we call $\mathbf{M}$ a triangulation. By a finite subdivision $\mathbf{M}$ we mean one where $\mathbf{M}$ is a finite collection. For a collection of cells $\mathbf{M}$ satisfying (b), the condition (c) is equivalent to: any two cells of $\mathbf{M}$ meet in a common face. For a collection of cells $\mathbf{M}$ satisfying (a) and (b), the condition (c) is equivalent to: the relative interiors of cells of $\mathbf{M}$ partition the carrier M. If the collection $\mathbf{M}$ satisfies (a), (b), and (c), then condition (d) is implied by $\dim \mathbf{M} = \dim M$. Note here that $\dim \mathbf{M}$ is the dimension of a collection of sets where $\dim M$ is the dimension of a subset of $\mathbb{C}^n$.

If $\mathbf{M}$ is a subdivision and locally finite, we refer to $\mathbf{M}$ as a locally finite subdivision. Without the local finiteness assumption the structure of a manifold can be weak for observe that

$$\{\{x\} \times \{0\} : x \in M\} \cup \{\{x\} \times \{1\} : x \in M\} \cup \{\{x\} \times [0,1] : x \in M\} \cup \{\phi\}$$

is a 1-subdivision with manifold $M \times [0,1]$ for any nonempty countable set M in $\mathbb{C}^n$.

For each subdivision we encounter we develop a representation set, a representation rule, a facet rule, a replacement rule, and sometimes a vertex correspondence rule. It is these notions which permits one to move about in the subdivision; let us discuss them.

If a set is in one-to-one correspondence with a subset of $\mathbf{M}$, for example, the subsets $\mathbf{M}$, $\mathbf{M}^k$, or $\mathbf{M}\backslash\phi$, we say the set is a representation set and the correspondence is a representation rule for that subset of $\mathbf{M}$. The notation $i\mathbf{M}^m$ is typically used to denote a representation set for $\mathbf{M}^m$, the m cells of $\mathbf{M}$. Given an m-subdivision $\mathbf{M}$, the representation set $i\mathbf{M}^m$ and the representation rule $\alpha \to \sigma_\alpha$ provide a unique indexing system for the m-cells of $\mathbf{M}$; given the element α in $i\mathbf{M}^m$ one applies the representation rule to generate the corresponding m-cell σ_α in $\mathbf{M}$. Given α in $i\mathbf{M}^m$ the facet rule uniquely indexes the facets of the cell σ_α indexed by α. Given α in $i\mathbf{M}^m$ and a facet index t for the cell σ_α, the replacement rule $(\alpha,t) \to \hat{\alpha}$ yields $\hat{\alpha}$ in $i\mathbf{M}^m|\alpha$ (if it exists) which indexes the unique cell $\sigma_{\hat{\alpha}}$ in $\mathbf{M}^m$ which shares the facet of σ_α indexed by t, see Figure 3.2. The m-cell $\sigma_{\hat{\alpha}}$ or index $\hat{\alpha}$ is referred to as the replacement of σ_α or α, respectively, given the index t. For triangulations treated, application of the representation rule generates the vertices of the simplex, and hence, orders the vertices. This ordering yields, de facto, a facet rule in a simplex, since there is a one-to-one correspondence between vertices and facets, namely, the facet opposite the vertex, see Figure 3.2. The exclusive vertices of σ_α and $\sigma_{\hat{\alpha}}$ are referred to as the dropped and adjoined vertices. For triangulations we shall develop as part of the replacement rule the vertex correspondence rule. Herein, as mentioned, for triangulations the representation rule orders the vertices of the indexed simplex; the vertex correspondence rule indicates how the vertices of adjacent simplices relate, see Figure 3.2; the squared indices indicate the

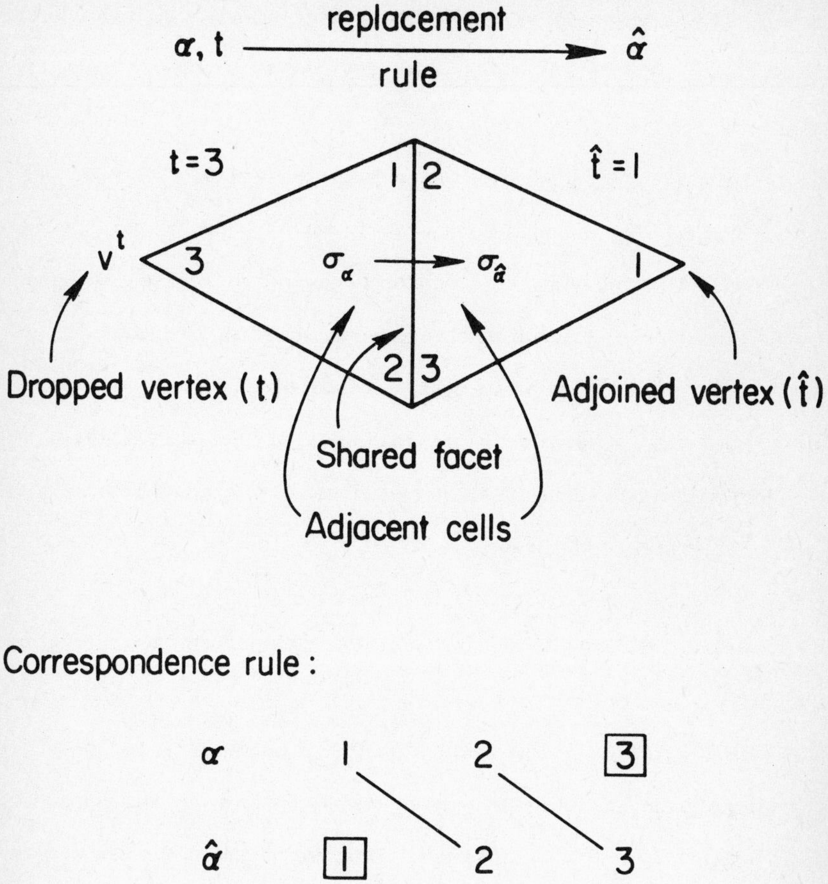

α, t
replacement
rule
α̂
t=3
t̂ =1
1
2
v^t
3
σ_α
σ_α̂
1
2
3
Dropped vertex (t)
Adjoined vertex (t̂)
Shared facet
Adjacent cells
Correspondence rule:
α
1
2
3
α̂
1
2
3

Figure 3.2

dropped and adjoined vertices. For subdivisions whose cells are simplicial cones one can proceed as this except one uses generators in lieu of vertices.

Repeated use of the replacement operations, as in Figure 3.2, is one-half of the path following procedures of the homotopy principle; the other half, namely, which vertex t to drop, is not discussed in this manuscript, that is, for present purposes we select t arbitrarily.

Let $\mathbf{M}$ be a subdivision with manifold M. We define a map $f : M \to R^k$ to be $\mathbf{M}$-PL (piecewise linear), if f is linear $(f(\lambda x + (1-\lambda)y) = \lambda f(x) + (1-\lambda) f(y)$ that is, of form $Ax+a)$ on each cell. If $\mathbf{M}$ is locally finite, an $\mathbf{M}$-PL map is continuous.

Let f be an $\mathbf{M}$-PL map, then we define $f(\mathbf{M})$ to be the collection $\{f(\sigma) : \sigma \in \mathbf{M}\}$. Suppose f is one-to-one, then $f(\mathbf{M})$ is a subdivision, and we say that f is an isomorphism and $\mathbf{M}$ is isomorphic or f-isomorphic to $f(\mathbf{M})$. Letting σ and τ be two cells of $\mathbf{M}$ and f be an isomorphism, then $f(\sigma \cap \tau) = f(\sigma) \cap f(\tau)$ and the dimension of σ and $f(\sigma)$ agree, however, $\mathbf{M}$ may be locally finite where $f(\mathbf{M})$ is not. If $\mathbf{M}$ lies in $\mathbb{C}^n$ and $f : \text{aff}(\mathbf{M}) \to \mathbb{C}^n$ is linear and one-to-one, of course, f restricted to M is an isomorphism and preserves local finiteness.

The $\|\,\|$ diameter of a cell σ is defined to be

$$\max_{x,y \in \sigma} \|x-y\| .$$

The $\|\,\|$ grid size of a subdivision is defined to be the maximum diameter of any of its cells. By a refining subdivision $\mathbf{M}$ of $\mathbb{C}^n \times [0,+\infty)$ we mean one such that for any $\varepsilon > 0$ there is a $\theta > 0$ such that if a cell of $\mathbf{M}$ meets $\mathbb{C}^n \times [\theta, +\infty)$ then the diameter of its projection to $\mathbb{C}^n \times 0$ does not exceed ε.

From a computational point of view the performance of a subdivision and an isomorphic copy could be radically different, but observe that if one has the representation, facet, replacement and/or vertex correspondence rules for a subdivision then one has them for all f-isomorphic subdivisions where one has f. If $\mathbf{M}$ and $\mathbf{N}$ are f-isomorphic and $i\mathbf{M}$ is a representation set for $\mathbf{M}$ then $i\mathbf{M}$ also is a representation set for $\mathbf{N}$ for consider

$$\begin{array}{ccccc} i\mathbf{M} & & \mathbf{M} & & \mathbf{N} \\ \alpha & \longrightarrow & \sigma & \longrightarrow & f(\sigma) \end{array} .$$

Namely, the representation rule for $\mathbf{N}$ is obtained by following the representation rule for $\mathbf{M}$ with f. Also one sees that the replacement rules of $\mathbf{M}$ and $\mathbf{N}$ are the same and that a facet rule for $\mathbf{M}$ is obtained from that of $\mathbf{M}$ by following it with f.

All representation sets, facet indicators, and replacement rules applied to representation sets herein are integral, and in computation there is no numerical error in the execution of the replacement operations. That is, before these integers become large enough to exceed machine capacity there would be binding deterioration of numerical operations of other portions of the algorithm, for example in application of the representation rule and subsequent function evaluation or in the determination of which vertex t to drop.

The following lemmas recount some of the properties of subdivisions.

3.1 Lemma: If $\mathbf{M}$ and $\mathbf{L}$ are locally finite m and ℓ-subdivisions with a common manifold M, then $m = \ell$. □

3.2 Lemma: If $\mathbf{M}$ is a locally finite subdivision with a convex manifold M, then $\dim \mathbf{M} = \dim M$. □

Given an m-subdivision $\mathbf{M}$ we let $\mathbf{M}^\partial$ be the collection of all cells of $\mathbf{M}$ that lie in an $(m-1)$-cell which, in turn, lies in exactly one m-cell; thus $(\mathbf{M}^\partial)^{m-1}$ is the collection of $(m-1)$-cells that lie in exactly one m-cell. We refer to elements of $\mathbf{M}^\partial$ as the boundary cells of $\mathbf{M}$. We define $\partial\mathbf{M}$ to be the union of all boundary cells of $\mathbf{M}$, that is, $\partial\mathbf{M} \equiv \cup(\mathbf{M}^\partial)$, and we refer to $\partial\mathbf{M}$ as the boundary of the subdivision, see Figure 3.3.

3.3 Lemma: If $\mathbf{M}$ and $\mathbf{N}$ are locally finite subdivisions with a common manifold M, then $\partial\mathbf{M} = \partial\mathbf{N}$. □

3.4 Lemma: Let $\mathbf{M}$ be a finite m-subdivision in $\mathbb{E}^m$. The manifold M of $\mathbf{M}$ is convex if and only if $\text{aff}(\sigma)$ is a supporting hyperplane of M for each boundary $(m-1)$-cell σ of $\mathbf{M}$. □

Let $\mathbf{M}$ be a collection of sets in $\mathbb{E}^n$, with carrier M, and let N be a subset of $\mathbb{E}^n$. We define the natural restriction $\mathbf{M}|N$ and the forced restriction $\mathbf{M}\|N$ of $\mathbf{M}$ to N by

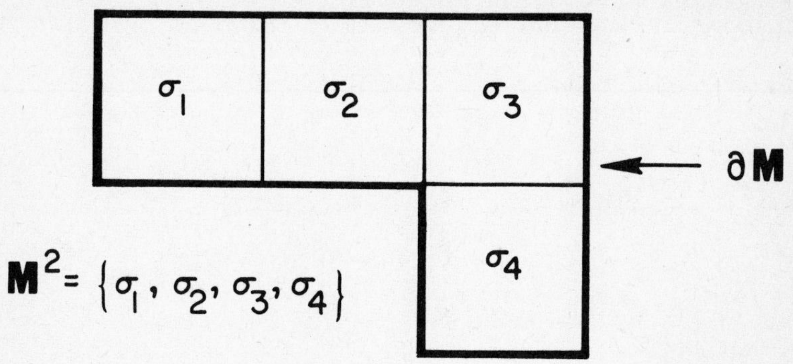

Figure 3.3

$$\mathbf{M}|N \triangleq \{\sigma \in \mathbf{M} : \sigma \subseteq N\} \quad ,$$

$$\mathbf{M}\|N = \{\sigma \cap N : \sigma \in \mathbf{M}\} \quad .$$

If $\mathbf{M}|N = \mathbf{M}\|N$ we speak of $\mathbf{M}|N$ as the restriction of $\mathbf{M}$ to N. Of course, we always have $\mathbf{M}\|N \supseteq \mathbf{M}|N$, and $\mathbf{M}\|N$ always covers $N \cap M$. If $A \supseteq B$, then $(\mathbf{M}|A)|B = \mathbf{M}|B$ and $(\mathbf{M}\|A)\|B = \mathbf{M}\|B$. As we shall see these restriction operations are important to our development. The following lemma is illustrated in Figure 3.4.

3.5 Lemma: Let $\mathbf{M}$ be a subdivision with manifold M. If N is a convex subset of M and $\mathbf{M}\|N$ is a locally finite collection of cells, then the collection of all faces of cells of $\mathbf{M}\|N$ is a subdivision with manifold N. □

3.6 Lemma: Let $\mathbf{M}$ be a subdivision with manifold M and N a subset of M. If N is convex and $\mathbf{M}\|N$ locally finite, then the following are equivalent.

(a) $\mathbf{M}|N$ subdivides N

(b) $\mathbf{M}|N$ covers N

(c) $\mathbf{M}\|N = \mathbf{M}|N$

(d) $\mathbf{M}\|N \subseteq \mathbf{M}$. □

3.7 Lemma: Let $\mathbf{M}$ and $\mathbf{N}$ subdivide M and N respectively. If $\mathbf{N}$ is a subset of $\mathbf{M}$, then the restriction of $\mathbf{M}$ to N is $\mathbf{N}$. □

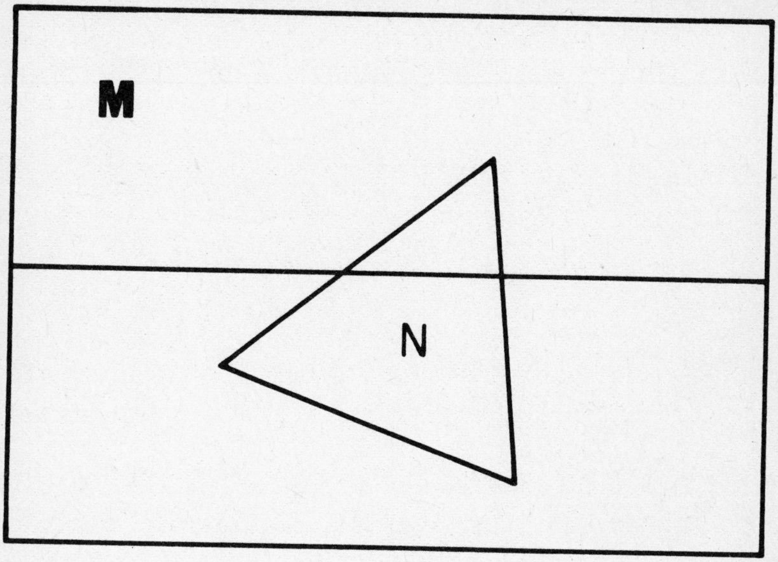

Figure 3.4

3.8 Lemma: Let $\mathbf{M}$ be a subdivision with convex manifold M. If N is a face of M and $\mathbf{M}|N$ is locally finite, then $\mathbf{M}|N$ subdivides N. □

3.9 Lemma: Let $\mathbf{M}$ and $\mathbf{N}$ be m-subdivisions with manifolds M and N. If $\mathbf{M}|N$ and $\mathbf{N}|M$ are equal and cover $M \cap N$, and if $M \cup N$ is of dimension m, then $\mathbf{M} \cup \mathbf{N}$ subdivides $M \cup N$.

Proof: We need only show that a cell σ of $\mathbf{M}$ and τ of $\mathbf{N}$ meet in a common face. Of course $\sigma \cap \tau$ is a cell in $M \cap N$. Since $\sigma \cap \tau$ is covered by cells $\sigma_i \in \mathbf{M}$ with $\sigma_i \subseteq N$, it is covered by cells $\sigma \cap \sigma_i \in \mathbf{M}$ with $\sigma \cap \sigma_i \subseteq N$. Thus $\sigma \cap \tau$ is covered by faces ρ_i of σ with $\rho_i \subseteq N$. But a union of faces of σ which is convex is a face of σ, and thus $\sigma \cap \tau$ is a face of σ, and similarly for τ. □

For the next lemma suppose $\mathbf{M}$ is an m-subdivision with manifold M, N is a convex subset of M, and $\mathbf{M}|N$ subdivides N.

3.10 Lemma: If σ and τ are two m-cells of $\mathbf{M}$ with $\sigma \subseteq N$ and $\tau \nsubseteq N$, then $\sigma \cap \tau$ is contained in a face of N other than N itself. □

This last lemma is illustrated in Figure 3.5.

Let $\mathbf{M}$ and $\mathbf{N}$ be two collections of sets with common carrier M. If each element σ of $\mathbf{M}$ is contained in some element τ of $\mathbf{N}$ we define $\mathbf{M}$ to be a refinement of $\mathbf{N}$ and $\mathbf{N}$ to be an encoarsement of $\mathbf{M}$.

If A is a nonsingular matrix, then $A(\mathbf{M})+a$ is a refinement of $\mathbf{M}$, if and only if $A^{-1}(\mathbf{M}) - A^{-1}a$ is an encoarsement of $\mathbf{M}$.

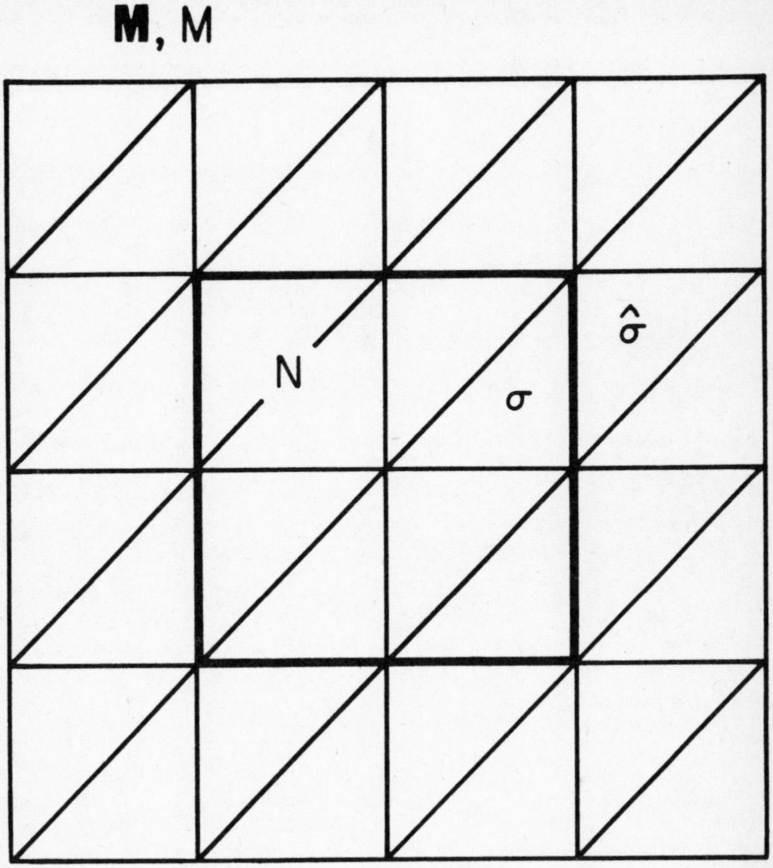

Figure 3.5

We shall from time to time be concerned with subdivisions of simplicial cones, see Section 2; we begin our study with a generator matrix and an index set.

Let $w = (w^1, \ldots, w^k)$ be an $n \times k$ matrix with $k \geq n$, and let $i\mathbf{W}$ be a collection of subsets of $\xi \triangleq \{1, \ldots, k\}$ with the following two properties,

(i) $\alpha \in i\mathbf{W}$ and $\beta \subseteq \alpha$ implies $\beta \in i\mathbf{W}$

(ii) $\alpha \in i\mathbf{W}$ implies there is a $\beta \in i\mathbf{W}$ with $\alpha \subseteq \beta$ and $\#\beta = n$.

Let $\mathbf{W}$ be the collection containing the empty set and all cones of form $W_\alpha \triangleq \{w^\alpha x_\alpha : x_\alpha \geqq 0\} = \{wx : x \geqq 0, x_{\xi\setminus\alpha} = 0\}$ where $\alpha \in i\mathbf{W}$ and let W be the union of all these cones. The question is, when is $\mathbf{W}$ a subdivision. The following lemma gives a result of this form: we say that a subdivision $\mathbf{W}$ is a subdivision with simplicial cones,if each element of $\mathbf{W}$ is a simplicial cone.

3.11 Lemma: $\mathbf{W}$ is an n-subdivision with simplicial cones with carrier W, if for each y in W there is a unique x and at least one α such that

(a) $wx = y$

(b) $x_{\xi\setminus\alpha} = 0$

(c) $x \geqq 0$

(d) $\alpha \in i\mathbf{W}$.

In this case, for any β and γ in $i\mathbf{W}$

(i) $\dim W_\beta = \#\beta$

(ii) $W_\beta \cap W_\gamma = W_{\beta\cap\gamma}$

(iii) $W_\beta \supseteq W_\gamma$ if and only if $\beta \supseteq \gamma$

(iv) $W_\beta = W_\gamma$ if and only if $\beta = \gamma$

(v) ρ is a proper face of W_β if and only if $\rho = W_\gamma$ for some $\gamma \subset \beta$.

Proof: As $w^\beta x_\beta = y$ has a unique solution for y in W_β we conclude that W_β is a simplicial cone and $\dim W_\beta = \#\beta$. If $w^\beta x_\beta = w^\gamma \bar{x}_\gamma$ then $x_i = \bar{x}_i$ for i in $\beta \cap \gamma$ and $x_i = \bar{x}_i = 0$ for i not in $\beta \cap \gamma$ since x is unique. Thus $W_\beta \cap W_\gamma \subseteq W_{\beta\cap\gamma}$ but trivially W_β and W_γ both contain $W_{\beta\cap\gamma}$. Therefore (ii) follows and from which (iii) and (iv) follow. As proper faces of W_β have a form W_γ for $\gamma \subset \beta$ and as the representation $\gamma \to W_\gamma$ is one to one; a proper face W_γ of W_β must have $\gamma \subset \beta$. Axioms (a), (b), and (c) in the definition of a subdivision follow immediately. The last axiom (d) follows from the fact that $\mathbf{W}$ is an n-subdivision in $\mathbb{C}^n$, see the remarks following the definition of a subdivision. □

In particular, we see that $i\mathbf{W}$ and $\alpha \to W_\alpha$ is a representation set and rule for $\mathbf{W}\backslash\phi$; note $\phi \to W_\phi = \{0\}$. Also $i\mathbf{W}^n \triangleq \{\alpha \in i\mathbf{W} : \#\alpha = n\}$ is a representation set for $\mathbf{W}^n$ and α in $i\mathbf{W}^n$ and t in α indexes the facets $W_{\alpha/t} = \{w^{\alpha\backslash t} x_{\alpha\backslash t} : x_{\alpha\backslash t} \geqq 0)\}$ of $W_\alpha = \{w^\alpha x_\alpha : x_\alpha \geqq 0\}$. For α in $i\mathbf{W}^n$ and $t \in \alpha$ the replacement $\hat{\alpha}$ for (α,t) is that $\hat{\alpha}$ of $i\mathbf{W}^n$, if it exists, such that $\#(\alpha \cap \hat{\alpha}) = n-1$ and $t \notin \hat{\alpha}$. We have now specified the representation set and rule, the facet rule, and the replacement rule for the subdivision $\mathbf{W}$ with simplicial cones.

We briefly consider two familiar subdivisions of $\mathbb{C}^n$, namely, $\mathbf{O}$, the orthants of $\mathbb{C}^n$ which is a subdivision with simplicial cones, and $\mathbf{C}$, the unit cubes of $\mathbb{C}^n$. First $\mathbf{O}$ is described, see Figure 3.6; we begin by defining its representation set $i\mathbf{O}$.

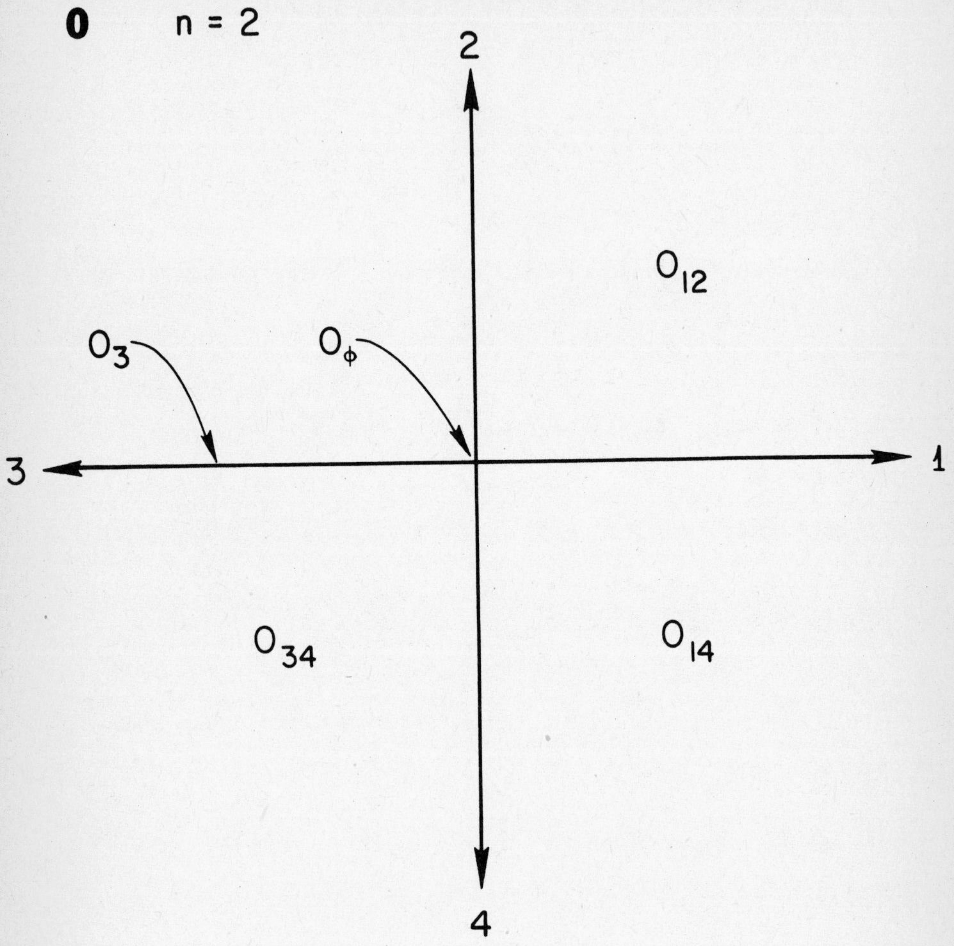

Figure 3.6

Let $\mathbf{i0}$ be the collection of all subsets α of $\xi \triangleq \nu \cup 2\nu = \{1,\ldots,2n\}$ such that α does not contain both i and $i+n$ for any $i = 1, \ldots, n$, that is, $(\alpha+n) \cap \alpha = \phi$. Define 0_α for $\alpha \in \mathbf{i0}$ to be the cone

$$\{(I, -I)^\alpha x_\alpha : x_\alpha \geqq 0\}$$

$$= \{(I, -I)x : x \geqq 0,\ x_{\xi\setminus\alpha} = 0\} .$$

That is, 0_α is the set of y in $\mathbb{C}^n$ such that

$$y_i \geqq 0 \qquad \text{if } i \in \alpha \cap \nu$$

$$y_i \leqq 0 \qquad \text{if } i \in \alpha \cap 2\nu$$

$$y_i = 0 \qquad \text{if } i \notin \alpha .$$

Now consider the system of equations

(a) $(I, -I)x = y$

(b) $x_{\xi\setminus\alpha} = 0$

(c) $x \geqq 0$

(d) $\alpha \in \mathbf{i0}$.

For each $y \in \mathbb{C}^n$ there is a solution (x,α) and the x is unique, and thus, from Lemma 3.11 we see that $\mathbf{0}$ is a subdivision of $\mathbb{C}^n$ with simplicial cones and that $\mathbf{i0}$ is a representation set for $\mathbf{0}\setminus\phi$.

Let $\mathbf{i0}^n$ be the set of elements α in $\mathbf{i0}$ with $\#\alpha = n$. Then $\mathbf{i0}^n$ indexes the n-cones or n-cells of $\mathbf{0}$. Given an n-cone 0_α its facets are of form $0_{\alpha\setminus t}$ for t in α. Given such α and t let

$$\hat{\alpha} = \begin{cases} (\alpha\backslash t) \cup (t+n) & \text{if } t \leqq n \\ (\alpha\backslash t) \cup (t-n) & \text{if } t > n \end{cases}$$

and we see that $0_{\hat{\alpha}}$ is an n-cone of **0** that shares the facet $0_{\alpha\backslash t}$ with 0_α. Thus, we have specified the representative set $\mathbf{i0}^n$ for $\mathbf{0}^n$, the representative rule $\alpha \to 0_\alpha$, the facet rule $(\alpha,t) \to 0_{\alpha\backslash t}$ for $t \in \alpha$, and the replacement rule $(\alpha,t) \to \hat{\alpha}$. We refer to **0** as the subdivision $\mathbb{C}^n$ with orthants. Observe that $r\mathbf{0} = \mathbf{0}$ for any nonzero element r in $\mathbb{C}^n$.

Next let us consider the subdivision **C** based on the unit cubes of $\mathbb{C}^n$, see Figure 3.7.

Define **iC** to be the set of all pairs (y,z) in $\mathbb{Z}^n \times \mathbb{Z}^n$ with $y \leqq z \leqq y+e$ where $e = (1, \ldots, 1)$. Define C_y^z to be the collection of all x in $\mathbb{C}^n$ with $y \leqq x \leqq z$. **C** is defined to be the collection of all such C_y^z with $(y,z) \in \mathbf{iC}$ and the empty set. It is easily verified that **C** is a locally finite subdivision of $\mathbb{C}^n$, that **iC** is a representation set of $\mathbf{C}\backslash\phi$, and that **C** is a refinement of **0**.

Of the elements in **C** observe that:

(a) $C_y^z = C_u^v$ if and only if $y = u$ and $z = v$

(b) ρ is a face of C_y^z, if and only if $\rho = C_u^v$ where $y \leqq u \leqq v \leqq z$ or $\rho = \phi$.

(c) $C_y^z \cap C_u^v = C_w^z$ where $w = \max(y,u)$ and $x = \min(z,v)$.

(d) $\dim C_z^y = e(z-y)$ where $e = (1, \ldots, 1)$ is $1 \times n$.

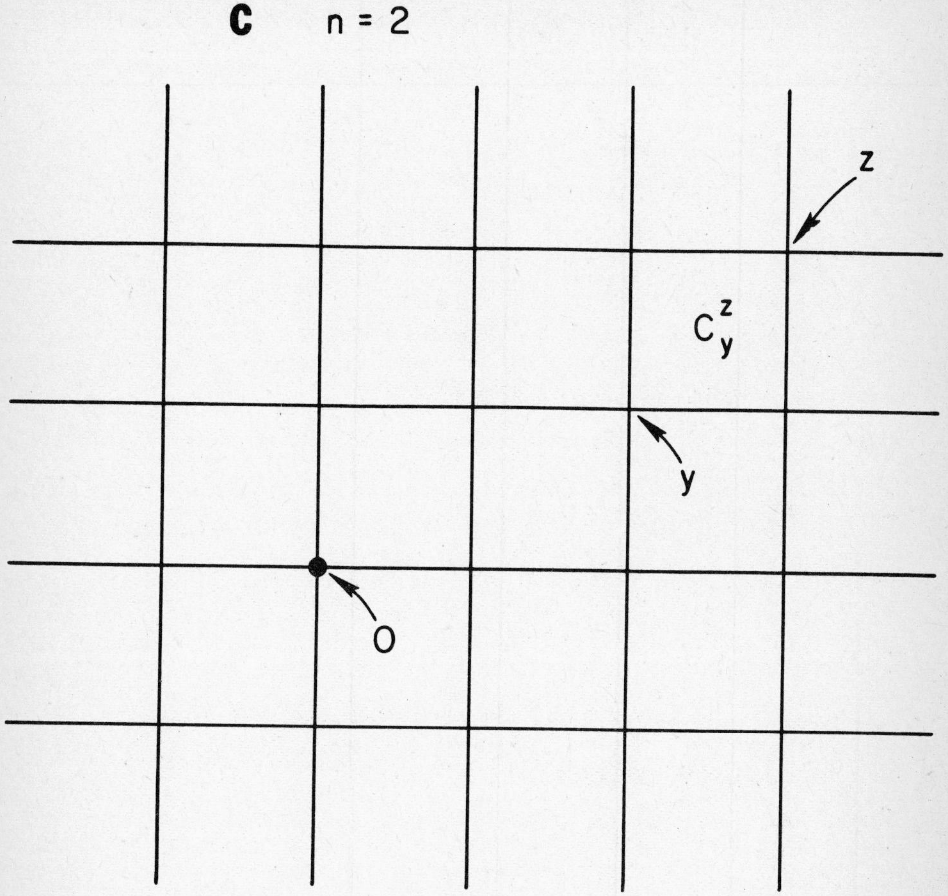

Figure 3.7

Let D be a diagonal matrix with positive integral diagonal and d be an integral vector. Observe that $\mathbf{DC} + d \triangleq \{D\sigma + d : \sigma \in \mathbf{C}\}$ is an encoursement of $\mathbf{C}$.

Let $i\mathbf{C}^n = \{(y,z) \in i\mathbf{C} : e(z-y) = n\}$ index the n-cells of $\mathbf{C}$. An n-cell C_y^z of $\mathbf{C}$ has $2n$ facets corresponding to $C_y^{z-e^t}$ for $t = 1, \ldots, n$ and $C_{y+e^{t-n}}^z$ for $t = n+1, \ldots, 2n$. The facet rule of $\mathbf{C}$ is thus specified and we consider the replacement rules.

Given the n-cell C_y^z and a facet indicator t then $C_{\hat{y}}^{\hat{z}}$ is the unique n-cell of $\mathbf{C} \backslash C_y^z$ sharing the facet of C_y^z indicated by t where

$$\hat{z} = z - e^t \qquad \hat{y} = y - e^t \qquad t = 1, \ldots, n$$

$$\hat{z} = z + e^{t-n} \qquad \hat{y} = y + e^{t-n} \qquad t = n+1, \ldots, 2n .$$

For $\mathbf{C}$ we have specified the representation set $i\mathbf{C}^n$ for the n-cells, the representation rule $(y,z) \to C_y^z$, the facet rule $t \to C_y^{z-e^t}$ for $t = 1, \ldots, n$, $t \to C_{y+e^{t-1}}^z$ for $t = n+1, \ldots, 2n$, and the replacement rule $(y,z,t) \to (\hat{y},\hat{z})$.

3.12 Bibliographical Remarks

The roots of representation and replacement rules are, probably, found in Hansen [1968] and Kuhn [1968].

Proofs of Lemmas 3.3 through 3.10 are not hard, but can be found in Eaves and Rothblum [in progress]. The definition of a subdivision used here is weaker than that used in PL topology, see for example, Rourke and Sandarson [1972]. □

4. STANDARD SIMPLEX S AND MATRIX OPERATIONS

The technical development of this paper begins with the selection of an n-simplex; different selections lead to subdivisions which are linear transformations of one another. There is no loss of generality in selecting a particular simplex; and, therefore, one selects that simplex which most enhances the development. In this regard we define the standard simplex S to be the n-simplex in $\mathbb{C}^n$ with vertices $s^1, \ldots, s^{n+1}$ where

$$s^1 = \begin{bmatrix} 0 \\ 0 \\ \vdots \\ 0 \\ 0 \end{bmatrix} \quad s^2 = \begin{bmatrix} 1 \\ 0 \\ \vdots \\ 0 \\ 0 \end{bmatrix} \quad s^3 = \begin{bmatrix} 1 \\ 1 \\ 0 \\ \vdots \\ 0 \\ 0 \end{bmatrix} \quad \cdots \quad s^{n+1} = \begin{bmatrix} 1 \\ 1 \\ \vdots \\ 1 \\ 1 \end{bmatrix}$$

or in general

$$s^i = \sum_{j=1}^{i-1} e^j$$

and $e^i = (0, \ldots, 0, 1, 0, \ldots, 0)$ is the ith unit vector, see Figure 4.1. We refer to S as the standard simplex.

Let s be the $n \times (n+1)$ matrix $(s^1, \ldots, s^{n+1})$ of vertices of S. For $n = 3$,

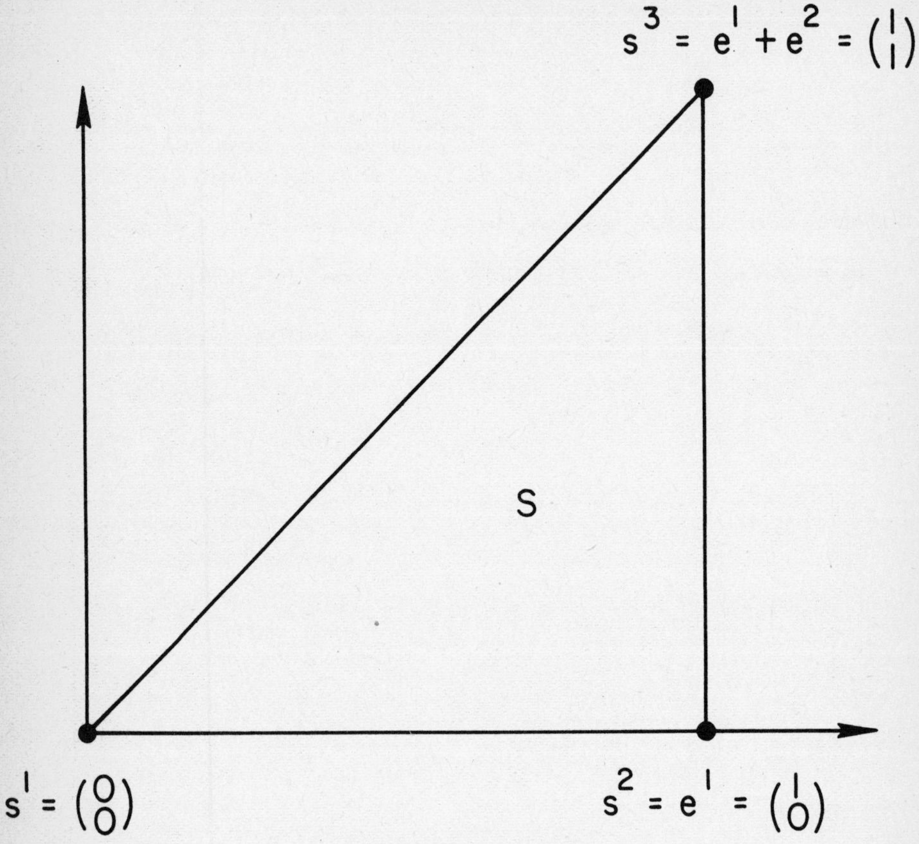

Figure 4.1

$$s = \begin{bmatrix} 0 & 1 & 1 & 1 \\ 0 & 0 & 1 & 1 \\ 0 & 0 & 0 & 1 \end{bmatrix} .$$

As S is the convex combination of its vertices, S is the set of $x = s\lambda$ where $e\lambda = 1$ and $\lambda \geqq 0$. Thus, we see that S is the set of all $x = (x_1, \ldots, x_n)$ such that $1 \geqq x_1 \geqq x_2 \geqq \cdots \geqq x_n \geqq 0$.

For $\alpha \subseteq \mu \triangleq \{1,\ldots,n+1\}$, $s^\alpha = (s^{\alpha 1}, \ldots, s^{\alpha k})$ is the submatrix of columns of s indexed by $\alpha = \{\alpha 1, \ldots, \alpha k\}$ where $\alpha i < \alpha(i+1)$ for $i = 1, \ldots, k-1$. For $\alpha \subseteq \mu$ define S_α to be the set $\{s^\alpha x : ex = 1, x \geqq 0\}$ where $e = (1, \ldots, 1)$ is a row; clearly S_α is the convex hull of the vectors s^i for $i \in \alpha$ and is a face of S of dimension $\#\alpha-1$.

Next we concern ourselves with the directions from one vertex of S to the next. For $i \in \mu$ define

$$q^i = s^{i+1} - s^i$$

where $s^{n+2} \triangleq s^1$. We have

$$q^i = e^i , \qquad i = 1, \ldots, n$$

$$q^{n+1} = -e$$

where $e = (1, \ldots, 1)$. Define q to be the $n \times (n+1)$ matrix $q \triangleq (q^1,\ldots,q^{n+1}) = (I,-e)$ and observe $\sum_\mu q^i = 0$. For $\alpha \subseteq \mu$ define Q_α to be the cone $\{q^\alpha x : x \geqq 0\}$ generated by the q^i with $i \in \alpha$.

By an ordering π of $\alpha \subseteq \mu = \{1,\ldots,n+1\}$ we mean a one-to-one map $\pi : \{1, \ldots, \#\alpha\} \to \alpha$ where $\#\alpha$ is the number of elements in α. We write πt for $\pi(t)$ and $\pi t+1$ for $\pi(t)+1$.

If π orders $\alpha \subseteq \mu$ then by q^{π} we denote the submatrix of columns of q ordered as π, that is,

$$q^{\pi} = (q^{\pi 1}, \ldots, q^{\pi k}) ,$$

where $\pi = (\pi 1, \ldots, \pi k)$ and $k = \#\alpha$. Throughout the paper we use the objects S, S_{α}, s, s^{α}, Q_{α}, q, q^{α}, and q^{π}, repeatedly.

Let π be an ordering (permutation) of μ and k an element of $\mu \triangleq \{1, \ldots, n+1\}$. By $\pi|k$ for $k = 0, 1, \ldots, n$ we mean the ordering $\gamma : \{1, \ldots, k\} \to \{\pi 1, \ldots, \pi k\}$ defined by $\gamma i = \pi i$ for $1 \leq i \leq k$. We have $(\pi|k)i = \pi i$ for $i = 1, \ldots, k$. In particular $q^{\pi|k}$ is the matrix $(q^{\pi 1}, \ldots, q^{\pi k})$.

In the following sections we are concerned from time to time with the nature of or computation of inverses and multiplication of matrices of form $q^{\pi|n}$ where π permutes μ. The remainder of this section is concerned with this matter; it seems best to treat these matrix operations all in one place even though their use is, for the most part, disperse. The reader might want only to peruse them now and return to them at the moment of application.

4.1 Lemma: If π orders μ and $q^{\pi|n}x = y$ then

$$y_i = x_{\gamma i} - x_{\gamma(n+1)}$$

if i in ν where $\gamma = \pi^{-1}$ and $x_{n+1} \triangleq 0$.

Proof: Examine the ith row, $q_i^{\pi|n}x = y_i$. There are three cases corresponding to

(a) $\pi(n+1) = n+1$

(b) $\pi(n+1) = i$

(c) $\pi(n+1) \notin \{i, n+1\}$.

In (a) we have $y_i = x_j$ where $j = \pi^{-1}(i)$ and $x_{\gamma(n+1)} = x_{n+1} = 0$. In (b) we have $y_i = -x_k$ where $k = \pi^{-1}(n+1)$ and $x_{\gamma i} = x_{n+1} = 0$. In (c) we have $y_i = x_j - x_k$ where $j = \pi^{-1}(i)$ and $k = \pi^{-1}(n+1)$. □

4.2 Lemma: If π permutes μ, then the inverse of $q^{\pi|n}$ is $q^{\gamma|n}$ where $\gamma = \pi^{-1}$.

Proof: Consider the ith column $q^{\gamma|n}\, q^{\pi i}$ of $q^{\gamma|n}\, q^{\pi|n}$. If $\pi i = n+1$ we have

$$q^{\gamma|n}\, q^{\pi i} = q^{\gamma|n}\, q^{n+1} = q^{\gamma|n}(-e) = q^{\gamma(n+1)} = q^i .$$

If $\pi i \neq n+1$ we have

$$q^{\gamma|n}\, q^{\pi i} = (q^{\gamma|n})^{\pi i} = q^{\gamma\pi i} = q^i .$$

Note $\pi\nu = \gamma\nu$ is not required. □

4.3 Lemma If π orders μ and $q^{\pi|n}x = y$ then

$$x_i = y_{\pi i} - y_{\pi(n+1)}$$

for i in ν where $y_{n+1} \triangleq 0$.

Proof: $x = (q^{\pi|n})^{-1}y = q^{\gamma|n}y$ where $\gamma = \pi^{-1}$, see Lemma 4.2. The result follows from Lemma 4.1. □

4.4 Lemma: Let π and γ order μ. Then $q^{\pi|n}\, q^{\gamma|n} = q^{\xi|n}$ where $\xi = \pi\gamma$.

Proof: Again we examine the ith column $q^{\pi|n}\, q^{\gamma i}$ of the product. If $\gamma i = n+1$ then $q^{\pi|n}\, q^{\gamma i} = q^{\pi|n}(-e) = q^{\pi(n+1)} = q^{\xi i}$. If $\gamma i \neq n+1$ then $q^{\pi|n}\, q^{\gamma i} = q^{\pi\gamma i} = q^{\xi i}$. □

4.5 Lemma: Let π and γ each order μ. Then

$$q^{\pi|n}\, q^{\xi|n} = q^{\gamma|n}$$

where $\xi = \pi^{-1}\gamma$.

Proof: Given π and ξ the result is $\gamma = \pi\xi$ from Lemma 4.4. □

Although we may use notation as $q^{\gamma|n}x$ and $q^{\gamma|n}\, q^{\pi|n}$ for conceptual clarity, it is understood that in actual computation one would proceed as above.

5. SUBDIVISION $\mathbf{Q}$ of $\mathbb{C}^n$

Our subdivision $\mathbf{Q}$ of $\mathbb{C}^n$ with simplicial cones, which we shortly describe, is of little interest in its own right, but shall be of considerable importance in our construction of triangulations. Nevertheless, because of its transparency $\mathbf{Q}$ is a pleasant place to begin. Our line of reasoning here follows Lemma 3.11 and will be employed again for the subdivision $\mathbf{P}$ in Section 13.

Define $\mathbf{iQ}$ to be the collection of all subsets α of $\mu \triangleq \{1, \ldots, n+1\}$ other than μ itself. For α in $\mathbf{iQ}$ define the cone $Q_\alpha = \{q^\alpha y : y \geqq 0\}$ where $q = (e^1, \ldots, e^n, -e)$, see Section 4. Define $\mathbf{Q}$ to be the collection of all such Q_α with $\alpha \in iQ$ and the empty set, see Figure 5.1.

From the fact that the system

$$qx = y \qquad 0 \nleq x \geqq 0 \tag{1}$$

has a unique solution for all y, that is, that the system

$$\text{(a)}\quad qx = y \qquad\qquad \text{(b)}\quad x_{\mu\setminus\alpha} = 0$$

$$\text{(c)}\quad x \geqq 0 \qquad\qquad \text{(d)}\quad \alpha \in \mathbf{iQ}$$

has a solution (x,α) for all y in $\mathbb{C}^n$ and x is unique, we see that $\mathbf{Q}$ is a finite subdivision of $\mathbb{C}^n$ with simplicial cones, see Lemma 3.11. Thus, for $\mathbf{Q}$, we have representation set $\mathbf{iQ}$, a representation rule $\alpha \to Q_\alpha$, a facet rule $(\alpha,t) \to Q_{\alpha\setminus t}$ and we proceed to develop the replacement rule.

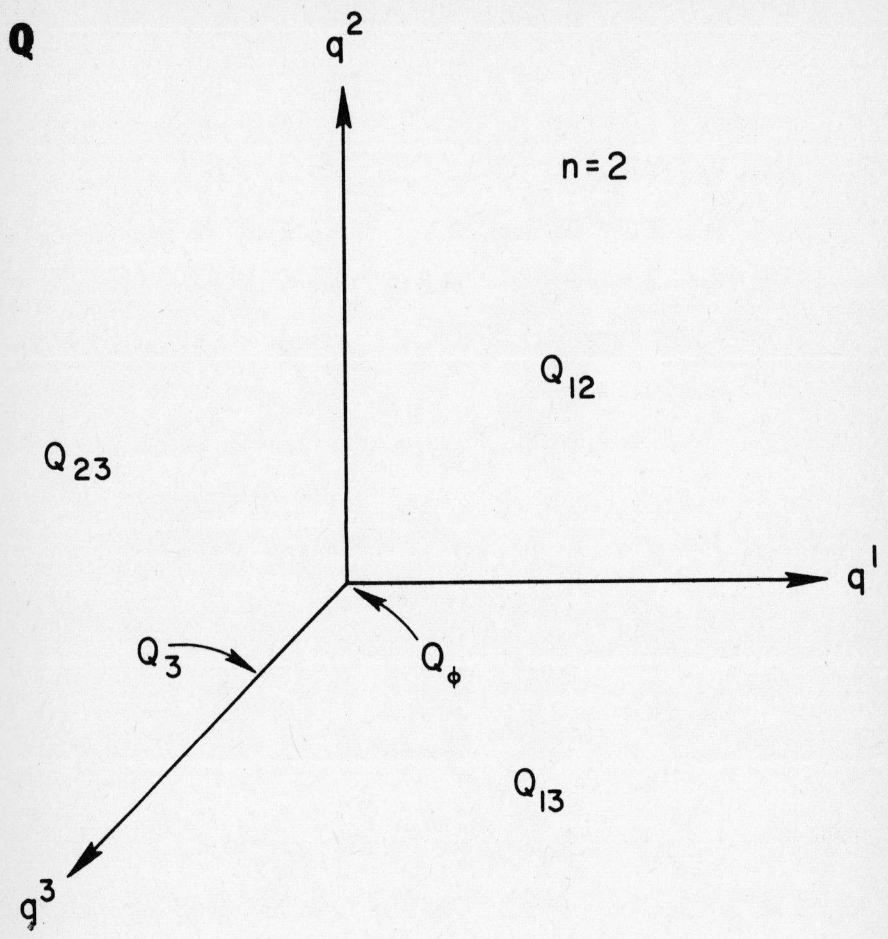

Figure 5.1

Let $i\mathbf{Q}^n = \{\alpha \in i\mathbf{Q} : \#\alpha = n\}$ index the n-cells of $\mathbf{Q}$. Let t be an element of α and $Q_{\alpha\backslash t}$ is a facet of Q_α. Let $\hat{\alpha} = \mu\backslash t$, then $Q_{\hat{\alpha}}$ is an n-cell of $\mathbf{Q}\backslash Q_\alpha$ which also has $Q_{\alpha\backslash t}$ as a facet. The representation and replacement rules of $\mathbf{Q}$ are complete.

From time to time given y we must compute an x solving $qx = y$ with $0 \nless x \geqq 0$ or $\alpha \subset \mu$ with $y \in Q_\alpha$. The procedure for this is: let j index the smallest component of y. If $y_j \geqq 0$ let $x = (y,0)$ and $\alpha = \nu$. If $y_j < 0$ let $x = (y-ey_j, -y_j)$, see Figure 5.2.

For r a positive scalar it is obvious that $r\mathbf{Q} = \mathbf{Q}$.

We now develop partial orderings based upon the cones of $\mathbf{Q}$. These results will be important in our study of Freudenthal's triangulation.

For α in $i\mathbf{Q}$ if $y^i = q^\alpha x^i$ for $i = 1, 2,$ we define $y^1 \leqq_\alpha y^2$, $y^1 \leq_\alpha y^2$, and $y^1 <_\alpha y^2$ to mean $x_1 \leqq x_2$, $x_1 \leq x_2$, and $x_1 < x_2$, respectively. Of course, $y^1 \leqq_\alpha y^2$, $y^1 \leq_\alpha y^2$, and $y^1 <_\alpha y^2$, if and only if $y^2 \in Q_\alpha + y^1$, $y^2 \in Q_\alpha + y^1$ and $y^1 \neq y^2$, and $y^2 \in riQ_\alpha + y^1$, respectively. If $\alpha = \nu \triangleq \{1, \ldots, n\}$ obviously $y^1 \leqq_\alpha y^2$, $y^1 \leq_\alpha y^2$, and $y^1 <_\alpha y^2$, if and only if $y^1 \leqq y^2$, $y^1 \leq y^2$, and $y^1 < y^2$, respectively.

It is routine to show that $\leqq_\alpha$ is a partial ordering; that is, it is reflexive ($y \leqq_\alpha y$), antisymetric ($y^1 \leqq_\alpha y^2$ and $y^2 \leqq_\alpha y^1$ implies $y^1 = y^2$), and transitive ($y^1 \leqq_\alpha y^2 \leqq_\alpha y^3$ implies $y^1 \leqq_\alpha y^3$).

If $\alpha \subseteq \beta$ are in $i\mathbf{Q}$ then clearly $y^1 \leqq_\alpha y^2$ and $y^1 \leq_\alpha y^2$ imply $y^1 \leqq_\beta y^2$ and $y^1 \leq_\beta y^2$, respectively. Note that $y^1 <_\alpha y^2$ or $y^1 \leqq_\beta y^2$ does not imply $y^1 <_\beta y^2$ or $y^1 \leqq_\alpha y^2$, respectively. The following lemma describes the feature of the ordering that we use later.

5.1 Lemma: If $v^1 \leq_\alpha v^2 \leq_\alpha \cdots \leq_\alpha v^k$, $w^1 \leq_\alpha w^2 \leq_\alpha \cdots \leq_\alpha w^k$, and $\{v^1, \ldots, v^k\} = \{w^1, \ldots, w^k\}$, then $v^i = w^i$ for $i = 1, \ldots, k$.

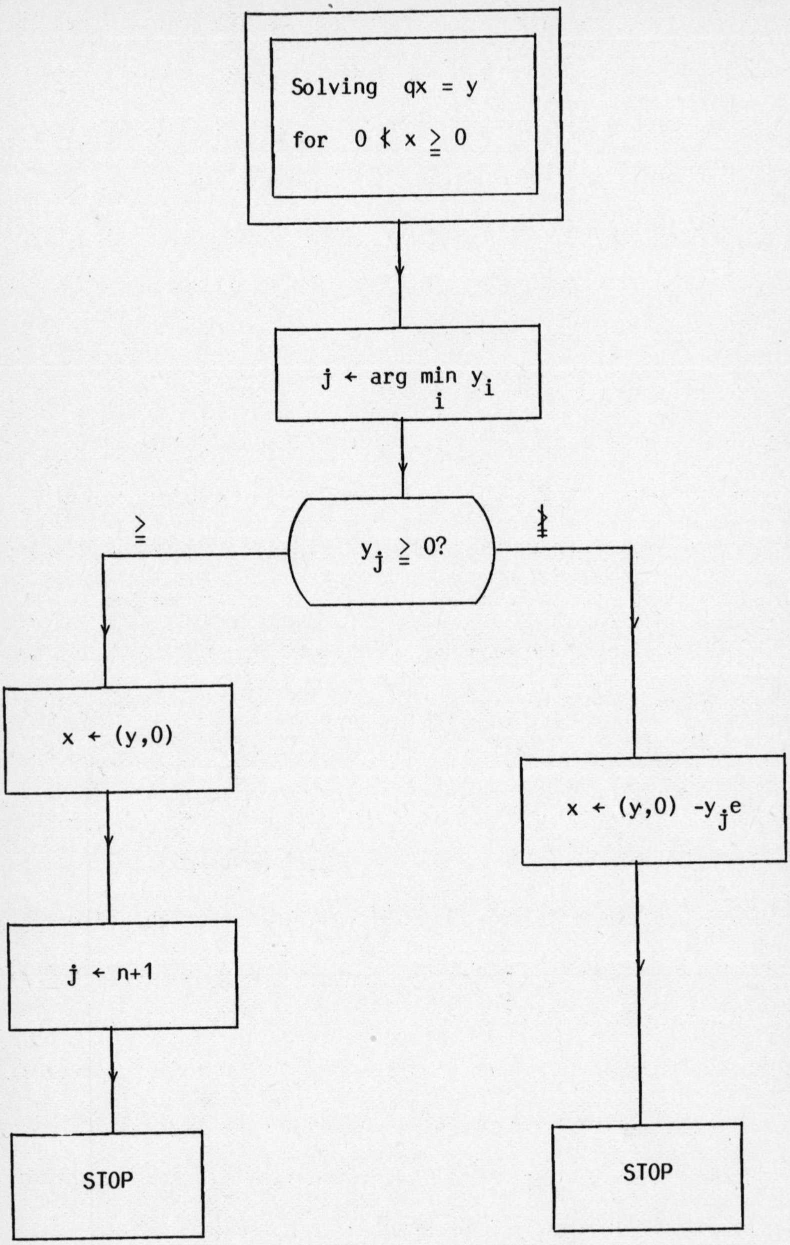
Solving qx = y
for 0 ≨ x ≧ 0
j ← arg min y_i (min over i)
y_j ≧ 0?
≧
≱
x ← (y,0)
j ← n+1
STOP
x ← (y,0) -y_j e
STOP

Figure 5.2

Proof: For $k = 1$ the result is obvious. If $v^1 = w^1$ the result follows by induction. Assuming $v^1 \neq w^1$, for some $i > 1$ and $j > 1$ we have

$$v^1 \leq_\alpha \cdots \leq_\alpha v^i = w^1 \leq_\alpha \cdots \leq_\alpha w^j = v^1$$

or $v^1 \leq_\alpha v^1$ which is a contradiction. □

In Section 10 we shall prove a stronger version of the following invariance theorem for $\mathbf{Q}$; recall the s^i are the vertices of S.

5.2 Lemma: Let j be an integer and $h : \mathbb{C}^n \to \mathbb{C}^n$ a linear map such that $h(s^i) = s^{i+j}$ for i in μ where $i+j$ is regarded mod $n+1$. Then $h(\mathbf{Q}) - h(0) = \mathbf{Q}$. □

5.3 Bibliographical Notes

The first use of the subdivision $\mathbf{Q}$ for solving equations was, probably, Lemke [1965], though implicitly. □

6. FREUDENTHAL TRIANGULATION **F** of $\mathbb{C}^n$, Part I

The triangulation **F** of $\mathbb{C}^n$ which we describe here is most fundamental to solving continuous equations with PL homotopies. Virtually every homotopy algorithm for solving equations (without special structure), has used the triangulation **F**.

For $n = 1$ we have **F** = **C** where **C** is the subdivision of $\mathbb{C}^n$ based on unit intervals, see Section 3. For all $n = 1, 2, \ldots,$ **F** is a refinement of **C**. For $n = 2$ one obtains **F** from **C** by subdividing each square along the diagonal $e = (1, \ldots, 1)$, see Figures 6.1 and 6.2.

For $n = 2$ each 2-simplex of **F** has vertices v, $v+e^1$, and $v+e^1+e^2$ and shape

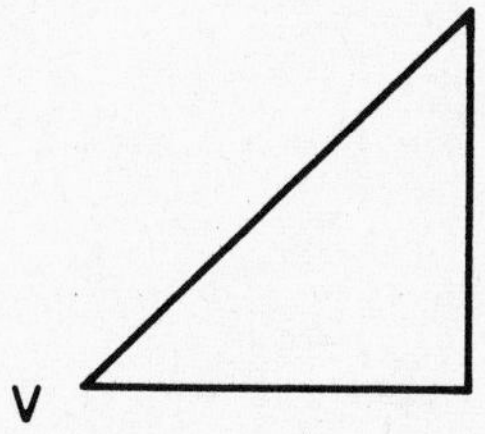

or vertices v, $v+e^2$, and $v+e^1+e^2$ and shape

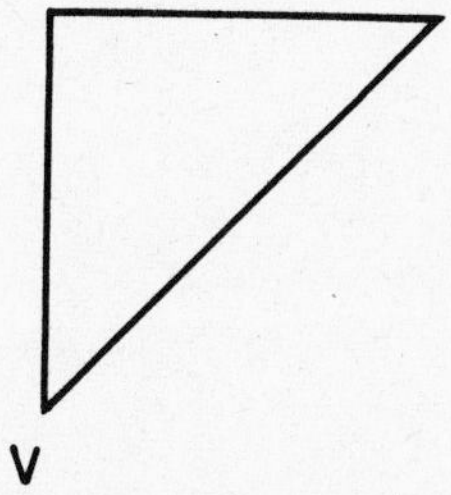

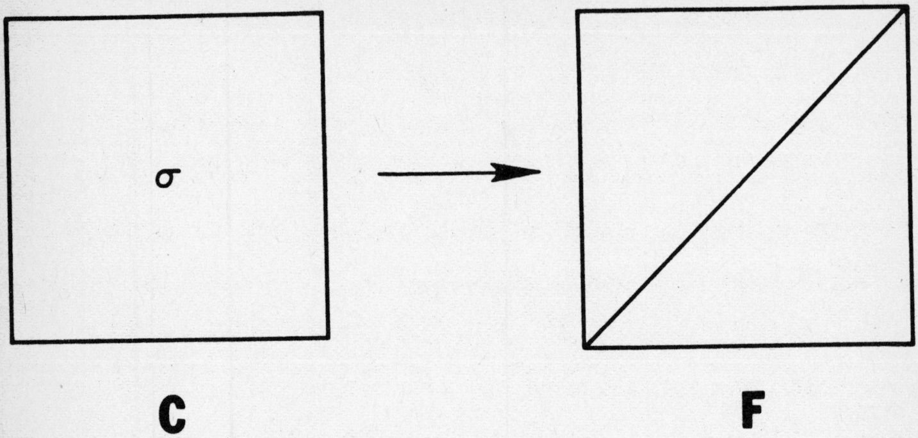

Figure 6.1

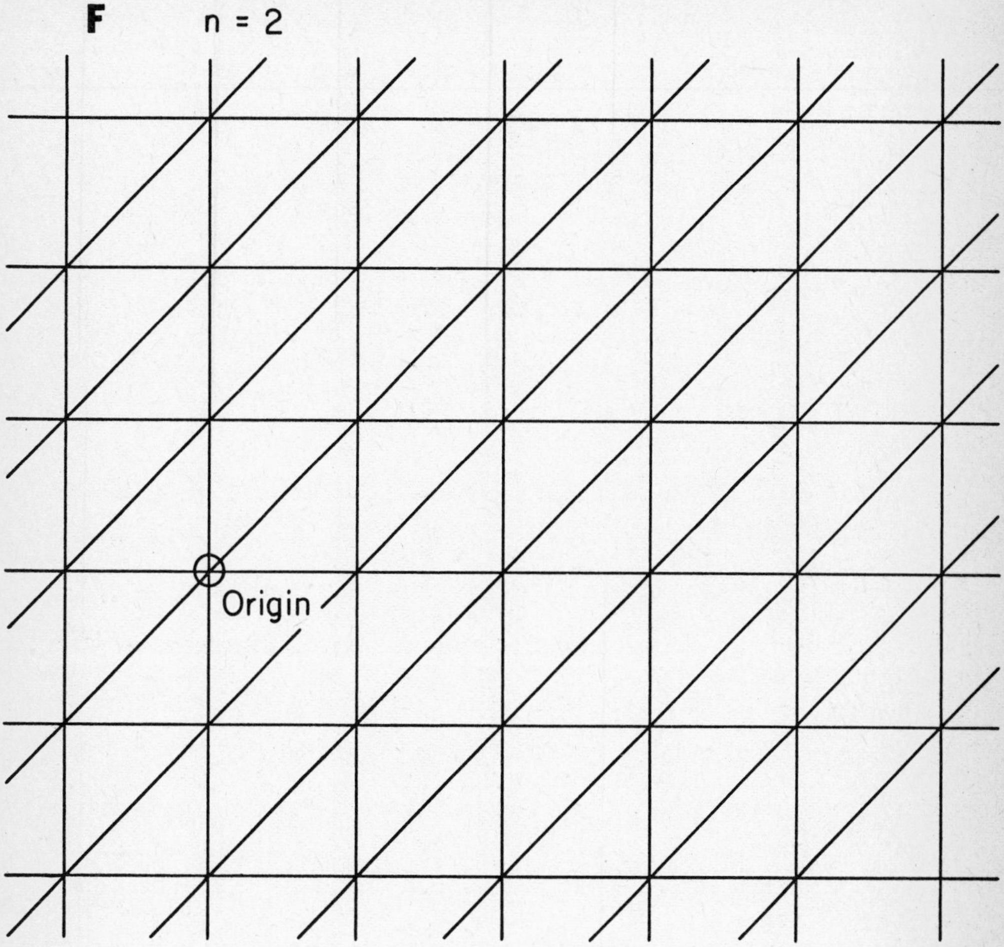

Figure 6.2

where v is integral, $e^1 = (1,0)$ and $e^2 = (0,1)$. Or, in general, each 2-simplex has vertices v, $v+e^{\pi 1}$, and $v+e^{\pi 1}+e^{\pi 2}$, where π is a permutation (ordering) of $\nu = \{1,2\}$, thus, given v and π the 2-simplex can be generated. In fact, define $(v;\pi;2)$ (with semicolon) as the 2-simplex with vertices

$$v^i = v + \sum_{j=1}^{i-1} e^{\pi i}$$

for $i = 1, 2, 3$.

Given a 2-simplex $(v; \pi; 2)$ the adjacent simplexes of $\mathbf{F}$ are shown in Figure 6.3 where two n-cells of an n-subdivision are defined to be adjacent if they share a common facet.

Given $(v; \pi; 2)$ there are three adjacent simplexes corresponding to the dropping each of the vertices. The three simplexes are $(v^2; \hat{\pi}; 2)$, $(v-e^{\pi 2}; \hat{\pi}; 2)$ and $(v; \hat{\pi}; 2)$ where $\hat{\pi} = (\pi 2, \pi 1)$ that is $\hat{\pi}1 = \pi 2$ and $\hat{\pi}2 = \pi 1$ as is easily verified in Figure 6.3 for $\pi = (1,2)$ and $\pi = (2,1)$.

With the foregoing as motivation let us now prepare for describing the Freudenthal triangulation $\mathbf{F}$ of $\mathbb{C}^n$ for arbitrary $n = 1, 2, \ldots$. We continue to use $\nu \triangleq \{1, \ldots, n\}$ and $\mu \triangleq \nu \cup \{n+1\}$.

Our formal development is begun by defining $i_*\mathbf{F}$ which is redundant and incomplete as a representation set for $\mathbf{F}$, nevertheless, it is helpful and from it we extract the representation set $i\mathbf{F}^n$ for the n-simplexes of $\mathbf{F}$.

Define $i_*\mathbf{F}$ to be the collection of all triples (v,π,k) where

(a) $v \in \mathbb{Z}^n$

(b) π permutes μ

(c) $k = 0, 1, \ldots, n$.

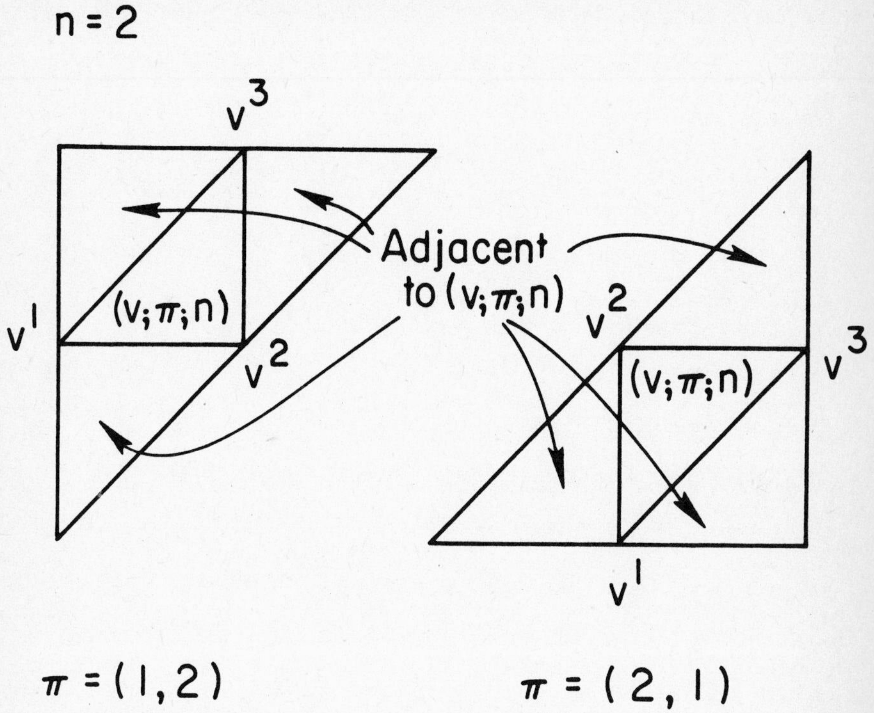

Figure 6.3

Given (v,π,k) in $i_*\mathbf{F}$ we define $(v;\pi;k)$ to be the convex hull of the points $v^1, \ldots, v^{k+1}$ where $v^1 = v$, $v^2 = v^1 + q^{\pi 1}$, $v^3 = v^2 + q^{\pi 2}, \ldots,$ and, in general,

$$v^{i+1} = v^i + q^{\pi i}$$

for $i = 1, \ldots, k$. Recall $q = (e^1, \ldots, e^n, -e)$. As the next lemma shows $(v;\pi;k)$ is a k-simplex. Note that the information $(\pi(k+1), \ldots, \pi(n+1))$ is not used in the definition of $(v;\pi;k)$; our choice of this notation stems from computational considerations and not mathematical elegance; on this matter, please be patient.

6.1 Lemma: For $(v,\pi,k) \in i_*\mathbf{F}$ the cell $(v;\pi;k)$ is a k-simplex.

Proof: $(v^2-v^1, \ldots, v^{k+1}-v^k) = q^{\pi|k}$ has rank k. □

Thus, it is seen that the points $v^1, \ldots, v^{k+1}$ are, in fact, vertices of the k-simplex $(v;\pi;k)$. Whenever we refer to the vertices $v^1, \ldots, v^{k+1}$ of $(v,\pi,k) \in i_*\mathbf{F}$ it is implicit that $v^1 = v$ and $v^{i+1} = v^i+q^{\pi i}$ for $i = 1, \ldots, k$.

As an example, let $v = 0$ and $\pi = (1, \ldots, n, n+1)$. Then $(v,\pi,n) \in i_*\mathbf{F}$ and $(v;\pi;n)$ is the standard n-simplex S of Section 4; note $v^i = s^i$ for i in μ, that is, $v^1 = 0$, $v^2 = e^1$, $v^3 = e^1 + e^2, \ldots,$ $v^{n+1} = e^1 + \cdots + e^n$.

The next lemma yields another form for $(v;\pi;n)$; let $v \circ e$ represent the $n \times (n+1)$ matrix $(v, v, \ldots, v)$.

6.2 Lemma: For $(v,\pi,n) \in i_*F$ let $(v^1, \ldots, v^{n+1})$ be the $n \times (n+1)$ matrix whose columns are the vertices of the n-simplex $(v;\pi;n)$. Then

$$(v^1, \ldots, v^{n+1}) = v \circ e + q^{\pi|n} s .$$

Proof: As s is the matrix of vertices of S, merely substitute $v = e^{\pi 1} + \cdots + e^{\pi i}$ for v^{i+1}. □

Thus, for all $\lambda = (\lambda_1, \ldots, \lambda_{n+1})$ with $e\lambda = 1$ we have

$$\sum_{\mu} v^i \lambda_i = v + q^{\pi|n} s\lambda .$$

Thus, the n-simplex $(v;\pi;n)$ is the set

$$\{v + q^{\pi|n} s\lambda : \lambda \geqq 0,\ e\lambda = 1\} .$$

and the interior of $(v;\pi;n)$ is

$$\{v + q^{\pi|s} \lambda : \lambda > 0,\ e\lambda = 1\} .$$

Of course, λ is the vector of barycentric coordinates for a point $\sum_{\mu} v^i \lambda_i$ in $(v;\pi;n)$.

The next lemma establishes some uniqueness of representation in i_*F. Let (v,π,k) and (u,γ,k) be elements of i_*F and let $(\pi|k)\mu = \{\pi 1, \ldots, \pi k\}$ and $(\gamma|k)\mu = \{\gamma 1, \ldots, \gamma k\}$, that is, $(\pi|k)\mu$ is the image of $\pi|k$.

6.3 Lemma: If the simplexes $(v;\pi;k)$ and $(u;\gamma;k)$ are equal and the sets $(\pi|k)\mu$ and $(\gamma|k)\mu$ are equal, then the pairs $(v, \pi|k)$ and $(u, \gamma|k)$ are equal.

Proof: We have $\{v^1, \ldots, v^{k+1}\} = \{u^1, \ldots, u^{k+1}\}$, that is, the vertex sets are equal. Let $\alpha = (\pi|k)\mu = (\gamma|k)\mu$. We have $v^i \leq_\alpha v^{i+1}$ and $u^i \leq_\alpha u^{i+1}$ for $i = 1, \ldots, k$ and so by Lemma 5.1 $v^i = u^i$ for i in μ. Thus, $\pi|k = \gamma|k$. □

The next lemma states that different representations in i_*F of the same simplex have different first vertices.

6.4 Lemma: If $(v;\pi;k) = (v;\gamma;\ell)$ then $k = \ell$ and $\pi|k = \gamma|\ell$.

Proof: Clearly the vertex sets $\{v^1, \ldots, v^{k+1}\} = \{w^1, \ldots, w^{\ell+1}\}$ are equal, so $k = \ell$. Select $h = 1, 2, \ldots$ maximum such that $v^i = w^i$ for $1 \leqq i \leqq h$. If $h = k$, then $\pi|k = \gamma|k$. Suppose $h < k$ and $v^{h+1} \neq w^{h+1}$. Some $w^{h+i} = v^{h+1}$ for $i > 1$. Thus

$$q^{\gamma h} + q^{\gamma(h+1)} + \cdots + q^{\gamma(h+j)} = q^{\pi h}$$

where $j \geqq 1$. But no subset of two or more columns of q sum to some other column of q and we have a contradiction. □

Henceforth, in this section we concern ourselves with the subset iF^n of i_*F where iF^n is defined to be the collection of all pairs (v,π,n)

in $i_*\mathbf{F}$ with $\pi(n+1) = n+1$. As we shall see, $i\mathbf{F}^n$ has a unique representative for each n-simplex of the Freudenthal triangulation $\mathbf{F}$. For the rest of this section mention of $(v;\pi;n)$ implies $(v,\pi,n) \in i\mathbf{F}^n$; in later sections we return to $i_*\mathbf{F}$. Observe for $(v,\pi,n) \in i\mathbf{F}^n$ we have $q^{\pi|n}$ as a permutation matrix.

Define $\mathbf{F}^n$ to be the collection of all n-simplexes $(v;\pi;n)$ where $(v,\pi,n) \in i\mathbf{F}^n$. As we shall see $\mathbf{F}$, the collection of all faces of simplexes of $\mathbf{F}^n$, is a triangulation of $\mathbb{C}^n$; we shall call $\mathbf{F}$ the Freudenthal triangulation. Our task is to show that the simplexes of $\mathbf{F}^n$ cover $\mathbb{C}^n$ and any two simplexes of $\mathbf{F}^n$ meet in a common face. Note that $\mathbf{F}^0$, the vertices of $\mathbf{F}$, is the set $\mathbb{Z}^n$ of integral vectors of $\mathbb{C}^n$.

6.5 Lemma: For any x in $\mathbb{C}^n$ there is a simplex $(v;\pi;n)$ of $\mathbf{F}^n$ containing x.

Proof: Let $v = \lfloor x \rfloor$, that is, component by component let v_i be the largest integer not exceeding x_i. Let $d = x-v$ and let π be any permutation of μ with $\pi(n+1) = n+1$ such that

$$1 \geqq d_{\pi 1} \geqq d_{\pi 2} \geqq \cdots \geqq d_{\pi n} \geqq 0 .$$

We argue that $x \in (v;\pi;n)$

$$
\begin{aligned}
x &= v + d \\
&= v + d_{\pi 1}\, e^{\pi 1} + \cdots + d_{\pi n}\, e^{\pi n} \\
&= v + (d_{\pi 1} - d_{\pi 2}) e^{\pi 1} \\
&\quad + (d_{\pi 2} - d_{\pi 3})\, (e^{\pi 1} + e^{\pi 2}) \\
&\quad + (d_{\pi 3} - d_{\pi 4})\, (e^{\pi 1} + e^{\pi 2} + e^{\pi 3}) \\
&\quad + \cdots + d_{\pi n} (e^{\pi 1} + \cdots + e^{\pi n}) \\
&= v + q^{\pi | n}\, s\lambda
\end{aligned}
$$

where $\lambda_1 = 1 - d_{\pi 1}$ and $\lambda_{i+1} = d_{\pi i} - d_{\pi(i+1)}$ for i in ν where $d_{n+1} \triangleq 0$ and $e\lambda = 1$. Thus, x is in $(v;\pi;n)$, see Lemma 6.2. □

Given two simplexes of $\mathbf{F}^n$ we show that they meet in a common face, thereby, showing that any two simplexes of $\mathbf{F}$ meet in a common face.

6.6 Lemma: Any two n-simplexes of $\mathbf{F}$ meet in a common face.

Proof: Let (v,π,n) and (u,γ,n) be elements of $\mathbf{iF}^n$. Following Lemma 2.4 we consider the system

$$
v + q^{\pi | n}\, s\lambda = u + q^{\gamma | n}\, s\xi
$$

$$
\lambda \geqq 0 \qquad e\lambda = 1 \qquad \xi \geqq 0 \qquad e\xi = 1
$$

or equivalently, to

$$\begin{bmatrix} q^{\pi|n}s & -q^{\gamma|n}s \\ e & 0 \\ 0 & e \end{bmatrix} \begin{bmatrix} \lambda \\ \xi \end{bmatrix} = \begin{bmatrix} u - v \\ 1 \\ 1 \end{bmatrix} \qquad (1)$$

$$\lambda \geqq 0 \qquad \xi \geqq 0$$

with $x = v + q^{\pi|n} s\lambda$. The matrix is unimodular. To see this, reverse the order of and sign of the last $n+1$ columns and reverse the sign of the last row, then each row has the form $(0, \ldots, 0, 1, \ldots, 1, 0, \ldots, 0)$, see Lemmas 2.1, 2.2, and 2.3. Thus, the extreme points of solutions to (1) are integral. Upon applying Lemma 2.5 twice and Lemma 2.4 twice, we see that $(v;\pi;n)$ and $(u;\gamma;n)$ meet in a common face. □

The n-simplex $(v;\pi;n)$ lies in the n-cube C_v^{v+e} of **C** and observe that $v^{n+1} = v+e$. Since there are $n!$ permutations of ν we have proved:

6.7 Lemma: There are $n!$ of the n-simplexes of **F** in each n-cube of **C**. □

From Lemma 6.5 we see that these $n!$ n-simplexes of **F** cover the cube. In the following theorem it is asserted that **F** is a triangulation of $\mathbb{C}^n$. We shall henceforth refer to **F** as the Freudenthal triangulation.

6.8 Lemma: **F** is a locally finite triangulation of $\mathbb{C}^n$.

Proof: $\mathbf{F}^n$ covers $\mathbb{C}^n$ and any two cells of $\mathbf{F}^n$ meet in a common face, see Lemmas 6.5 and 6.6. For local finiteness see Lemma 6.7. That each (n-1)-simplex of **F** lies in exactly two n-simplexes of **F** follows from $\dim \mathbf{F} = \dim \mathbb{C}^n$ and $\partial\mathbb{C}^n = \phi$, see the comments following the definition of a subdivision and Lemma 3.2. □

The following lemma gives one some sense of the overall regularity of F.

6.9 Lemma: **F** is a refinement of **O** and **C**.

Proof: The n-simplex $(v;\pi;n)$ is contained in the cube $C_v^{v+e} = \{x : v \leqq x \leqq v + e\}$ thus **F** refines **C** which, in turn refines **O**. □

The next lemma is obvious, but helpful.

6.10 Lemma: The nonzero differences $v^k - v^j$ between vertices of $(v;\pi;n)$ in $\mathbf{F}^n$ are all distinct.

Proof: If $k > j$ then $v^k - v^j = \sum_{i=j}^{k-1} q^{\pi i} \geq 0$, but the sum of columns of q^ν is uniquely composed. □

From Lemma 6.3 we see that each element of σ of $\mathbf{F}^n$ is uniquely represented by $(v;\pi;n)$ for some (v,π,n) in $\mathbf{iF}^n$. Having established the representation of $\mathbf{F}^n$ we turn to explore the replacement rules.

Given (v,π,n) in $\mathbf{iF}^n$ what are the adjacent simplexes $(\hat{v};\hat{\pi};n)$ of $(v;\pi;n)$, that is, what n-simplexes $(\hat{v};\hat{\pi};n) \neq (v;\pi;n)$ share an (n-1)-face with $(v;\pi;n)$. There are $n+1$ such simplexes corresponding to dropping each of the vertices v^t. We have three cases corresponding to $t = 1$, $2 \leqq t \leqq n$, and $t = n+1$ where v^t is the vertex to be dropped.

Case $t = 1$: Let $\hat{v} = v^2$ and $\hat{\pi} = (\pi 2, \ldots, \pi n, \pi 1, \pi(n+1))$, from Lemma 6.3 we see that $(\hat{v};\hat{\pi};n) \neq (v;\pi;n)$. Further, it is clear that $(\hat{v}^1, \ldots, \hat{v}^n) = (v^2, \ldots, v^{n+1})$. Thus, $(\hat{v};\hat{\pi};n)$ does not contain v^1 and $(\hat{v};\hat{\pi};n)$ and $(v;\pi;n)$ share a common (n-1)-face, namely, the (n-1)-simplex with vertices $(v^2, \ldots, v^{n+1})$.

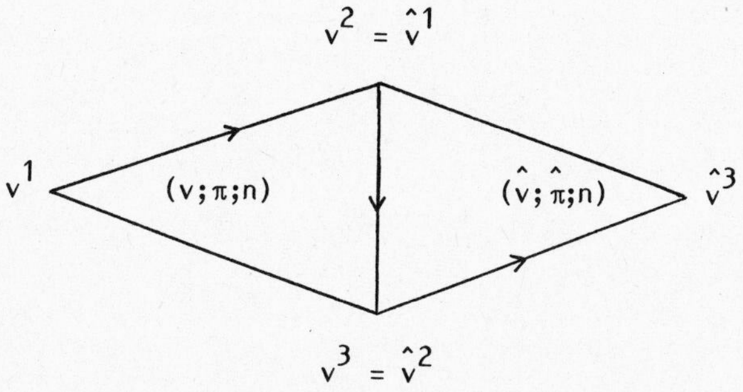

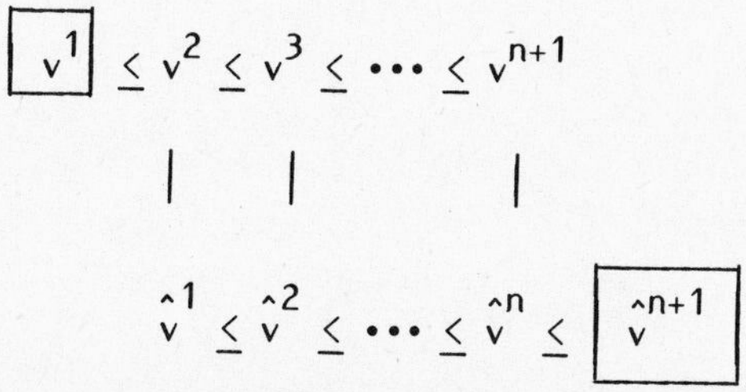

Note that since $\mathbf{F}$ is a subdivision of $\mathbb{C}^n$ we have $(\hat{v};\hat{\pi};n)$ as the unique element of $\mathbf{F}^n$ other than $(v;\pi;n)$ that has among its vertices $v^2, \ldots, v^{n+1}$. □

Case $t = 2, \ldots, n$: Let $\hat{v} = v$ and $\hat{\pi} = (\pi 1, \ldots, \pi(t-2), \pi t, \pi(t-1), \pi(t+1), \ldots, \pi n, \pi(n+1))$. From Lemma 6.3 we see that $(\hat{v};\hat{\pi};n) \neq (v;\pi;n)$. Further, it is clear that $(\hat{v}^1, \ldots, \hat{v}^{t-1}, \hat{v}^{t+1}, \ldots, \hat{v}^{n+1}) = (v^1, \ldots, v^{t-1}, v^{t+1}, \ldots, v^{n+1})$. Thus, $(\hat{v};\hat{\pi};n)$ does not contain v^t and $(\hat{v};\hat{\pi};n)$ and $(v;\pi;n)$ share a common $(n-1)$-face, namely, the $(n-1)$-simplex with vertices $(v^1, \ldots, v^{t-1}, v^{t+1}, \ldots, v^n)$.

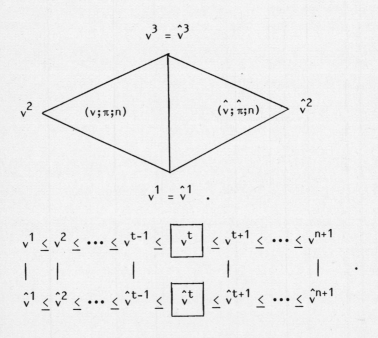

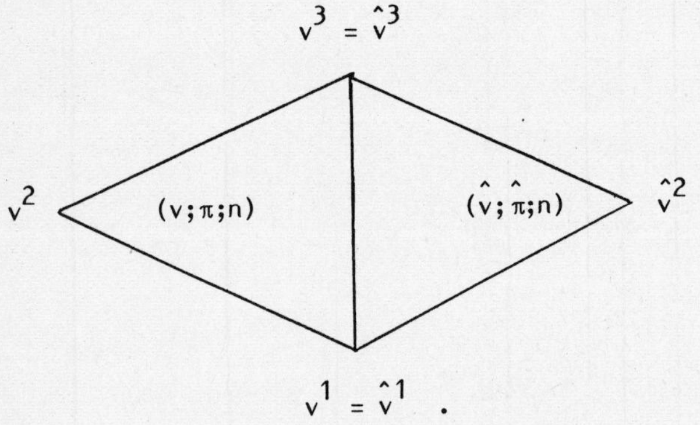

$$\begin{array}{ccccccccccccc}
v^1 & \le & v^2 & \le & \cdots & \le & v^{t-1} & \le & \boxed{v^t} & \le & v^{t+1} & \le \cdots \le & v^{n+1} \\
| & & | & & & & | & & & & | & & | \\
\hat{v}^1 & \le & \hat{v}^2 & \le & \cdots & \le & \hat{v}^{t-1} & \le & \boxed{\hat{v}^t} & \le & \hat{v}^{t+1} & \le \cdots \le & \hat{v}^{n+1}
\end{array}.$$

Again, since $\mathbf{F}$ is a subdivision of $\mathbb{C}^n$ we have $(\hat{v};\hat{\pi};n)$ as the unique element of $\mathbf{F}^n \backslash (v;\pi;n)$ that has among its vertices $(v^1, \ldots, v^{t-1}, v^{t+1}, \ldots, v^{n+1})$. □

Case $t = n+1$: Let $\hat{v} = v - e^{\pi n}$ and $\hat{\pi} = (\pi n, \pi 1, \ldots, \pi(n-1), \pi(n+1))$. From Lemma 6.3 we see that $(\hat{v};\hat{\pi};n) \neq (v;\pi;n)$. Further, it is clear that $(\hat{v}^2, \ldots, \hat{v}^{n+1}) = (v^1, \ldots, v^n)$. Thus, $(\hat{v};\hat{\pi};n)$ does not contain v^{n+1} and $(\hat{v};\hat{\pi};n)$ and $(v;\pi;n)$ share a common $(n-1)$-face, namely, the $(n-1)$-simplex with vertices $(v^1, \ldots, v^{n-1})$.

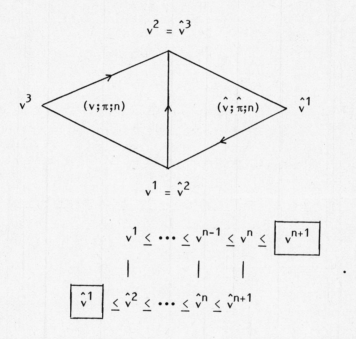

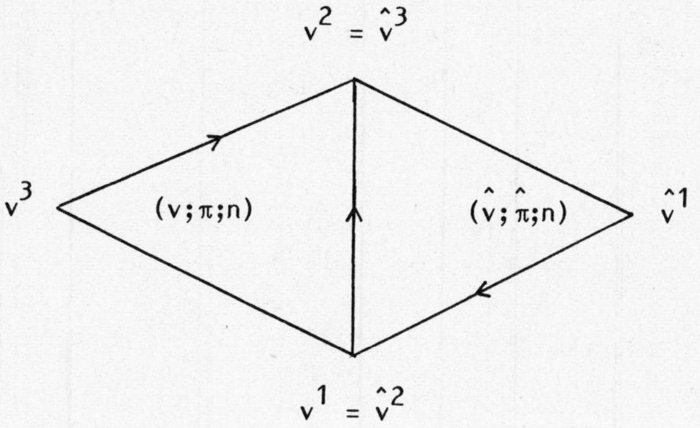

$$\begin{array}{cccccccc} & v^1 & \le \cdots \le & v^{n-1} & \le & v^n & \le & \boxed{v^{n+1}} \\ & | & & | & & | & & \\ \boxed{\hat{v}^1} & \le \hat{v}^2 & \le \cdots \le & \hat{v}^n & \le & \hat{v}^{n+1} & & \end{array} .$$

Finally, since $\mathbf{F}$ is a subdivision of $\mathbb{C}^n$ we have $(\hat{v};\hat{\pi};n)$ as the unique element of $\mathbf{F}^n \backslash (v;\pi;n)$ with vertices $(v^1, \ldots, v^n)$.

We summarize the foregoing replacement rules for $\mathbf{F}$ in the chart of Figure 6.4. At STOP, $(\hat{v};\hat{\pi};n)$ is adjacent to $(v;\pi;n)$ but does not contain v^t. The adjoined vertex $\{\hat{v}^1, \ldots, \hat{v}^{n+1}\} \backslash \{v^1, \ldots, v^{n+1}\}$ and the correspondence between vertices of $(v;\pi;n)$ and of $(\hat{v};\hat{\pi};n)$ are indicated in the notes. □

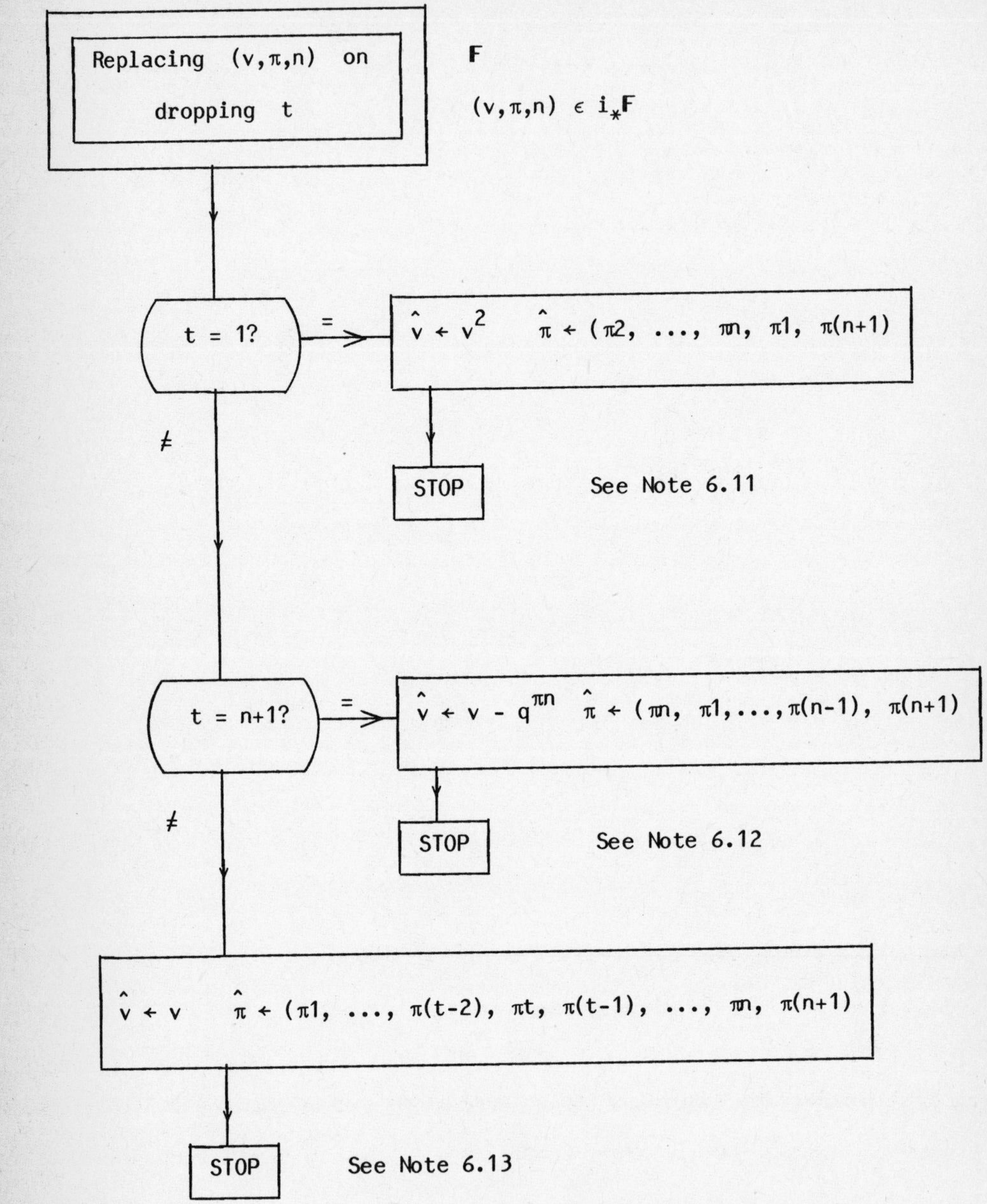
Replacing (v,π,n) on dropping t
F
(v,π,n) ∈ i*F
t = 1?
=
v̂ ← v²   π̂ ← (π2, ..., πn, π1, π(n+1))
≠
STOP
See Note 6.11
t = n+1?
=
v̂ ← v - q^πn   π̂ ← (πn, π1,...,π(n-1), π(n+1))
≠
STOP
See Note 6.12
v̂ ← v   π̂ ← (π1, ..., π(t-2), πt, π(t-1), ..., πn, π(n+1))
STOP
See Note 6.13

Figure 6.4

6.11 Note (for Figure 6.4): The adjoined vertex is $\hat{v}^{n+1} = v^{n+1} + e^{\pi 1}$. The correspondence between vertices of $(v;\pi;n)$ and $(\hat{v};\hat{\pi};n)$ is

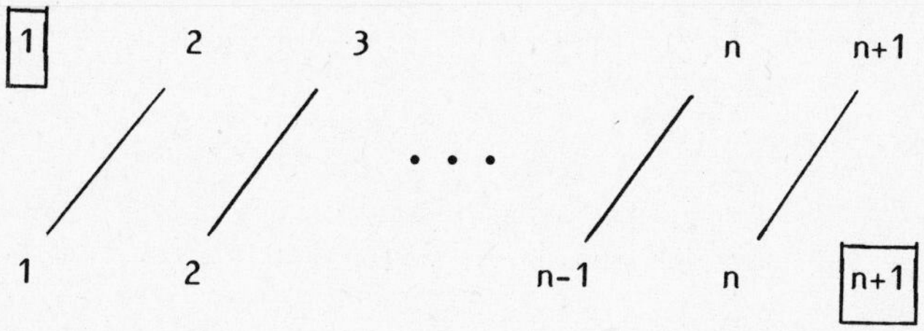

That is, $v^1 = \hat{v}^{i-1}$ for $i = 2, \ldots, n+1$. The squares indicate the dropped and adjoined vertices. □

6.12 Note (for Figure 6.4): The adjoined vertex is $\hat{v}^1 = v - q^{\pi n}$. The correspondence between vertices of $(v;\pi;n)$ and $(\hat{v};\hat{\pi};n)$ is

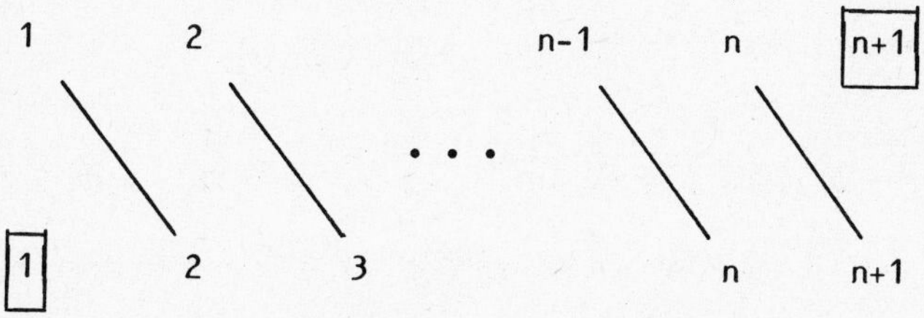

. □

6.13 Note (for Figure 6.4): The adjoined vertex is $\hat{v}^t = v^{t-1} + q^{\pi t}$. The correspondence between vertices of $(v;\pi;n)$ and $(\hat{v};\hat{\pi};n)$ is

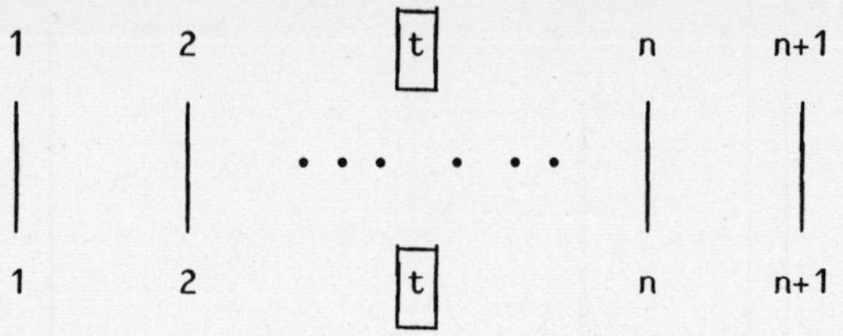

. □

6.14 Example: Suppose for $n = 5$, $v = (1,0,3,-1,2)$, $\pi = (1,5,3,2,4,6)$, and that vertex v^6 is dropped. What is the replacement simplex? Following the chart we see that $\hat{v} = (1,0,3,-2,2)$ and $\hat{\pi} = (4,1,5,3,2,6)$. It is quite trivial to execute! □

6.15 Example: Beginning with simplex 1, $(v;\pi;2)$, where $v = (-2,-1)$ and $\pi = (1,2,3)$ in Figure 6.5, the chart of Figure 6.4 is used to generate the n-simplexes along the path. The vertex v^t to be dropped in order to follow the path must be determined by inspection. This manuscript does not consider the problem of which vertex to drop. □

6.16 Exercise: Given the 5-simplex $(v;\pi;5)$ in $\mathbb{C}^5$ with $v = (1,10,-1,0,1)$ and $\pi = (1,5,4,2,3,6)$, what is the n-simplex $(\hat{v};\hat{\pi};n)$ adjacent to (v,π,n) but does not contain v^3. □

6.17 Exercise: Draw a path in Figure 6.5 that does not meet any vertices. Then generate the $(v;\pi;2)$ encountered along the path. Use the chart the first time and without assistance the second time. □

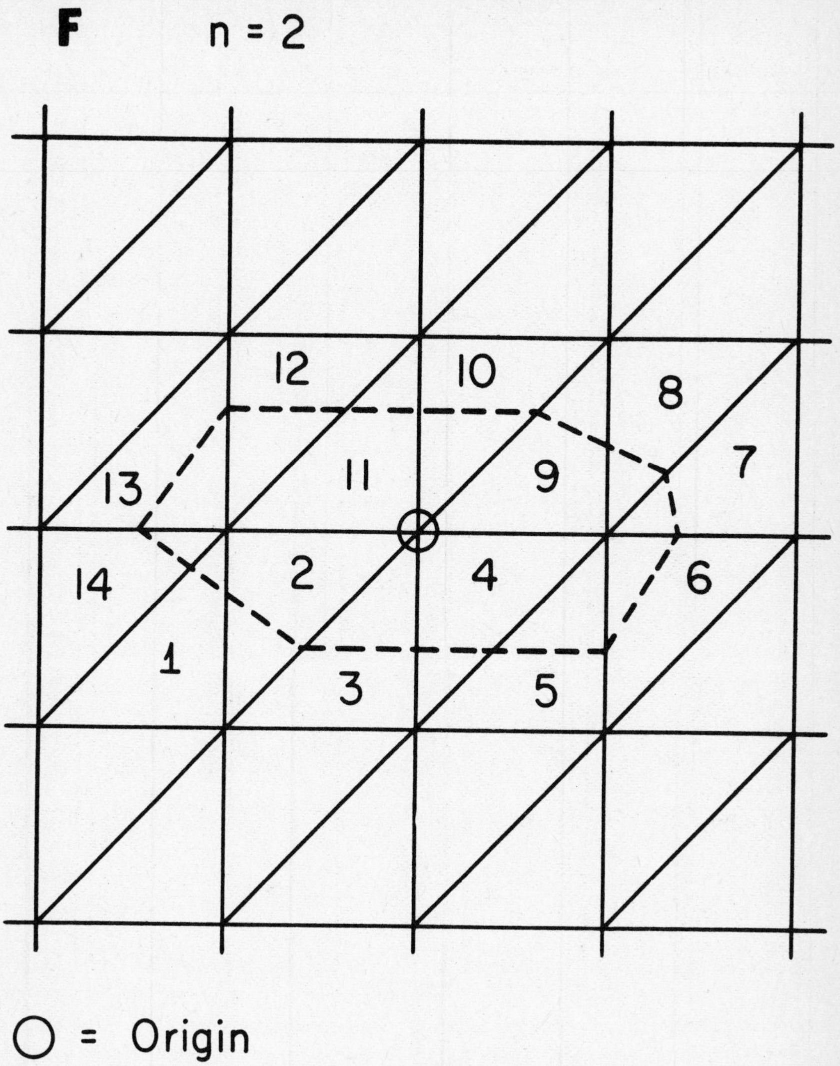

Figure 6.5

Simplex i	v_1	v_2	π_1	π_2	t
1	-2	-1	1	2	1
2	-1	-1	2	1	2
3	-1	-1	1	2	1
4	0	-1	2	1	2
5	0	-1	1	2	1
6	1	-1	2	1	1
7	1	0	1	2	2
8	1	0	2	1	3
9	0	0	1	2	2
10	0	0	2	1	3
11	-1	0	1	2	2
12	-1	0	2	1	3
13	-2	0	1	2	3
14	-2	-1	2	1	3

Figure 6.6

When we must distinguish between the Freudenthal triangulation of $\mathbb{C}^n$ and $\mathbb{C}^{n+1}$, etc., we shall employ the notation $\mathbf{F}_n$ and $\mathbf{F}_{n+1}$, etc. The next lemma states a relationship between $\mathbf{F}_n$ and $\mathbf{F}_{n+1}$.

6.18 Lemma: The natural restriction of $\mathbf{F}_{n+1}$ to $\mathbb{C}^n \times h$ is $\mathbf{F}_n \times h$ for $h = 0, \pm 1, \pm 2, \ldots$.

Proof: Clearly, $\mathbf{F}_n \times h$ subdivides $\mathbb{C}^n \times h$. For $(v;\pi;n) \in \mathbf{F}_n$ let $\bar{v} = (v,h)$ and $\bar{\pi} = (\pi, n+2)$, and we have $(\bar{v}, \bar{\pi}, n+1) \in i\mathbf{F}_{n+1}^{n+1}$ and $(\bar{v}; \bar{\pi}; n+1) \cap (\mathbb{C}^n \times h) = (v;\pi;n) \times h$. As $\mathbf{F}_{n+1}|\mathbb{C}^n \times h \supseteq \mathbf{F}_n \times h$ we see that $\mathbf{F}_{n+1}|\mathbb{C}^n \times h = \mathbf{F}_n \times h$, see Lemma 3.7. □

Although the previous lemma focuses on the $(n+1)^{th}$ coordinate, the result clearly can be extended to any coordinate $i = 1, \ldots, n+1$.

The $\|\|_1$, $\|\|_2$, and $\|\|_\infty$ grid sizes of $\mathbf{F}$ are easily seen to be n, $\sqrt{n}$, and 1 respectively; note that these grid sizes obtain for $\mathbf{C}$ as well.

For a one-to-one and onto $\mathbf{F}$-PL map $h : \mathbb{C}^n \to \mathbb{C}^n$, of course, $h(\mathbf{F})$ is also a triangulation of $\mathbb{C}^n$. Direct representation and replacement rules of $h(\mathbf{F})$ are quite easy to establish, however, this is not the approach one should take for $h(\mathbf{F})$. Instead, one should continue to use the representation set, facet rule, and replacement rules of $\mathbf{F}$ and when an actual vertex in $h(\mathbf{F})$ is needed one uses $h(v^i)$ where v^i is a vertex of a simplex in $\mathbf{F}$. There is a very important advantage in this line of reasoning. In Chapter 9 this concept is expanded and termed "squeeze and shear". For the moment, however, we point out that since $\mathbf{F}^0 = \mathbb{Z}^n$ there are no numerical errors in applying the representation and replacement rules of $\mathbf{F}$. Of course, the evaluation $h(v^i)$ may suffer numerical errors, but this is unavoidable.

The next sequence of lemmas deals with linear transformations of $\mathbf{F}$ that refine or encoarse $\mathbf{F}$.

6.19 Lemma: Let r be -1 or $+1$ and p be integral, then $r\mathbf{F}+p = \mathbf{F}$.

Proof: As $-(v,\pi,n) = (w,\gamma,n)$ where $w = -v^{n+1}$ and $\gamma = (\pi n, \ldots, \pi 1(n+1))$ we have $-\mathbf{F} = \mathbf{F}$. Since $(v,\pi,n)+p = (v+p,\pi,n)$ we have $\mathbf{F}+p = \mathbf{F}$. □

6.20 Lemma: Let r be a nonzero integer and p an integral vector. Then $\mathbf{F}$ is a refinement of $r\mathbf{F}+p$.

Proof: Since $\mathbf{F}-p = \mathbf{F}$ and $-\mathbf{F} = \mathbf{F}$ we need only show that $\mathbf{F}$ refines $r\mathbf{F}$ where r is positive. Thus, we need to show that there is a simplex $r(u;\gamma;n)$ containing $(v;\pi;n)$. Let y be any interior point of $(v;\pi;n)$ and let $r(u;\gamma;n)$ be a simplex of $r\mathbf{F}$ containing y.

A point x is in $(v;\pi;n) \cap r(u;\gamma;n)$ if and only if there is a solution (λ,η) to

$$\begin{bmatrix} q^{\pi|n}s & -q^{\gamma|n}s \\ e & 0 \\ 0 & e \end{bmatrix} \begin{bmatrix} \lambda \\ \eta \end{bmatrix} = \begin{bmatrix} ru - v \\ 1 \\ r \end{bmatrix}$$

$$\lambda \geqq 0 \qquad \eta \geqq 0$$

with $x = v+q^{\pi|n} s\lambda = ru+q^{\gamma|n} s\eta$, see Lemma 6.2. But as in Lemma 6.6 the matrix is unimodular. In view of y there is a solution (λ,η) to the system with $\lambda > 0$. Consequently, there is a solution (λ^i,η^i) to the system with $\lambda^i_i = 1$ for i in μ, and we have $(v;\pi;n)$ in $r(u;\gamma;n)$, see Lemmas 2.1 through 2.5. □

Later in this section we use Lemma 6.20 for constructing an algorithm for generating the $(w;\gamma;n)$ of the previous proof. This procedure is required in Section 17.

As a consequence of Lemma 6.20 we have that $r^{-1}\mathbf{F} - r^{-1}p$, a refinement of $\mathbf{F}$ where r is a nonzero integer and p is integral.

6.21 Lemma: Let r be a positive integer and p and integral vector in $\mathbb{G}^n$, then $\mathbf{F}_{n+1}|((rS+p) \times [0,1])$ subdivides $(rS+p) \times [0,1]$.

Proof: Let (x,t) be an interior point of $(rS+p) \times [0,1]$ and let $(v;\pi;n+1) \in \mathbf{F}_{n+1}$ be an $(n+1)$-simplex containing (x,t). Let $u = (v_1,\ldots,v_n)$ and $\gamma = (\pi 1, \ldots, \pi(k-1), \pi(k+1), \ldots, \pi(n+1), \pi k)$ where $\pi k = n+1$. Then $(u;\gamma;n) \in \mathbf{F}_n$ and $(u;\gamma;n)$ contains x which is interior to $rS+p$ in

$\mathbb{C}^n$. Thus, $(u;\gamma;n) \subseteq rS+p$ by Lemma 6.20, and then $(v;\pi;n+1) \subseteq (u;\gamma;n) \times [0,1] \subseteq (rS+p) \times [0,1]$, and the result follows. □

The next lemma will play an important role in the construction of subdivision **V** of Section 15. The subdivision **Q** is that of Section 5. Recall $q = (I, -e)$ and $Q_\alpha = \{q^\alpha x : x \geqq 0)\}$ for $\alpha \subseteq \mu$.

6.22 Lemma: Let r be a nonzero scalar and p an integral vector. Then $r\mathbf{F} + rp$ is a refinement of **Q**.

Proof: Since $-\mathbf{F} = \mathbf{F}$ we may assume $r > 0$. Since $r^{-1}\mathbf{Q} = \mathbf{Q}$ and $\mathbf{F}+p = \mathbf{F}$ it is sufficient to prove that **F** is a refinement of **Q**. We must show that each n-simplex $(v;\pi;n)$ in **F** lies entirely in some cone Q_α of **Q**. Let y be an interior point of $(v;\pi;n)$ and let Q_α be an n-cone of **Q** containing y. A point x is in $(v;\pi;n) \cap Q_\alpha$ if and only if there is a solution (λ,ξ) to

$$\begin{bmatrix} q^{\pi|n}s & -q^\alpha \\ e & 0 \end{bmatrix} \begin{bmatrix} \lambda \\ \xi \end{bmatrix} = \begin{bmatrix} -v \\ 1 \end{bmatrix}$$

$$\lambda \geqq 0 \qquad \xi \geqq 0 \quad .$$

However, after reversing the order and changing the sign of the columns of $-q^\alpha$, the matrix of the system has form

$$\begin{bmatrix} & I & 0 \\ A\ , & & \\ & 0 & 0 \end{bmatrix}$$

which is unimodular since A is unimodular, see Lemma 2.1, 2.2, and 2.3. In view of y there is a solution with $\lambda > 0$ and, hence, a solution λ^i with $\lambda^i_i = 1$ for $i \in \mu$. Consequently, $(v;\pi;n)$ is in Q_α, see Lemmas 2.4 and 2.5. □

Later we shall develop an algorithm for generating α with Q_α containing $(v;\pi;n)$; the task is much easier knowing that such an α exists. This procedure will be required in Section 17.

6.23 Lemma: Let σ and τ be two n-simplexes of **F** and $h : \mathbb{C}^n \to \mathbb{C}^n$ a linear map with $h(\sigma) = \tau$. Then $h(\mathbb{Z}^n) = \mathbb{Z}^n$, that is, $h(x) = Ax + \alpha$ where A is integral and has a determinant of ± 1 and α is integral.

Proof: Let $\sigma = (v;\pi;n)$ and $\tau = (u;\gamma;n)$. For any z in $\mathbb{Z}^n$ we have

$$z = v^1 + \sum_{i \in \nu} (v^{i+1} - v^i)x_i$$

where the x_i are integral. Summing the x_i we obtain integral λ with $z = V\lambda$, $e\lambda = 1$, and $V = (v^1, \ldots, v^{n+1})$. Now $h(V\lambda) = h(V)\lambda$ where

$h(V) = (h(v^1), \ldots, h(v^{n+1}))$ is integral and, thus, $h(V)\lambda = h(z)$ is integral. On the other hand, for z in $\mathbb{Z}^n$ we have $z = U\lambda$ where $U = (u^1, \ldots, u^{n+1})$, λ is integral, $e\lambda = 1$. Thus, $h^{-1}(z) = h^{-1}(U\lambda) = h^{-1}(U)\lambda$ which is integral since $h^{-1}(U) \triangleq (h^{-1}(u^1), \ldots,. h^{-1}(u^{n+1}))$ is integral. □

6.24 Remark: One might suspect from the above lemma that $h(\mathbf{F}) = \mathbf{F}$ would also be a proper conclusion, however, as the example below shows this would be quite improper. Let $h(x) = Ax$ where

$$A \triangleq \begin{bmatrix} 1 & 0 & 0 \\ 1 & -1 & 1 \\ 0 & 0 & 1 \end{bmatrix}$$

and observe that $A(v;\pi;n) = (v;\pi;n)$ and $A(u;\gamma;n) \notin \mathbf{F}$ for $v = u = 0$, $\pi = (1,2,3,4)$ and $\gamma = (3,1,2,4)$. □

After introducing alternative representation rules for $\mathbf{F}$ in Section 10 we shall state additional conditions in h for which $h(\mathbf{F}) = \mathbf{F}$ and $h(\mathbf{Q}) - h(0) = \mathbf{Q}$. For the present we continue by describing efficient procedures for finding an element of $\mathbf{F}$ and $\mathbf{Q}$ containing $r^{-1}(v;\pi;n)$ where r is a positive integer. These procedures will become part of the replacement rules in the triangulation $\mathbf{S}$ of Section 17. Lemmas 6.20 and 6.22 insure that such an element of $\mathbf{F}$ and $\mathbf{Q}$ exists; they also indicate that step one in computing such an element is to obtain an interior point of $r^{-1}(v;\pi;n)$. Although an arbitrary interior point would suffice, the next lemma indicates an especially convenient one. The entity ω is a a positive infinitesimal (that is, positive and arbitrarily small); let $\omega^{\nu} = (\omega^1,\ldots,\omega^n)$ where $\nu = \{1, \ldots, n\}$ and ω^i is the ith power of ω.

6.25 Lemma: For positive scalar r the point

$$r^{-1}v + q^{\pi|n}\ \omega^{\nu}$$

is interior to $r^{-1}(v;\pi;n)$.

Proof: We show that the barycentric coordinates are positive. Consider

$$r^{-1}v + q^{\pi|n}\ \omega^{\nu} = r^{-1}\sum_{i\epsilon\mu} v^i\lambda_i$$

with $\sum_{i\epsilon\mu}\lambda_i = 1$, or $rq^{\pi|n}\ \omega^{\nu} = q^{\pi|n}\ s\lambda$ with $e\lambda = 1$. Thus, $r\omega^{\nu} = s\lambda$ or $\lambda_{n+1} = r\omega^n$, $\lambda_i = r(\omega^{i-1} - \omega^i)$ for $i = 2, \ldots, n$, $\lambda_1 = 1 - \sum_{i=2}^{n+1}\lambda_i$, and $\lambda > 0$. □

Equivalently, one can think of $r^{-1}v + q^{\pi|n}\ \varepsilon^{\nu}$ as being interior to $r^{-1}(v;\pi;n)$ for all small positive ε.

To find a simplex $(u;\gamma;n)$ containing $r^{-1}(v;\pi;n)$ where r is a positive integer we need only find a $(u;\gamma;n)$ which contains the point $r^{-1}v + q^{\pi|n}\ \omega^{\nu}$ for consider Lemma 6.20. To find $(u;\gamma;n)$ containing the point $r^{-1}v + q^{\pi|n}\ \omega^{\nu}$ we apply the reasoning of Lemma 6.5. First let

$$u = \lfloor r^{-1}v + q^{\pi|n}\ \omega^{\nu} \rfloor$$

which is equivalent to letting $u = [r^{-1}v]$. Next for

$$d = r^{-1}v - u + q^{\pi|n}\ \omega^{\nu}$$

we construct γ with $d_{\gamma i} > d_{\gamma(i+1)}$ for $i = 1,\ldots,n-1$ and $\gamma(n+1) = n+1$. From Lemma 4.2 we have

$$(q^{\pi|n}\ \omega^{\nu})_i = \omega^{\pi^{-1}(i)}$$

for i in ν. Thus, equivalently, conceptually, but more efficiently, we may obtain γ by permuting $(v_i - ru_i, - \pi^{-1}(i))$ for i in ν into a lexico decreasing order. The chart summarizing this process is given in Figure 6.7.

In view of Lemma 6.22, to find $Q_{\mu|\ell}$ with contains $(v;\pi;n)$ we again need only find $Q_{\mu|\ell}$ which contains the interior point $v + q^{\pi|n}\ \omega^{\nu}$, or equivalently, to find a solution x to

$$q^{x} = v + q^{\pi|n}\ \omega^{\nu}$$

$$0 \nleq x \geqq 0 \ .$$

Following Figure 5.2, first select the index j of the smallest element of $v + q^{\pi|n}\ \omega^{\nu}$, or equivalently, the index $j \in \nu$ of the lexico smallest element of $(v_i, -\pi^{-1}i)$. If $v_j \geqq 0$ let $\ell = n+1$ and if $v_j < 0$ let $\ell = j$ and we have $Q_{\mu|\ell}$ containing $(v;\pi;n)$, see Figure 6.8.

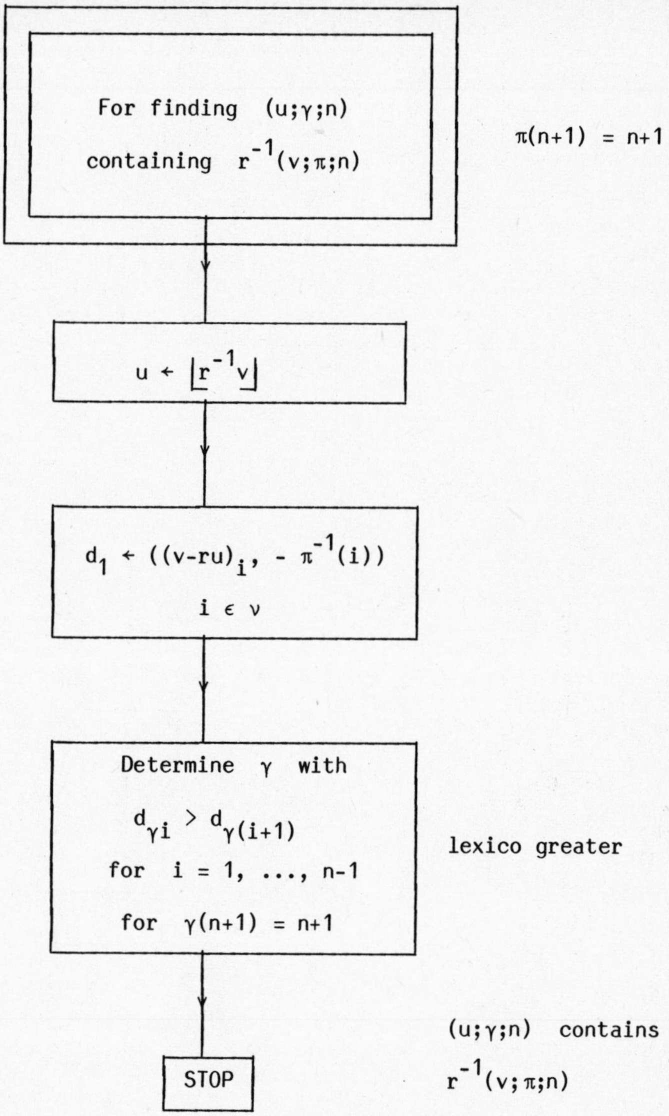
For finding (u;γ;n)
containing r⁻¹(v;π;n)
π(n+1) = n+1
u ← ⌊r⁻¹v⌋
d₁ ← ((v-ru)ᵢ, - π⁻¹(i))
i ∈ ν
Determine γ with
d_γi > d_γ(i+1)
for i = 1, ..., n-1
for γ(n+1) = n+1
lexico greater
(u;γ;n) contains
r⁻¹(v;π;n)
STOP

Figure 6.7

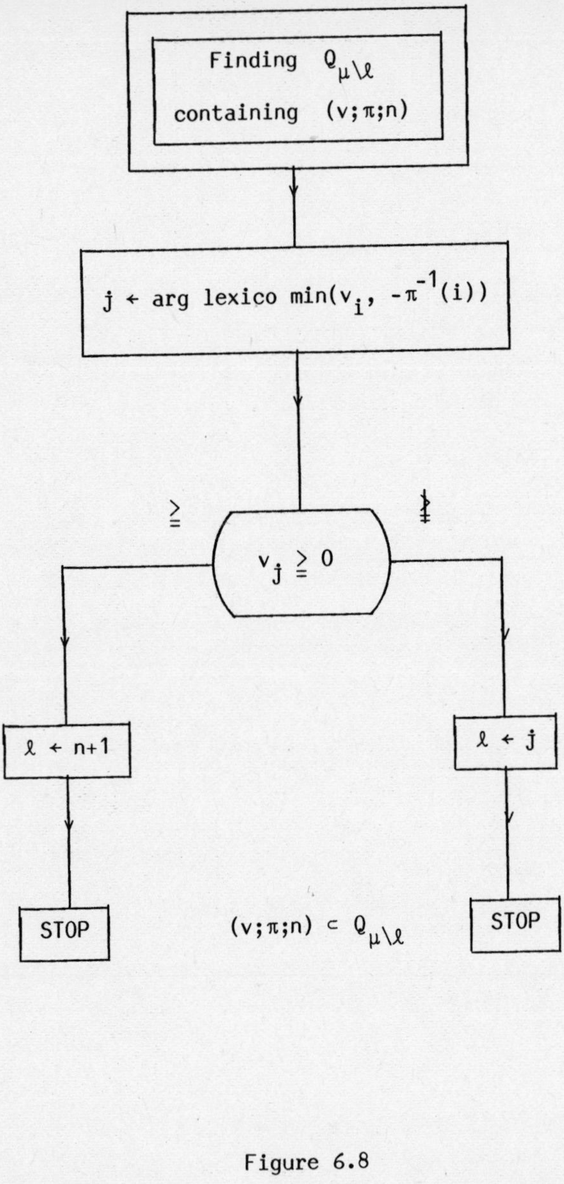
Finding $Q_{\mu\backslash\ell}$
containing $(v;\pi;n)$
$j \leftarrow \arg \text{lexico} \min(v_i, -\pi^{-1}(i))$
$\geqq$
$v_j \geqq 0$
$\ngeqq$
$\ell \leftarrow n+1$
$\ell \leftarrow j$
STOP
$(v;\pi;n) \subset Q_{\mu\backslash\ell}$
STOP

Figure 6.8

Recall that a matrix P is orthogonal if $PP^T = I$, the identity. Action as $x \to Px+p$ where P is orthogonal is like "rigid" motion. One view of this is that

$$\|x-y\|_2 = \|(Px+p) - (Py+p)\|_2$$

for all x and y, which is, say, that distances are univariant under the action.

Two cells σ and τ are said to be congruent or permutation congruent if there is an orthogonal matrix or permutation matrix P and vector p such that $P\sigma+p = \tau$, respectively. We define a subdivision of $\mathbb{C}^n$ to be congruent or permutation congruent if any two n-cells are congruent or permutation congruent, respectively.

6.26 Proposition: The triangulation $\mathbf{F}$ of $\mathbb{C}^n$ is permutation congruent.

Proof: Consider $\sigma = (v;\pi;n)$ and $\tau = (u;\gamma;n)$. The orthogonal transformation

$$x \to q^{\pi|n}(q^{\gamma|n})^{-1}\,(x-u) + v$$

is seen to carry the vertices $u + q^{\gamma|n}\,s^i$ of τ to the vertices $v + q^{\pi|n}\,s^i$ of σ. □

Obviously $A\mathbf{F} + a$ is congruent if A is orthogonal, however, in general $A\mathbf{F} + a$ is not.

6.27 Exercise: Show $A(0;\pi;4)$ and $A(0,\gamma;4)$ are not congruent where $\pi = (1,2,3,4)$ and $\gamma = (1,3,2,4)$ and

$$A = \begin{bmatrix} 1 & -1 & 0 \\ 0 & 1 & -1 \\ 0 & 0 & 1 \end{bmatrix}.$$

□

6.28 Bibliographical Notes

The triangulation **F** was introduced in Freudenthal [1942]; see Bibliographical Notes 1.1. □

7. SANDWICH TRIANGULATION $\mathbf{F}|\mathbb{C}^{n-1} \times [0,1]$

Let $\mathbf{C}$ and $\mathbf{F}$ be the cubical subdivision and the Freudenthal triangulation of $\mathbb{C}^n$, respectively. Obviously, the natural restriction of $\mathbf{C}$ to $\mathbb{C}^{n-1} \times [0,1]$ subdivides $\mathbb{C}^{n-1} \times [0,1]$. As $\mathbf{F}$ is a refinement of $\mathbf{C}$, see Lemma 6.9, we see that the natural restriction $\mathbf{F}|(\mathbb{C}^{n-1} \times [0,1]) \triangleq \mathbf{F}|\mathbb{C}^{n-1} \times [0,1]$ of $\mathbf{F}$ to $\mathbb{C}^{n-1} \times [0,1]$ triangulates $\mathbb{C}^{n-1} \times [0,1]$, see Figure 7.1. We briefly develop the representation and replacement rules for $(\mathbf{F}|\mathbb{C}^{n-1} \times [0,1])^n$; these rules, as one would expect, are very closely related to those of $\mathbf{F}^n$ as treated in the previous section.

The representation set $i(\mathbf{F}|\mathbb{C}^{n-1} \times [0,1])^n$ is a subset of the representation set $i\mathbf{F}^n$ of $\mathbf{F}^n$, namely, $i(\mathbf{F}|\mathbb{C}^{n-1} \times [0,1])^n$ is the set of all $(v,\pi,n) \in i\mathbf{F}^n$ such that $(v;\pi;n) \subseteq \mathbb{C}^{n-1} \times [0,1]$. Uniqueness of representation follows from that of $\mathbf{F}$. For (v,π,n) in $i\mathbf{F}^n$ we have $v^1 \leq v^2 \leq \cdots \leq v^{n+1}$ and thus, we see that $(v;\pi;n) \subseteq \mathbb{C}^{n-1} \times [0,1]$ if and only if $v_n^1 = 0$ and $v_n^{n+1} = 1$. The representation rules for $\mathbf{F}|\mathbb{C}^{n-1} \times [0,1]$ are exactly as those for $\mathbf{F}$. The replacement rules, however, require a check for top and bottom exits, see Figure 7.1. In the case of a top or bottom exit, no replacement for (v,π,n) exists in $\mathbf{F}|\mathbb{C}^{n-1} \times [0,1]$.

If $(v;\pi;n)$ is an n-simplex in $(\mathbf{F}|\mathbb{C}^{n-1} \times [0,1])^n$ and vertex t is dropped, let $(\hat{v};\hat{\pi};n)$ be the replacement n-simplex in $\mathbf{F}$. The n-simplex $(\hat{v};\hat{\pi};n)$ is the replacement in $\mathbf{F}|\mathbb{C}^{n-1} \times [0,1]$ if and only if $(\hat{v};\hat{\pi};n)$ is in $\mathbb{C}^{n-1} \times [0,1]$, which is to say, $(\hat{v};\hat{\pi};n)$ is in $\mathbb{C}^{n-1} \times [0,1]$, if and only if the adjoined vertex v^* is in $\mathbb{C}^{n-1} \times [0,1]$.

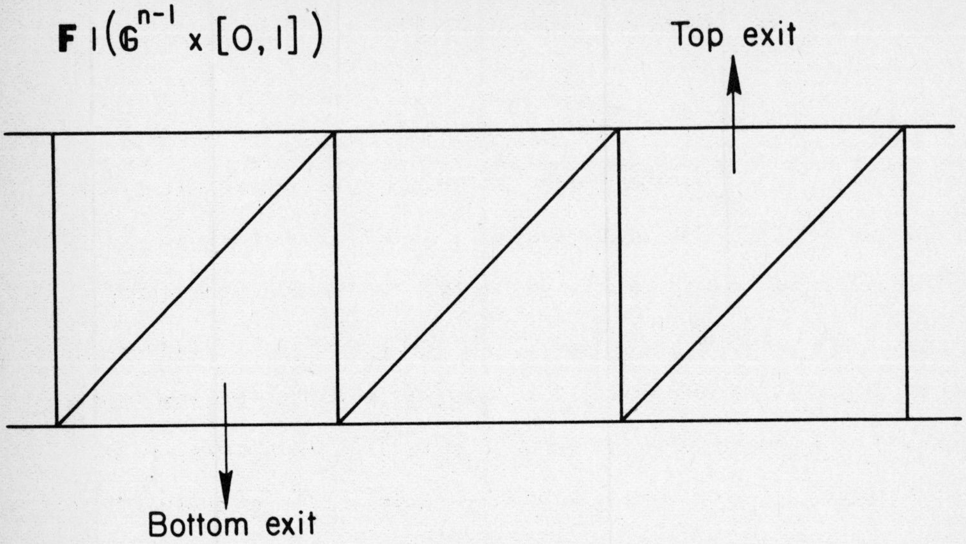

Figure 7.1

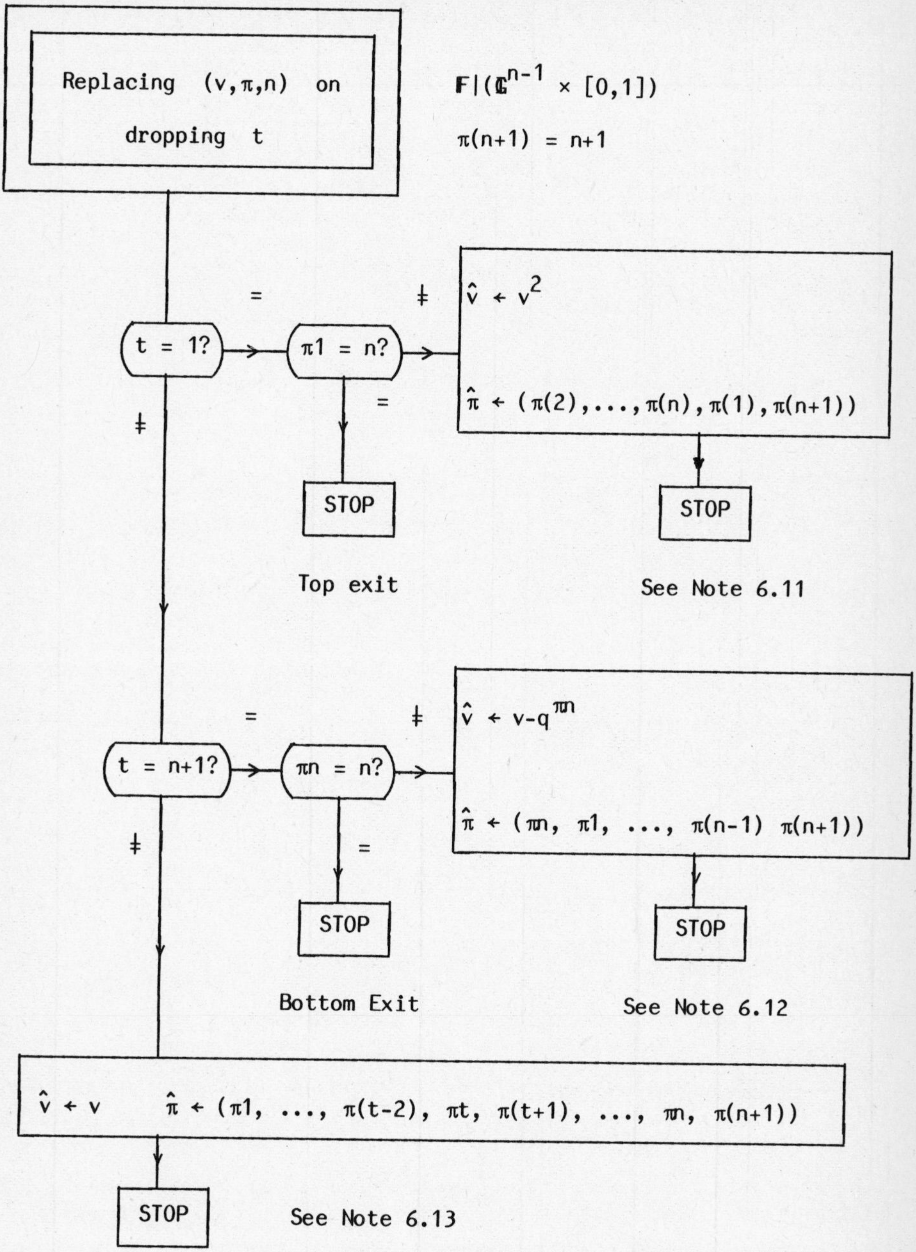

Replacing (v,π,n) on dropping t
F|(G^{n-1} × [0,1])
π(n+1) = n+1
t = 1?
=
π1 = n?
≠
v̂ ← v²
π̂ ← (π(2),...,π(n),π(1),π(n+1))
=
≠
STOP
STOP
Top exit
See Note 6.11
t = n+1?
=
πn = n?
≠
v̂ ← v-q^{πn}
π̂ ← (πn, π1, ..., π(n-1) π(n+1))
=
≠
STOP
STOP
Bottom Exit
See Note 6.12
v̂ ← v
π̂ ← (π1, ..., π(t-2), πt, π(t+1), ..., πn, π(n+1))
STOP
See Note 6.13

Figure 7.2

For $t = 1$, $2 \leqq t \leqq n$, and $t = n+1$ let us consider the adjoined vertex v^*. Recall $q^i = e^i$ for i in ν.

Case $t = 1$: The adjoined vertex is $\hat{v}^{n+1} = v^{n+1} + q^{\pi 1}$. Since $v_n^{n+1} \geqq 1$, clearly $\hat{v}_n^{n+1} \geqq 1$ and a bottom exit does not occur. Thus, we need only check if $\hat{v}_n^{n+1} = 2$, or equivalently, if $\pi 1 = n$. Assuming $\pi 1 = n$ we have a top exit and we observe that all vertices of $(v;\pi;n)$ except $v^t = v^1$ lie in $\mathbb{C}^{n-1} \times 1$, see Lemma 3.10. □

Case $2 \leqq t \leqq n$: Since $(v;\pi;n)$ is in $\mathbb{C}^n$, $v^1 = \hat{v}^1$, $v^{n+1} = \hat{v}^{n+1}$, and $\hat{v}^1 \underline{<} \cdots \underline{<} \hat{v}^{n+1}$ we see that $(\hat{v};\hat{\pi};n)$ is in $\mathbb{C}^{n-1} \times [0,1]$. Thus, in this case a top or bottom exit cannot occur. □

Case $t = n+1$: The adjoined vertex is $\hat{v}^1 = v^1 - q^{\pi n}$. As $v_n^1 = 0$, clearly $\hat{v}_n^1 \leqq 0$ and it is observed that a top exit cannot occur in this case. A bottom exit occurs if and only if $\hat{v}_n^1 = -1$, or equivalently, if $\pi n = n$. In the case of a bottom exit, all vertices of $(v;\pi;n)$ except v^t lie in $\mathbb{C}^{n-1} \times 0$. □

The replacement rules are summarized in Figure 7.2.

7.1 Bibliographical Notes: Triangulations as $F|\mathbb{C}^{n-1} \times [0,1]$ were used in Merrill [1972] and Kuhn and McKinnon [1975]. □

8. TRIANGULATION $\mathbf{F}|rS$

Letting r be a positive integer we know that the natural restriction $\mathbf{F}|rS$ of Freudenthal's triangulation $\mathbf{F}$ to rS triangulates rS where S is the standard simplex, see Lemma 6.20 and Figure 8.1. Our purpose here is simply to develop the representation and replacement rules for $\mathbf{F}|rS$.

Define $i(\mathbf{F}|rS)^n$ to be the collection of triples (v,π,n) in $i\mathbf{F}^n$ such that $(v;\pi;n)$ is in rS. Uniqueness of representation in $i(\mathbf{F}|rS)^n$ follows from the uniqueness in $i\mathbf{F}^n$. The representation rules are as in $\mathbf{F}$; let us proceed to the replacement rules.

Given the triple (v,π,n) in $i(\mathbf{F}|rS)^n$ suppose vertex v^t is to be dropped. In $\mathbf{F}$ let $(\hat{v};\hat{\pi};n)$ be the replacement for $(v;\pi;n)$ upon dropping v^t. In $\mathbf{F}|rS$ the replacement is also $(\hat{v};\hat{\pi};n)$ if and only if $(\hat{v};\hat{\pi};n)$ is in rS. If $(\hat{v};\hat{\pi};n)$ is not in rS, a replacement does not exist and we say that we have exited a side or facet of S, see Figure 8.2.

In the event of a side exit, all vertices of $(v;\pi;n)$ except v^t lie in some facet $S_{\mu\backslash i}$ of S where $S_\alpha \triangleq \text{cvx}\{s^i : i \in \alpha\}$ and we say that the replacement attempt has lead to a side exit through $rS_{\mu\backslash i}$, see Lemma 3.10 and Figure 8.2.

In order to test whether or not $(\hat{v};\hat{\pi};n)$ is in rS, or equivalently, whether or not an exit has occurred we employ barycentric coordinates.

Given a point x in $\mathbb{C}^n$, the barycentric coordinates λ of x with respect to S are defined by

$$\binom{s}{e}\lambda = \binom{x}{1}$$

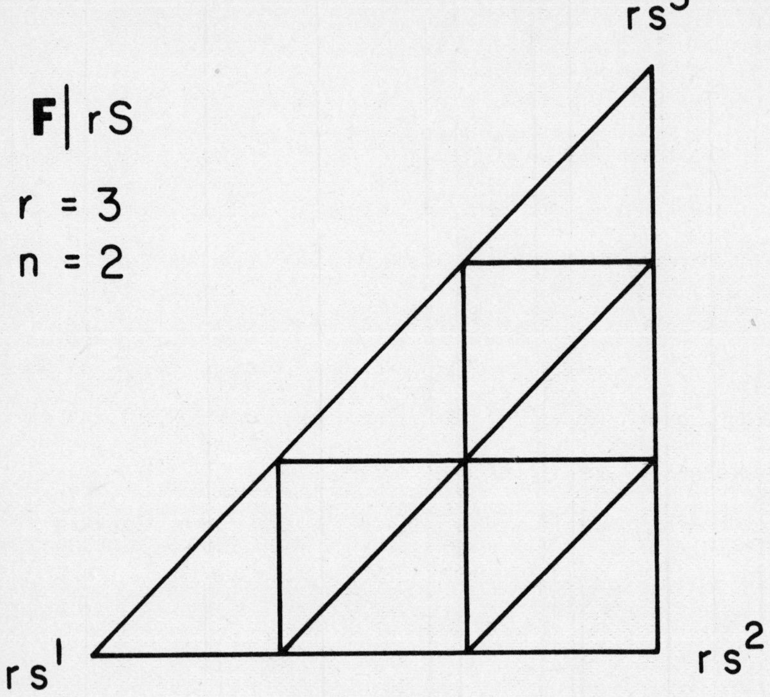

Figure 8.1

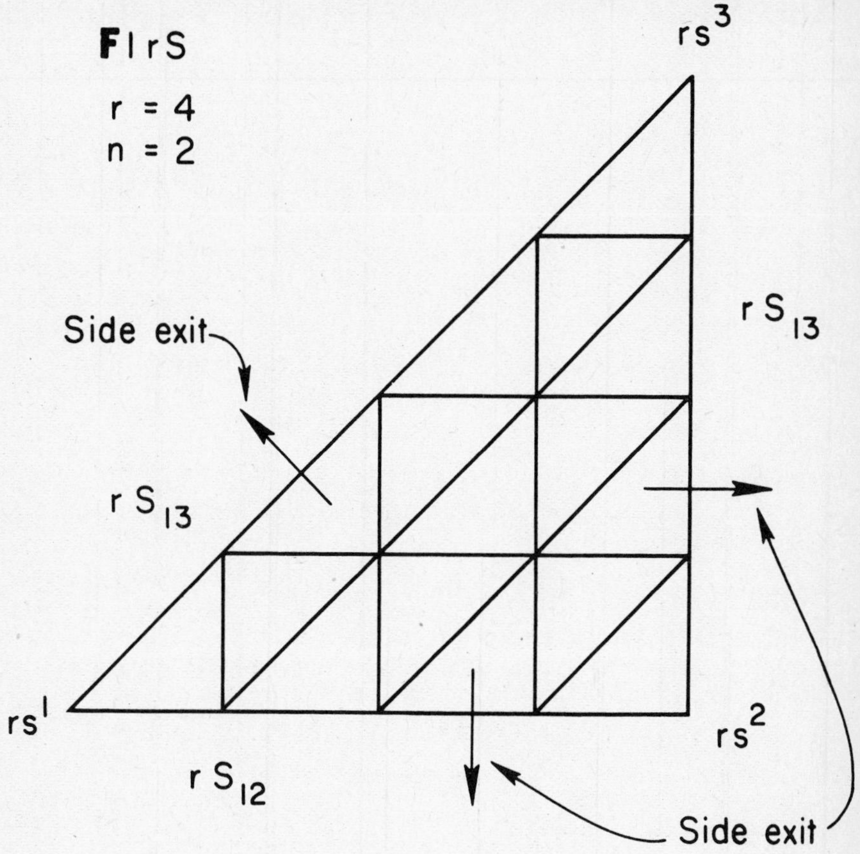

Figure 8.2

and, therefore, we have

$$\lambda_1 = 1 - x_1$$

$$\lambda_i = x_{i-1} - x_i , \qquad 2 \leqq i \leqq n$$

$$\lambda_{n+1} = x_n .$$

Let us define $\Lambda(x,i)$ to be the ith barycentric coordinates λ_i of x with respect to S for i in μ. This x is in rS if and only if $r^{-1}x$ is in S or if

$$r\Lambda(r^{-1}x, i) = x_{i-1} - x_i \geqq 0$$

for i in μ where $x_0 \triangleq r$ and $x_{n+1} \triangleq 0$. In particular, rS is the collection of $(x_1, \ldots, x_n)$ such that $r \geqq x_1 \geqq x_2 \geqq \cdots \geqq x_n \geqq 0$.

If one begins at v in $\mathbb{G}^n$ and moves in the direction q^i, the next lemma indicates the behavior of the barycentric coordinates. Note $q^i = s^{i+1} - s^i$ for i in μ where superscripts are regarded modulo $n+1$, that is $0 = n+1$, $1 = n+2$, etc.

8.1 Lemma: For all i in μ, j in $\mu\backslash\{i, i+1\}$, v in $\mathbb{G}^n$, and θ in $\mathbb{G}$ we have

$$\Lambda(v + \theta q^i, i) = \Lambda(v, i) - \theta$$

$$\Lambda(v + \theta q^i, i+1) = \Lambda(v, i+1) + \theta$$

$$\Lambda(v + \theta q^i, j) = \Lambda(v, j)$$

where $\Lambda(\bullet, n+2) \triangleq \Lambda(\bullet, 1)$.

Proof: Apply the definition of q^i and $\Lambda(v,i)$. □

Mutatis mutandis, the result of Lemma 8.1, is true for any simplex and not merely S. That is, if one moves in the direction of some edge of a simplex, exactly two barycentric coordinates change; one gets larger and the other smaller, etc.

Now using Lemma 8.1 and assuming v is integral and in rS, we see that

(a) for i in μ, $v+q^i$ is in rS if and only if $r\Lambda(r^{-1}v, i)$ $= v_{i-1} - v_i > 0$

(b) for i in ν, $v-q^i$ is in rS if and only if $r\Lambda(r^{-1}v, i+1)$ $= v_i - v_{i+1} > 0$

(c) $v - q^{n+1}$ is in rS if and only if $r\Lambda(r^{-1}v, n+2) = r\Lambda(r^{-1}v, 1)$ $= v_0 - v_1 > 0$.

Note we require $v_0 \triangleq r$ and $v_{n+1} \triangleq 0$.

Given (v,π,n) in $i(\mathbf{F}|rS)^n$ assume we are dropping vertex v^t. Let $(\hat{v};\hat{\pi};n)$ be the replacement for $(v;\pi;n)$ in $\mathbf{F}$ if v^t is dropped, see Figure 6.4. Let v^* be the adjoined vertex. Then $(\hat{v};\hat{\pi};n)$ is the replacement in $F|rS$ if and only if $v^* \in rS$, that is, if and only if $r\Lambda(r^{-1}v^*, i) \geqq 0$ for $i \in \mu$. Let us now consider the three cases of $t = 1$, $2 \leqq t \leqq n$, and $t = n+1$.

Case $t = 1$: The adjoined vertex v^* is $v^{n+1} + q^{\pi 1} = \hat{v}^{n+1}$. Thus, v^* is in rS, if and only if $r\Lambda(r^{-1} v^{n+1}, \pi 1) = v^{n+1}_{\pi 1 - 1} - v^{n+1}_{\pi 1}$ is positive. If the quantity is zero, there is a side exit through $rS_{\mu \backslash \pi 1}$. □

Case $2 \leqq t \leqq n$: The adjoined vertex v^* of $(\hat{v}, \hat{\pi}, n)$ is $v^{t-1} + q^{\pi t} = \hat{v}^{t}$. Thus, v^* is in rS, if and only if $r\Lambda(r^{-1} v^{t-1}, \pi t) = v^{t-1}_{\pi t - 1} - v^{t-1}_{\pi t}$ is positive. If the quantity is zero, there is a side exit through $rS_{\mu \backslash \pi t}$. Note $\pi t - 1 = (\pi(t)) - 1$. □

Case $t = n+1$: The adjoined vertex v^* of $(\hat{v}; \hat{\pi}; n)$ is $v^1 - q^{\pi n} = \hat{v}^1$. Thus, v^* is in S, if and only if $r\Lambda(r^{-1}v^1, \pi n + 1 = v^1_{\pi n} - v^1_{\pi n + 1}$ is positive. If the quantity is zero, there is a side exit through $rS_{\mu \backslash (\pi n + 1)}$. □

The replacement rules are encapsulated in the chart of Figure 8.3. Notes for each terminus describe the new vertex or face exited.

Note 8.2: Later we observe that the chart applies for any permutation of μ, namely, $\pi(n+1) = n+1$ is not required. If $\pi n = n+1$ in Case $t = n+1$, then $r\Lambda(r^{-1}v^1, \pi n + 1) = r\Lambda(r^{-1}v^1, n+2) = r\Lambda(r^{-1}v^1, 1) = v^1_0 - v^1_1$ and not $v^1_{n+1} - v^1_{n+2}$, see Lemma 8.1. □

8.3 Bibliographical Notes: Triangulations as $\mathbf{F}|rS$ were used in the early algorithms, see Scarf [1967], Kuhn [1968], and Eaves [1971]. □

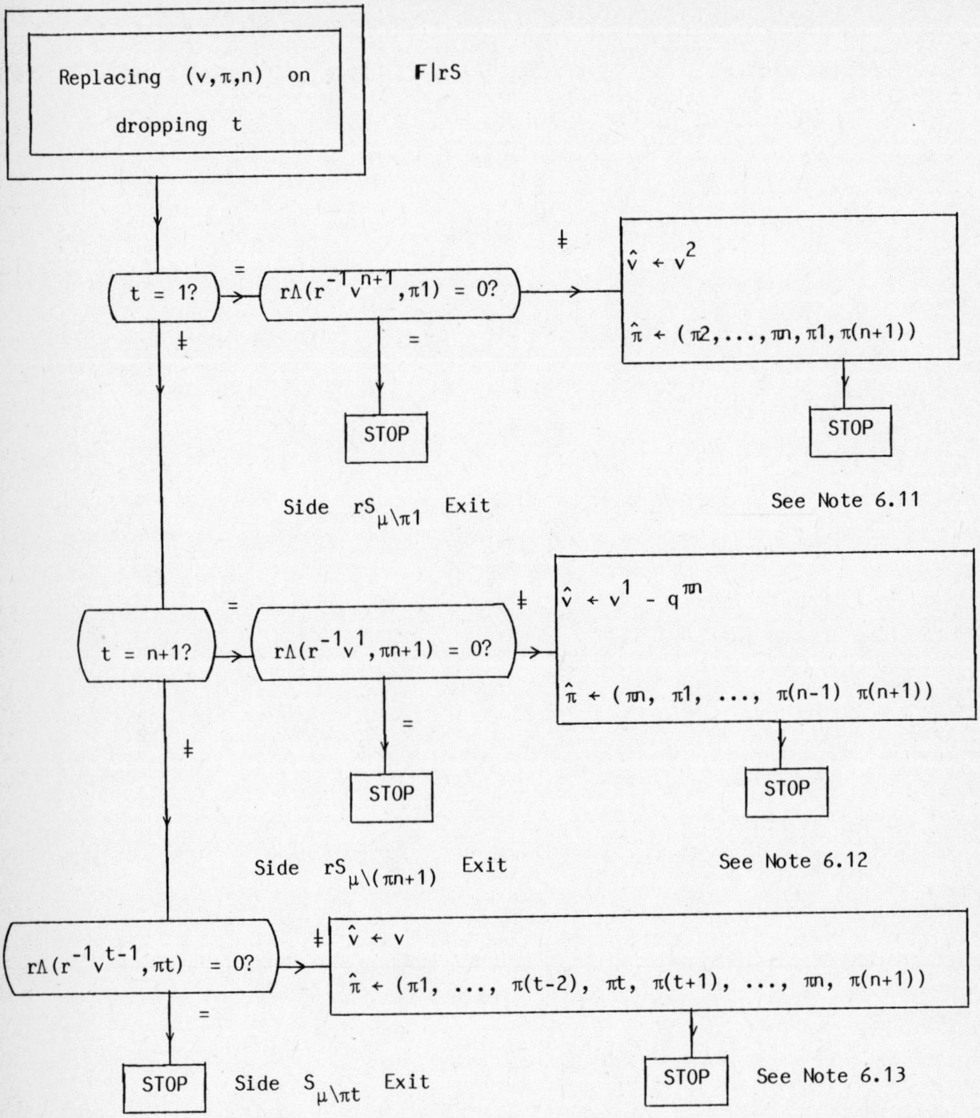
Replacing (v,π,n) on
dropping t
F|rS
t = 1?
=
rΛ(r^-1 v^n+1, π1) = 0?
≠
v̂ ← v^2
π̂ ← (π2, ..., πn, π1, π(n+1))
=
STOP
STOP
Side rS_μ\π1 Exit
See Note 6.11
≠
=
≠
v̂ ← v^1 - q^πn
t = n+1?
rΛ(r^-1 v^1, πn+1) = 0?
π̂ ← (πn, π1, ..., π(n-1) π(n+1))
=
≠
STOP
STOP
Side rS_μ\(πn+1) Exit
See Note 6.12
≠
v̂ ← v
rΛ(r^-1 v^t-1, πt) = 0?
π̂ ← (π1, ..., π(t-2), πt, π(t+1), ..., πn, π(n+1))
=
STOP
Side S_μ\πt Exit
STOP
See Note 6.13

Figure 8.3

9. SQUEEZE AND SHEAR

Given a triangulation $\mathbf{M}$ with vertices in $\mathbb{C}^{n+1}$ all of whose vertices lie in $\mathbb{C}^n \times 0$ or $\mathbb{C}^n \times 1$ we "squeeze" and "shear" $\mathbf{M}$ to yield a second triangulation $\mathbf{N}$ isomorphic to $\mathbf{M}$ and whose vertices also lie in $\mathbb{C}^n \times 0$ or $\mathbb{C}^n \times 1$, see Figure 9.1 and Section 3. After describing the general procedure of sqeeze and shear, we shall illustrate it by constructing a "refining" triangulation from $\mathbf{F}$; even so, we have more important roles in store for squeeze and shear, among them, the construction of $\mathbf{V}[r,p]$ in Section 16.

Once again assume the vertices of the triangulation $\mathbf{M}$ lie in $\mathbb{C}^n \times 0$ and $\mathbb{C}^n \times 1$ and that M is the manifold.

Let r be any positive scalar and p be any vector. Define the $\mathbf{M}$-PL map $g : M \to \mathbb{C}^n \times [0,1]$ by

$$g(x,1) = (rx+p,\ 1)$$

$$g(x,0) = (x,0)\ .$$

As g is defined on the vertices of $\mathbf{M}$ it is defined on all of M. Consider $g(\mathbf{M})$ and see Figure 9.1.

We shall say that $g(\mathbf{M})$ has been obtained from $\mathbf{M}$ by squeeze and shear where r is the squeeze and p is the shear. In applications typically $0 < r < 1$, and hence, for the term "squeeze".

We proceed to show that g is one-to-one on M and hence that $g(\mathbf{M})$ is a triangulation which is isomorphic to $\mathbf{M}$.

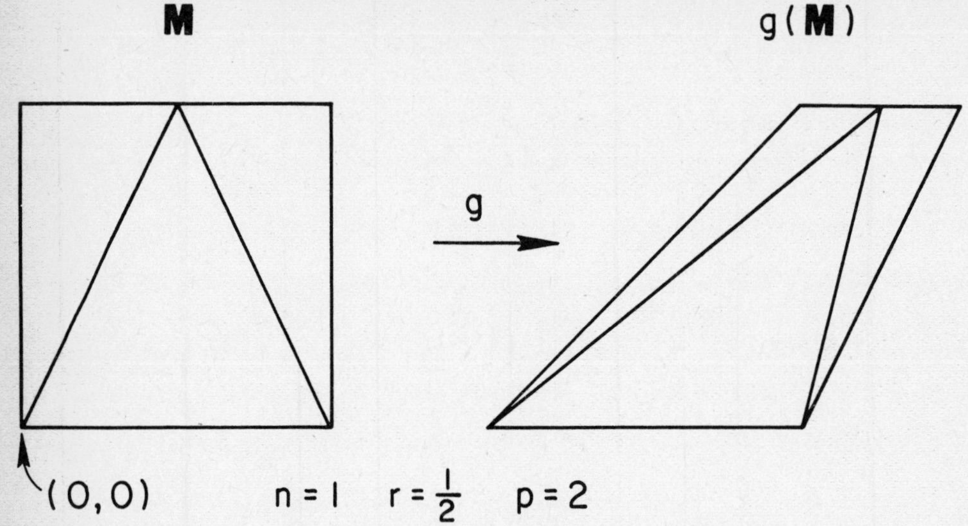

Figure 9.1

9.1 Theorem: **M** and g(**M**) are g-isomorphic.

Proof: We need only show that g is one-to-one. Suppose $x \in \sigma \in$ **M** and $y \in \tau \in$ **M**, and $g(x) = g(y)$. Then

$$\Sigma_i g(v^i,1)\lambda_i + \Sigma_j g(u^j,0)\zeta_j = \Sigma_i g(\tilde{v}^i,1)\tilde{\lambda}_i + \Sigma_j g(\tilde{u}^j,0)\tilde{\zeta}_j$$

$$x = \Sigma_i (v_i,1)\lambda_i + \Sigma_j (u_j,0)\zeta_j$$

$$y = \Sigma_i (\tilde{v}_i,1)\tilde{\lambda}_i + \Sigma_j (\tilde{u}_j,0)\tilde{\zeta}_j$$

$$\Sigma_i \lambda_i + \Sigma_j \zeta_j = 1 \qquad \lambda_i > 0 \qquad \zeta_j > 0$$

$$\Sigma_i \tilde{\lambda}_i + \Sigma_j \tilde{\zeta}_j = 1 \qquad \tilde{\lambda}_i > 0 \qquad \tilde{\zeta}_j > 0$$

where the $(v^i,1)$ and $(u^j,0)$ are vertices of σ and the $(\tilde{v}^i,1)$ and $(\tilde{u}^j,0)$ are vertices of τ. Therefore,

$$\Sigma_i (rv^i + p,\ 1)\lambda_i + \Sigma_j (u^j,0)\zeta_j = \Sigma_i (r\tilde{v}^i + p,\ 1)\tilde{\lambda}_i + \Sigma_j (\tilde{u}^j,0)\tilde{\zeta}_j$$

$$\Sigma_i \lambda_i = \Sigma_i \tilde{\lambda}_i \qquad \Sigma_j \zeta_j = \Sigma_j \tilde{\zeta}_j$$

and, consequently,

$$r\Sigma_i v^i \lambda_i + \Sigma_j u^j \zeta_j = r\Sigma_i \tilde{v}^i \tilde{\lambda}_i + \Sigma_j \tilde{u}^j \tilde{\zeta}_j \quad .$$

Let $t = r\Sigma_i \lambda_i + \Sigma_j \zeta_j = r\Sigma_i \tilde{\lambda}_i + \Sigma_j \tilde{\zeta}_j$ and

$$\theta_i = r\lambda_i t^{-1} \qquad \gamma_j = \zeta_j t^{-1} \ ,$$

$$\tilde{\theta}_i = r\tilde{\lambda}_i t^{-1} \qquad \tilde{\gamma}_j = \tilde{\zeta}_j t^{-1} \ .$$

Then

$$\Sigma_i (v^i,1)\theta_i + \Sigma_j (u^j,0)\gamma_j = \Sigma_i (\tilde{v}^i,1)\tilde{\theta}_i + \Sigma_j (\tilde{u}^j,0)\tilde{\gamma}_j$$

$$\Sigma_i \theta_i + \Sigma_j \gamma_j = 1 = \Sigma_i \tilde{\theta}_i + \Sigma_j \tilde{\gamma}_j$$

$$\theta_i \geqq 0 \qquad \gamma_j \geqq 0 \qquad \tilde{\theta}_i \geqq 0 \qquad \tilde{\gamma}_j \geqq 0 \ .$$

But since each point in M can be expressed in only one way as the convex combination of vertices all of which are in one simplex we have, perhaps, after re-indexing,

$$\theta_i = \tilde{\theta}_i \qquad v^i = \tilde{v}^i \qquad \text{all } i$$

$$\gamma_j = \tilde{\gamma}_j \qquad u^j = \tilde{u}^j \qquad \text{all } j$$

and, consequently, that

$$\lambda_i = \tilde{\lambda}_i \qquad \zeta_j = \tilde{\zeta}_j$$

or that $x = y$. □

It is worth noting that if one first squeezes and then shears, the shear is merely the linear transformation $(x,\theta) \to (x+\theta p, \theta)$; of course, squeeze is not a linear transformation, only PL.

Our next step is to show that if $\mathbf{M}$ is finite and M is convex then $g(M)$ is convex. We shall say that a hyperplane in $\mathbb{C}^n \times \mathbb{C}$ is not horizontal, if it meets both $\mathbb{C}^n \times 0$ and $\mathbb{C}^n \times 1$.

Let $\mathbf{H}$ be the collection of hyperplanes in $\mathbb{C}^n \times \mathbb{C}$ that are not horizontal. It is easy to see that η is in $\mathbf{H}$, if and only if η is the affine hull of two sets $\eta_0 \times 0$ and $\eta_1 \times 1$ where η_0 and η_1 are parallel hyperplanes in $\mathbb{C}^n$, see Figure 9.2. Indeed, given η in $\mathbf{H}$ define η_0 and η_1 by $\eta_i \times i = \eta \cap (\mathbb{C}^n \times i)$ for $i = 0, 1$. Clearly, η is the affine hull of the two sets $\eta_0 \times 0$ and $\eta_1 \times 1$, see Figure 9.2.

Given a hyperplane η_1 in $\mathbb{C}^n$ it is a simple matter to show that η_1 and $r\eta_1+p$ are parallel hyperplanes. With this in mind, we define the map $h : \mathbf{H} \to \mathbf{H}$ by setting $h(\eta)$ to be the affine hull of the two sets $\eta_0 \times 0$ and $(r\eta_1+p) \times 1$ where $\eta_i \times i = \eta \cap (\mathbb{C}^n \times i)$ for $i = 0, 1$. For the next lemma we continue to regard M as the manifold of $\mathbf{M}$ all of whose vertices lie in $\mathbb{C}^n \times 0$ and $\mathbb{C}^n \times 1$.

9.2 Lemma: The map $h : \mathbf{H} \to \mathbf{H}$ is one-to-one and onto. Furthermore, $\eta \in \mathbf{H}$ is a supporting hyperplane of M if and only if $h(\eta) \in \mathbf{H}$ is a supporting hyperplane of $g(M)$, and in this case,

$$g(M \cap \eta) = g(M) \cap h(\eta) .$$

Proof: For $\eta \in \mathbf{H}$, $h^{-1}(\eta)$ is the affine hull of $\eta_0 \times 0$ and $(r^{-1}\eta_1 - r^{-1}p) \times 1$ where $\eta \cap (\mathbb{C}^n \times i) = \eta_i \times i$. Let $M_i \times i = M \cap (\mathbb{C}^n \times i)$. If η_1 supports M_1 in $\mathbb{C}^n$, then clearly $r\eta_1+p$ supports rM_1+p. As η_0 supports M_0, r is positive, and vertices of $\mathbf{M}$ are in $\mathbb{C}^n \times [0,1]$,

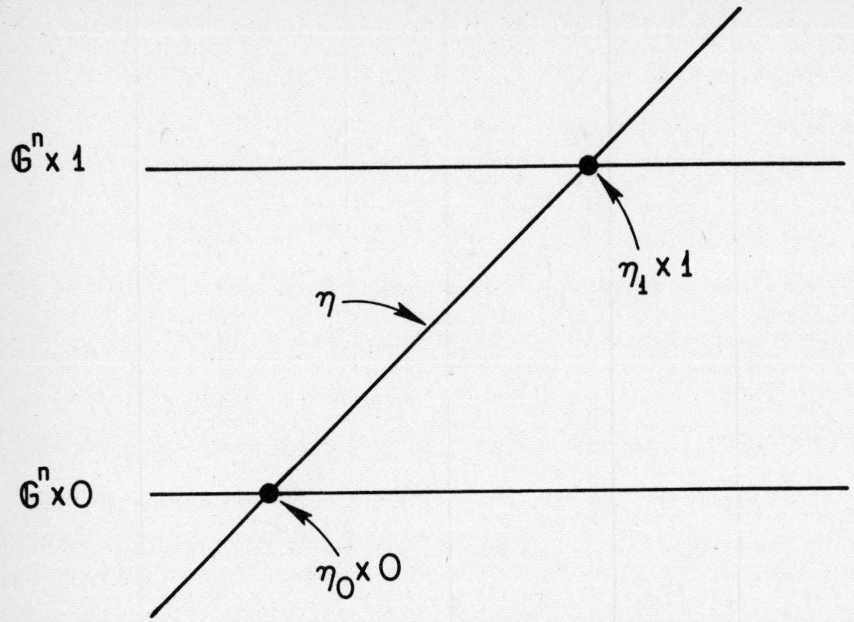
$\mathbb{G}^n \times 1$
$\eta_1 \times 1$
η
$\mathbb{G}^n \times 0$
$\eta_0 \times 0$

Figure 9.2

it follows that the affine hull of $\eta_0 \times 0$ and $(r\eta_1+p) \times 1$ supports $g(M)$ and that $g(M \cap \eta) = g(M) \cap h(\eta)$. □

As g is linear on $\mathbb{C}^n \times 0$ and $\mathbb{C}^n \times 1$, it is clear that $g(M_i \times i)$ is convex where $M_i \times i \triangleq M \cap (\mathbb{C}^n \times i)$ for $i = 0, 1$. As all vertices of $g(\mathbf{M})$ lie in $g(M_0 \times 0)$ and $g(M_1 \times 1)$, it is evident that $g(M)$ is contained in the convex hull of $g(M_0 \times 0)$ and $g(M_1 \times 1)$. From the next lemma it follows that $g(M)$ is the convex hull of $g(M_0 \times 0)$ and $g(M_1 \times 1)$.

9.3 Theorem: If $\mathbf{M}$ is finite and M is convex, then $g(M)$ is convex.

Proof: Assume $\mathbf{M}$ is an m-triangulation. If σ is a boundary $(m-1)$-simplex of $\mathbf{M}$, then $\text{aff}(\sigma)$ is a supporting hyperplane of M, and furthermore, $\text{aff}(g(\sigma)) = h(\text{aff}(\sigma))$ is a supporting hyperplane of $g(M)$, see Lemmas 9.2 and 3.4. If $g(M)$ is not convex, there is a boundary $(m-1)$-cell τ of $g(\mathbf{M})$ such that $\text{aff}(\tau)$ is not a supporting hyperplane. Let $\sigma = g^{-1}(\tau)$ which is a boundary $(m-1)$-simplex of $\mathbf{M}$. But then $h(\text{aff}(\sigma)) = \text{aff}\, g(\sigma) = \text{aff}(\tau)$ is a supporting hyperplane of $g(M)$ which is a contradiction. □

9.4 Exercise: Given y in $g(M)$, how can one solve $g(x) = y$? Hint: Use a PL-homotopy. This exercise involves notions outside the scope of this manuscript. □

In some of our applications we show $g(M)$ is convex where $\mathbf{M}$ is infinite. The following result is our vehicle. We say that a sequence of subsets M_i, $i = 1, 2, \ldots$ of M increase to M; if $M_1 \subseteq M_2 \subseteq \cdots$ and $\cup_i M_i = M$; note if each M_i is convex, then M is convex.

9.5 Lemma: Let M_i for $i = 1, 2, \ldots$ be a sequence of convex subsets of M which increases to M. If each $\mathbf{M}|M_i$ is finite and triangulates M_i, then $g(M)$ is convex.

Proof: As $g(\cup_i M_i) = \cup_i g(M_i)$ and using Theorem 9.3 we see that the $g(M_i)$ are convex and increase to $g(M)$. Let y_1 and y_2 be two points in $g(M)$. For some i we have y^1 and y^2 in $g(M_i)$ and thus the convex hull of y^1 and y^2 lies in $g(M_i) \subseteq g(M)$. □

Using squeeze and shear and the Freudenthal triangulation $\mathbf{F}$ of $\mathbb{C}^n$ we construct a refining triangulation $\mathbf{R}$ of $\mathbb{C}^n \times [0,+\infty)$ where $[0, +\infty) = \{x \in \mathbb{C} : x \geqq 0\}$. $\mathbf{R}$ depends upon an infinite sequence (r_h, p^h) in $\mathbb{C} \times \mathbb{C}^n$ with r_h positive and $h = 0, 1, 2, \ldots$; given such a sequence, the triangulation $\mathbf{R}$ of $\mathbb{C}^n \times [0, +\infty)$ will have the property that its natural restriction of $\mathbb{C}^n \times h$ for $h = 0, 1, 2$ is $(r_h \mathbf{F}_n + p^h, h)$ where $\mathbf{F}_n$ is the Freudenthal triangulation of $\mathbb{C}^n$. Thus in particular if the r_h sequence tends to zero as h tends to infinity, then the grid size of $\mathbf{R}|(\mathbb{C}^n \times h)$ tends to zero.

As a preview of such a triangulation consider Figure 9.3. As we shall see $\mathbf{R}$ restricted to $\mathbb{C}^n \times [0,k]$ depends only on $r_0, \ldots, r_k$ and $p_0, \ldots, p_k$.

Let $\mathbf{F}_{n+1}$ be the Freudenthal triangulation of $\mathbb{C}^{n+1}$. Of course, the natural restriction of $\mathbf{F}_{n+1}$ to $\mathbb{C}^n \times [0, +\infty)$ subdivides $\mathbb{C}^n \times [0, +\infty)$, and we are only interested in this portion of $\mathbf{F}$. Define $i\mathbf{R}^{n+1}$ to be the collection of all $(v,\pi,n+1)$ in $i\mathbf{F}^{n+1}_{n+1}$ with $v_{n+1} \geq 0$, that is, with

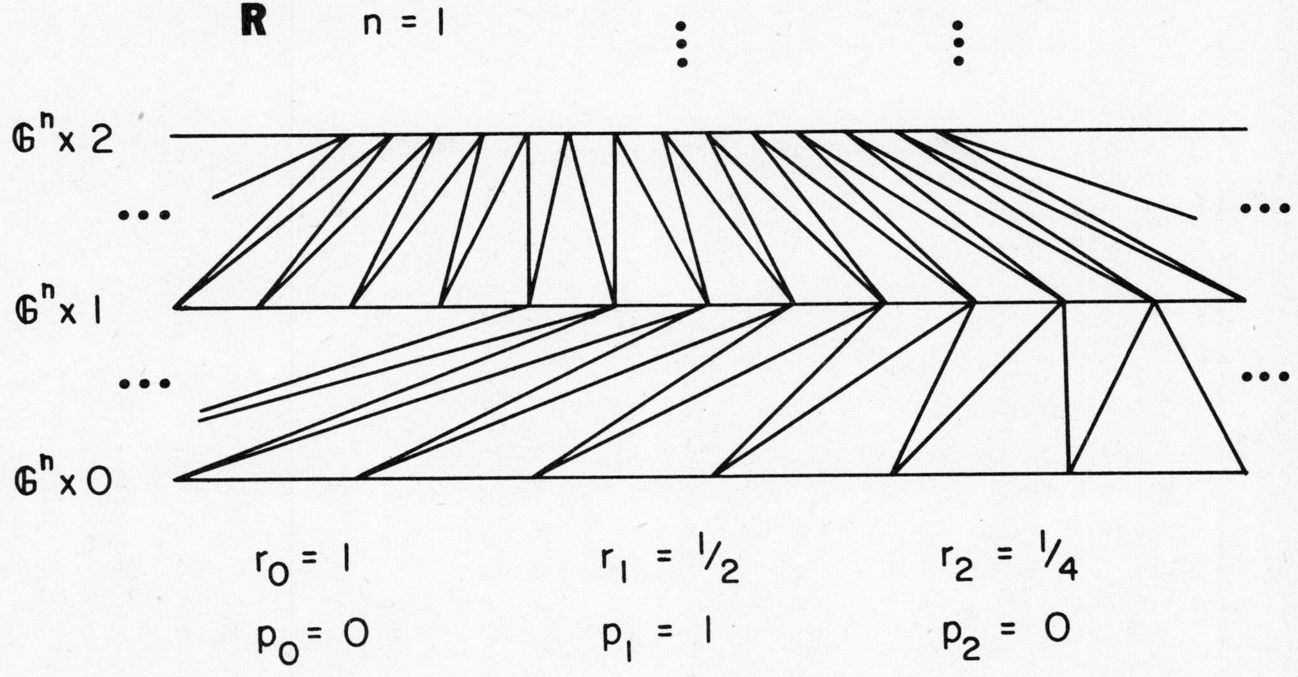
R
n = 1
$\mathbb{G}^n \times 2$
$\mathbb{G}^n \times 1$
$\mathbb{G}^n \times 0$
$r_0 = 1$
$p_0 = 0$
$r_1 = 1/2$
$p_1 = 1$
$r_2 = 1/4$
$p_2 = 0$

Figure 9.3

$(v;\pi;n+1) \in \mathbb{C}^n \times [0, +\infty)$. Notice that $v = (v_1, \ldots, v_{n+1})$ and that $\pi = (\pi 1, \ldots, \pi(n+2))$.

Given $(v,\pi,n+1)$ in $i\mathbf{R}^{n+1}$ define $[v;\pi;n+1]$ to be the convex hull of the points $w^j = (r_h u + p^h, h)$ for $j = 1, \ldots, n+2$ where

(a) $v^1 = v$

(b) $v^{i+1} = v^i + q^{\pi i}$, $i = 1, \ldots, n+1$

(c) $u = (v_1^j, \ldots, v_n^j)$

(d) $h = v_{n+1}^j$.

Define $\mathbf{R}$ to be the collection of all $[v;\pi;n+1]$ and their faces where $(v,\pi,n+1)$ ranges over $i\mathbf{R}^{n+1}$. As we shall prove with repeated application of the squeeze and shear device, $\mathbf{R}$ is a triangulation of $\mathbb{C}^n \times [0, +\infty)$ with the property that $\mathbf{R}$ naturally restricted to $\mathbb{C}^n \times h$ is $(r_h \mathbf{F}_n + p^h) \times h$ for $h = 0, 1, \ldots$.

9.6 Proposition: $\mathbf{R}$ triangulates $\mathbb{C}^n \times [0,+\infty)$ and the natural restriction of $\mathbf{R}$ to $\mathbb{C}^n \times h$ is $(r_h \mathbf{F}_h + p^h, h)$ for $h = 0, 1, 2, \ldots$.

Proof: We shall only prove that $\mathbf{R}$ naturally restricted to $\mathbb{C}^n \times [0,1]$ triangulates $\mathbb{C}^n \times [0,1]$ and $\mathbf{R}$ restricted to $\mathbb{C}^n \times h$ is over $(r_h \mathbf{F}_n + p^h) \times i$ for $h = 0, 1$, respectively. It is then clear that the complete result can be obtained by induction. Let $\mathbf{T} = \mathbf{F}_{n+1} | (\mathbb{C}^n \times [0,1])$ and define the linear function ℓ by $\ell(x,\theta) = (r_0 x + p^0, \theta)$ for (x,θ) in $\mathbb{C}^n \times [0,1]$. Clearly $\ell(\mathbf{T})$ triangulates $\mathbb{C}^n \times [0,1]$ and $\ell(\mathbf{T}) | (\mathbb{C}^n \times i) = (r_0 \mathbf{F}_n + p^0) \times i$ for $i = 0, 1$. Now we squeeze and shear $\ell(\mathbf{T})$ to obtain $\mathbf{R} | \mathbb{C}^n \times [0,1]$.

Define the $\ell(T)$ - PL map $g : \mathbb{C}^n \times [0,1] \to \mathbb{C}^n \times [0,1]$ by $g(x,_0) = (x,0)$ and $g(x,1) = (r_1 r_0^{-1} x + p^1 - r_1 r_0^{-1} p^0, 1)$. Clearly by squeeze and shear $g(\ell(T)) = \mathbf{R}|(\mathbb{C}^n \times [0,1])$ and the restriction to $\mathbb{C}^n \times i$ is $(r_i \mathbf{F}_n + p^i, i)$ for $i = 0, 1$. Let $c = (n, \ldots, 1)(n+1)^{-1}$ be the barycenter of S, the standard n-simplex with vertices $s^0, s^1, \ldots, s^n$. Let $M_k = (kS-kc) \times [0,1]$ for $k = n+1, n+2, \ldots$ and we have a sequence of convex sets increasing to $\mathbb{C}^n \times [0,1]$. From Theorem 9.3 and Lemma 9.5 we see that $g(\mathbb{C}^n \times [0,1]) = \mathbb{C}^n \times [0,1] = \mathrm{cvx}((\mathbb{C}^n \times 0) \cup (\mathbb{C}^n \times 1))$. □

The representation set and replacement rules for $\mathbf{R}$ are essentially those of $\mathbf{F}$; one must merely test for a bottom exit, see Figure 9.4. As for the representation and facet rules of $\mathbf{R}$ they are merely those of $\mathbf{F}_{n+1}$ followed by a linear translation.

Let the $w^1, \ldots, w^{n+2}$ be the $n+2$ vertices of $[v;\pi;n+1]$ in $\mathbf{R}$. Suppose we want to drop vertex w^t and compute the replacement. As mentioned, the rules are exactly as in $\mathbf{F}_{n+1}$ except that a check must be included for a bottom exit. This case occurs when all vertices of $[v;\pi;n+1]$ are in $\mathbb{C}^n \times 0$ except w^t. As $v^i \leqq v^{i+1}$ we see that this can only occur if $t = n+2$, $\pi(n+1) = n+1$ and $v_{n+1} = 0$. The replacement rules are summarized in the chart of Figure 9.5.

9.7 Note (for Figure 9.5): The adjoined vertex is $\hat{w}^1 = (r_h u + ph, h)$ where $u = (\hat{v}_1, \ldots, \hat{v}_n)$ and $h = \hat{v}_{n+1}$. The correspondence between vertices of $[v;\pi;n+1]$ and $[\hat{v};\hat{\pi};n+2]$ are given by

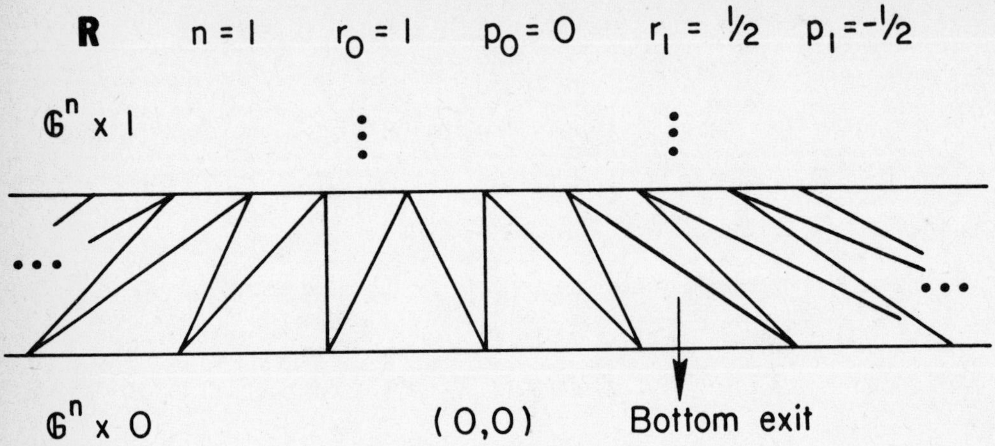

Figure 9.4

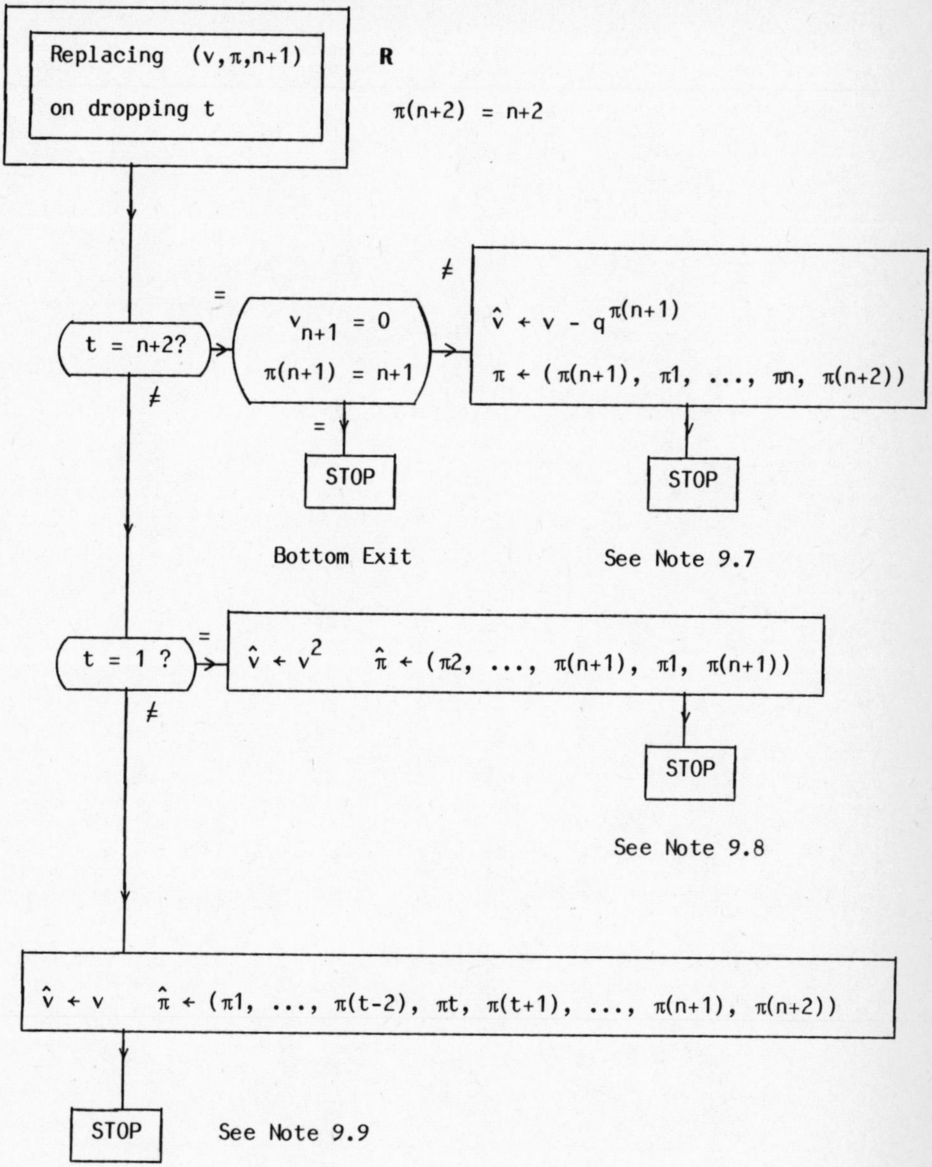
Replacing (v,π,n+1)
on dropping t
R
π(n+2) = n+2
t = n+2?
=
≠
v_{n+1} = 0
π(n+1) = n+1
v̂ ← v - q^{π(n+1)}
π ← (π(n+1), π1, ..., πn, π(n+2))
STOP
STOP
Bottom Exit
See Note 9.7
t = 1 ?
v̂ ← v^2
π̂ ← (π2, ..., π(n+1), π1, π(n+1))
STOP
See Note 9.8
v̂ ← v
π̂ ← (π1, ..., π(t-2), πt, π(t+1), ..., π(n+1), π(n+2))
STOP
See Note 9.9

Figure 9.5

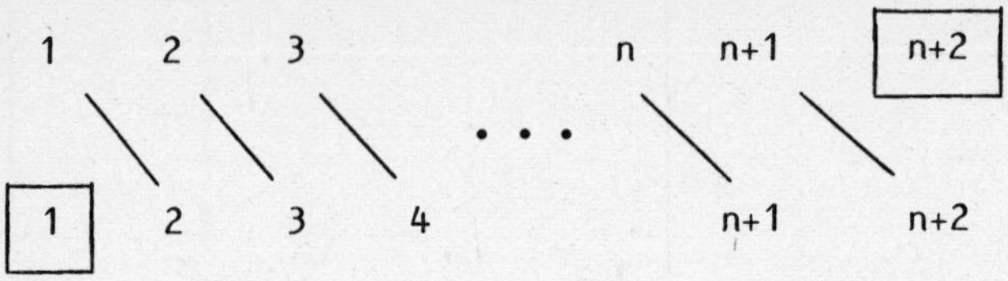

that is, $w^j = \hat{w}^{j+1}$ for $j = 1, \ldots, n+1$; w^{n+2} was dropped and $\hat{w}^1$ is adjoined. $\square$

9.8 Note (for Figure 9.5): The adjoined vertex is $\hat{w}^{n+2} = (r_h u + p_h, h)$ where $u = (\hat{v}_1^{n+2}, \ldots, \hat{v}_n^{n+2})$ and $h = \hat{v}_{n+1}^{n+1}$. The correspondence between vertices of $[v;\pi;n+1]$ and $[\hat{v};\hat{\pi};n+1]$ is

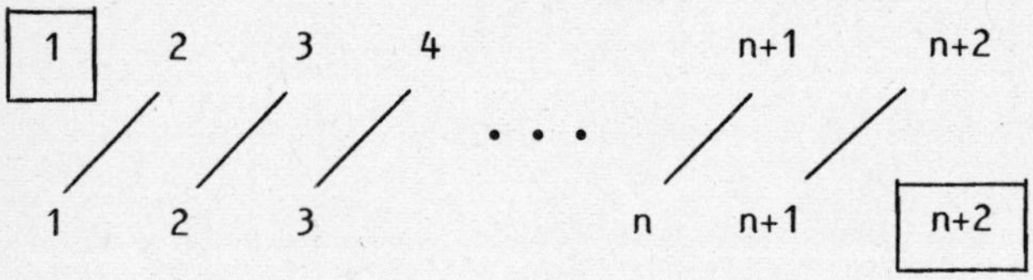

9.9 Note (for Figure 9.5): The adjoined vertex is $\hat{w}^t = (r_h u + p_h, h)$ where $u = (\hat{v}_1^t, \ldots, \hat{v}_n^t)$ and $h = \hat{v}_{n+1}^t$. The correspondence between vertices of $[v;\pi;n+1]$ and $[\hat{v};\hat{\pi};n+1]$ is

$$
\begin{array}{ccccccc}
1 & \cdots & t-1 & \boxed{t} & t+1 & \cdots & n+1 \\
| & & | & | & | & & | \\
1 & & t-1 & \boxed{t} & t+1 & & n+1
\end{array}
\qquad \square
$$

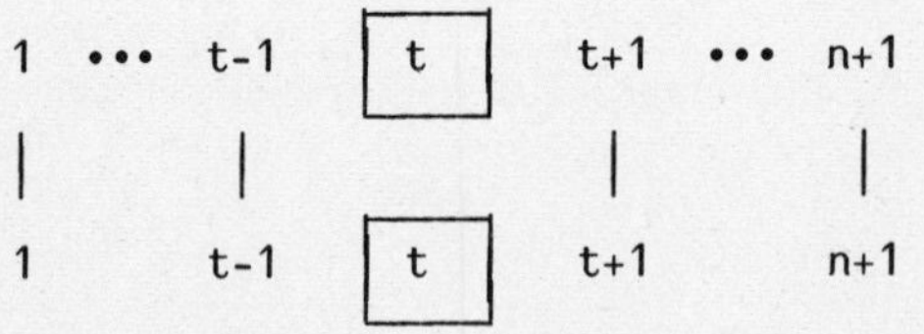

Typically in the usage of a triangulation like **R** the scale of the vertical coordinate x_n is irrelevant. That is, given a simplex σ in **R**, it is the diameter of σ projected to $\mathbb{C}^n \times 0$ that is of interest and not the diameter of σ itself. A triangulation **T** of $\mathbb{C}^n \times [0,+\infty)$ is called refining if for any $\varepsilon > 0$ there is an $\theta = 1, 2, \ldots$ such that if σ in **T** meets $\mathbb{C}^n \times [\theta,+\infty)$, then the diameter of the projection of σ to $\mathbb{C}^n \times 0$ does not exceed ε. In this sense, **R** is not a refining triangulation even though $\mathbf{R}|(\mathbb{C}^n \times h) = (r_h\mathbf{F}_n + p_h) \times h$, however, the triangulation S of Section 17 will be. □

9.10 Bibliographical Notes: Squeeze was first used in Eaves [1972] and shear was first used in Saigal and Todd [1978]. Shamir [1979] used squeeze and shear to construct the configuration "**R**", however neither Eaves nor Shamir were unaware that squeeze yielded a "geometric triangulation." Garcia and Zangwill [1980], Solow [1981], and Zangwill [1977] were indirectly concerned with the notion of distorting triangulation. □

10. Freudenthal Triangulation F of $\mathbb{C}^n$, Part II

The present section is a continuation of the study of the Freudenthal triangulation F which was initiated in Section 6. We have already developed a set of representation and replacement rules for F. Nevertheless, herein, we shall develop others, in fact, for each n-simplex we shall have n+1 representations, only one of which is from iF^n. Although it might seem that one is enough, there is no doubt that the overall development is immensely enhanced with the additional representations. In this section we study features of F that are more conveniently examined in the presence of the alternative representations.

Recall that i_*F is the collection of triples (v,π,n) where $v \in \mathbb{Z}^n$, π orders $u = \{1, \ldots, n+1\}$ and $k = 0, 1, \ldots, n$. The vertices of the k-simplex $(v;\pi;k)$ are $v^i = v + \Sigma_{j=1}^{i-1} q^{\pi j}$ for $i = 1, \ldots, k+1$.

For (v,π,n) in iF^n Figure 10.1 illustrates that one can begin with vertex v^1 and travel through all vertices $v^2, \ldots, v^{n+1}$ and return to v^1 by traveling to the directions of $q \triangleq (I,-e)$. Any link in the path can be omitted and, yet, one can still travel the path and visit all vertices by beginning on the correct vertex. Algebraically, this amounts to little more than $qe = 0$ where $e = (1, \ldots, 1)$.

The next lemma shows that in i_*F there are n+1 representations for each n -simplex of F. Recall $\mu \triangleq \{1, \ldots, n+1\}$.

10.1 Lemma: Let ℓ be an element of μ, then F^n is the set of simplexes $(v;\pi;n)$ with $(v,\pi,n) \in i_*F$ and $\pi(n+1) = \ell$.

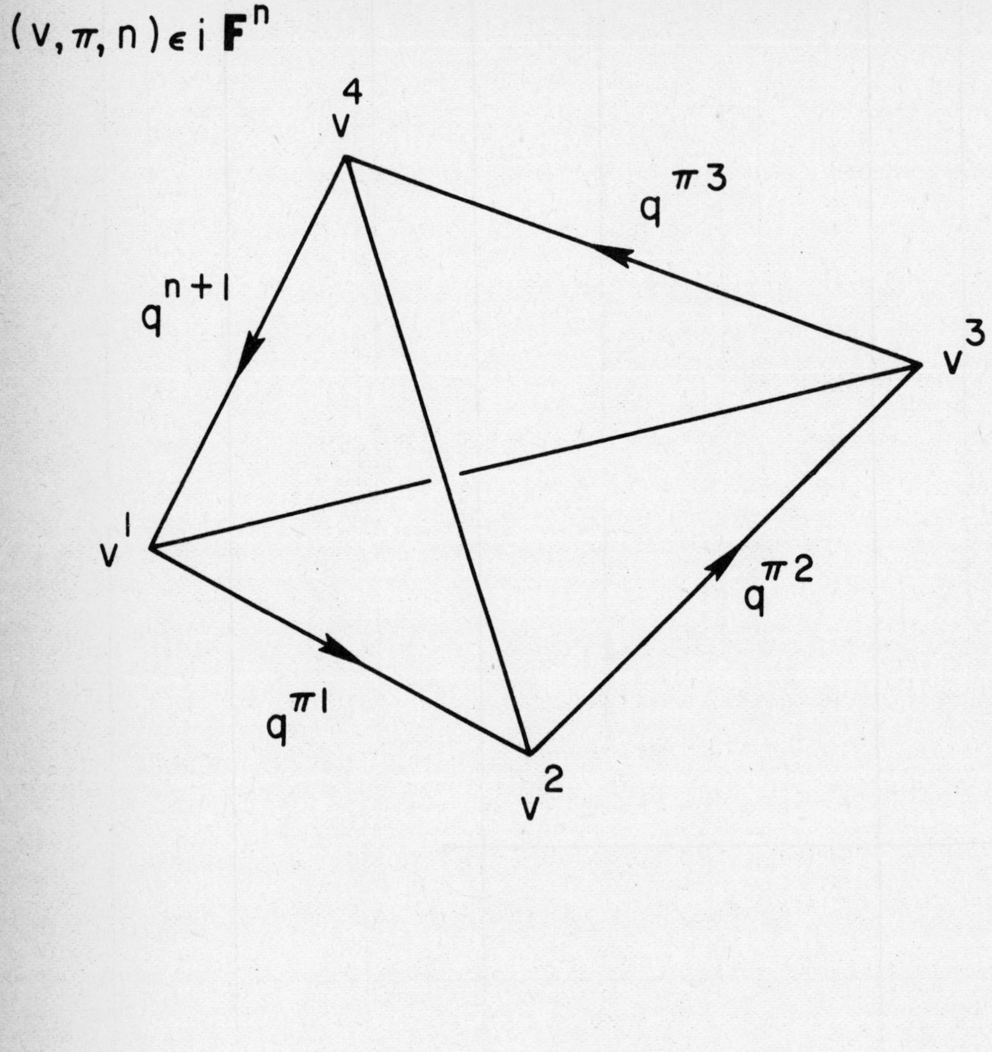

Figure 10.1

Proof: If $\ell = n+1$ there is nothing to prove as we are then examining the collection of simplexes $(v;\pi;n)$ with $(v;\pi;n)$ in $i\mathbf{F}^n$. Assume $\ell \neq n+1$. Let $(v,\pi,n) \in i\mathbf{F}^n$. Let $k = \pi^{-1}\ell$, $w = v^{k+1}$, and $\gamma = (\pi(k+1), \ldots, \pi(n+1), \pi 1, \ldots, \pi k)$. Then $(w;\gamma;n) = (v;\pi;n)$ for $(w^1, \ldots, w^{n+1}) = (v^{k+1}, \ldots, v^{n+1}, v^1, \ldots, v^k)$. For any $(w,\gamma,n) \in i_*\mathbf{F}$ with $\gamma(n+1) = \ell$, let $k = \gamma^{-1}(n+1)$, $v = w^{k+1}$, and $\pi = (\gamma(k+1), \ldots, \gamma(n+1), \gamma 1, \ldots, \gamma k)$. Then $(v;\pi;n) = (w;\gamma;n)$ for $(v^1, \ldots, v^{n+1}) = (w^{k+1}, \ldots, w^{n+1}, w^1, \ldots, w^k)$. □

10.2 Lemma: If σ is in $\mathbf{F}^n$ there are exactly $n+1$ distinct triples $(v,\pi,n) \in i_*\mathbf{F}$ with $\sigma = (v;\pi;n)$.

Proof: If $(v;\pi;n) = (w;\gamma;n) = \sigma$ and $\pi(n+1) \neq \gamma(n+1)$, then $(v,\pi) \neq (w,\gamma)$. Thus, there are at least $n+1$ triples, one corresponding to each $\pi(n+1) \in \mu$. From Lemma 6.3 for each $\ell \in \mu$ there is but one triple (v,π,n) in $i_*\mathbf{F}$ with $(v;\pi;n) = \sigma$ and $\pi(n+1) = \ell$. □

The next lemma states that every triple in $i_*\mathbf{F}$ indexes a simplex of $\mathbf{F}$, but it is a simple matter for $n \geq 3$ to find an $(n-2)$-simplex in $\mathbf{F}$ that cannot be represented by a triple in $i_*\mathbf{F}$.

10.3 Lemma: For any (v,π,k) in $i_*\mathbf{F}$, the simplex $(v;\pi;k)$ is in $\mathbf{F}$.

Proof: $(v;\pi;k)$ is a face of $(v;\pi;n)$. □

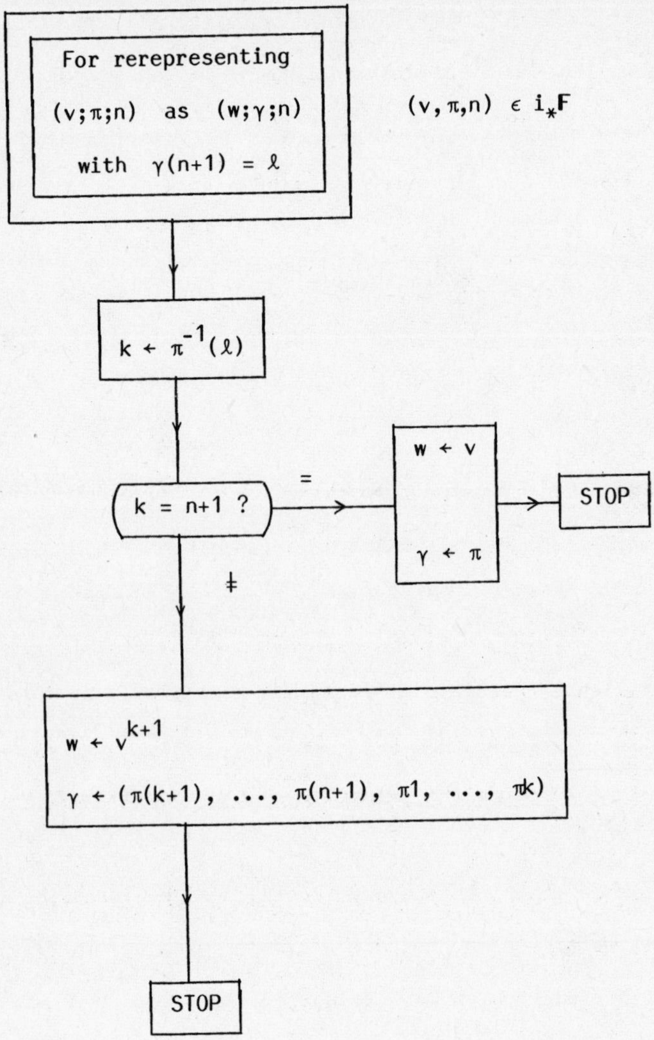
For rerepresenting
(v;π;n) as (w;γ;n)
with γ(n+1) = ℓ
(v,π,n) ∈ i*F
k ← π⁻¹(ℓ)
k = n+1 ?
=
w ← v
γ ← π
STOP
≠
w ← v^{k+1}
γ ← (π(k+1), ..., π(n+1), π1, ..., πk)
STOP

Figure 10.2

Let $(v;\pi;k)$ be a triple of $i_*\mathbf{F}$ and ℓ an element of μ. We often need to compute (w,γ,n) in $i_*\mathbf{F}$ with $(w;\gamma;n) = (v;\pi;n)$ where $\gamma(n+1) = \ell$, that is, we need to represent $(v;\pi;n)$. As the proof of Lemma 10.1 indicates, this is a simple matter. This procedure is given in Figure 10.2.

10.4 Note (for Figure 10.2): $(w,\gamma,n) \in i_*\mathbf{F}$ and $(w;\gamma;n) = (v;\pi;n)$. □

In Figure 6.4 the replacement rules for $\mathbf{F}$ using the representation $i\mathbf{F}^n$ are given. The replacement rules for $\{(v,\pi,n) \in i_*\mathbf{F} : \pi(n+1) = \ell\}$ and a representation set for $\mathbf{F}^n$ for any ℓ in μ are exactly the same as for $i\mathbf{F}^n$. Thus, if one follows the chart of Figure 6.4 with (v,π,n) in $i_*\mathbf{F}$ where $\pi(n+1) = \ell$, one obtains $(\hat{v},\hat{\pi},n)$ in $i_*\mathbf{F}$ with $\hat{\pi}(n+1) = \ell$ where $(\hat{v};\hat{\pi};n)$ is the replacement for $(v;\pi;n)$ upon dropping v^t. Similarly in the chart of Figure 8.3 for $\mathbf{F}|rS$ one requires only $(v,\pi,n) \in i_*\mathbf{F}$.

We return to the matter of invariance of $\mathbf{F}$ and $\mathbf{Q}$ under linear maps. In the next lemma let (v,π,n) and (w,γ,n) be elements of $i_*\mathbf{F}$; $\pi(n+1) = \gamma(n+1)$ is not required. The superscripts are regarded modulo $n+2$.

10.5 Theorem: Let $h : \mathbb{C}^n \to \mathbb{C}^n$ be a linear map such that for some i in μ

$$h(v^j) = w^{j+i}$$

for all j in μ where v^j and w^j are the vertices of $(v;\pi;n)$ and $(w;\gamma;n)$. Then $h(\mathbf{F}) = \mathbf{F}$ and $h(\mathbf{Q}) - h(0) = \mathbf{Q}$.

Proof: As h is linear one-to-one and onto it is clear that $h(\mathbf{F})$ is a triangulation of $\mathbb{C}^n$; we need only show that $h(u;\delta;n)$ is in $\mathbf{F}$ for each $(u;\delta;n)$ in $\mathbf{F}$. We have

$$u^i = u + \Sigma_{j=1}^{i-1} q^{\delta j}$$

for i in μ and regarding indices modulo $n+2$ we have

$$q^{\delta j} = v^{\beta j+1} - v^{\beta j}$$

for j in ν for some ordering β of n elements of μ. Thus

$$u^i = u + \Sigma_{j=1}^{i-1}(v^{\beta j+1} - v^{\beta j})$$

and

$$h(u^i) = h(u) + \Sigma_{j=1}^{i-1}(h(v^{\beta j+1}) - h(v^{\beta j}))$$

for i in μ. By Lemma 6.23 $z = h(w)$ is integral. So

$$h(u^i) = z + \Sigma_{j=1}^{i-1}(w^{\beta j+1+i}) - w^{\beta j+i})$$

for i in μ. Therefore, $h(u;\delta;n) = (z;\zeta;n)$ where ζ is defined by

$$q^{\zeta i} = w^{\beta j+1+i} - w^{\beta j+i} .$$

See Lemma 6.8. Now we consider the assertion of $\mathbf{Q}$. For $\alpha \subset \mu$

$$h(q^{\alpha}y) - h(0) = h(q^{\alpha})y$$

which equals, for some ordering β of $k = \#\alpha$ elements of μ, the quantity

$$(h(v^{\beta 1+1} - v^{\beta 1}), \ldots, h(v^{\beta k+1} - v^{\beta k}))y$$

$$= ((w^{\beta 1+1+i} - w^{\beta 1+i}), \ldots, (w^{\beta k+1+i} - w^{\beta k+i}))y$$

which equals $q^{\delta}y$ for the obvious choice of $\delta \subset \mu$. □

If $h(\mathbf{F}) = \mathbf{F}$, does it follows that $h(r^{-1}\mathbf{F}) = r^{-1}\mathbf{F}$? The next lemma indicates that the answer is, yes.

10.6 Lemma: Let $h : \mathbb{C}^n \to \mathbb{C}^n$ be a linear map such that $h(\mathbf{F}) = \mathbf{F}$ and r be a positive integer. Then $h(r^{-1}\mathbf{F}) = r^{-1}\mathbf{F}$.

Proof: Assume h has the form $Ax+a$. Since $h(0)$ is in $\mathbf{F}^0$ the vector a is integral.

$$A(r^{-1}\mathbf{F}) + a = r^{-1}A\mathbf{F} + r^{-1}a + (1-r^{-1})a$$

$$= r^{-1}(A\mathbf{F} + a) + (1-r^{-1})a$$

$$= r^{-1}\mathbf{F} + (1-r^{-1})a$$

$$= r^{-1}(\mathbf{F} + (r-1)a) = r^{-1}\mathbf{F}$$

since $\mathbf{F}+z = \mathbf{F}$ for z integral. □

If in the lemma above $h(0) = 0$, then, clearly, $h(r\mathbf{F}) = r\mathbf{F}$ for any r in $\mathbb{C}$.

Our attention is transferred to the convenient property of $\mathbf{F}$ that each vertex can be assigned to an integer in $\mu \triangleq \{1, \ldots, n+1\}$ in such a way that each h-simplex $(v;\pi)$ of $\mathbf{F}$ has a complete set of labels, namely, μ, on its vertices, see Figure 10.3.

Define $\underline{\ell} : \mathbb{C}^n \to \mathbb{C}$ by $\underline{\ell}(v) = ev \bmod(n+1)$. Formally, $\underline{\ell}(v)$ is the collection of scalars ℓ such that

$$\ell + k(n+1) = \Sigma_{i \in \nu} v_i$$

where $v = (v_1, \ldots, v_n)$, k is an integer, and $\nu \triangleq \{1, \ldots, n\}$. However, as is customary, we shall represent the class $\underline{\ell}(v)$ with any member of it.

10.7 Example: $\underline{\ell}(2,0,9,-1) = 5$ or 0 or 15 etc.; note $n = 4$. □

For each integral point v of $\mathbb{C}^n$ we can represent $\underline{\ell}(v)$ as an element of $\mu \triangleq \{1, \ldots, n+1\}$. Note that $\underline{\ell}$ is additive, that is, $\underline{\ell}(u+v) = \underline{\ell}(u) + \underline{\ell}(v) \bmod n+1$.

10.8 Example: For $n = 2$, we have $\underline{\ell}(5,1) = 6 = 3 = \ell(6,6) + (-1,-5))$ $= 12-6 = 6 = 6 = 3 = 0$ modulo 3. □

10.9 Lemma: For x in $\mathbb{Z}^n$ with $0 \leqq x \leqq e$ or $-e \leqq x \leqq 0$ we have $\underline{\ell}(x) = 1$, if and only if $x = q^i$ for some i in μ.

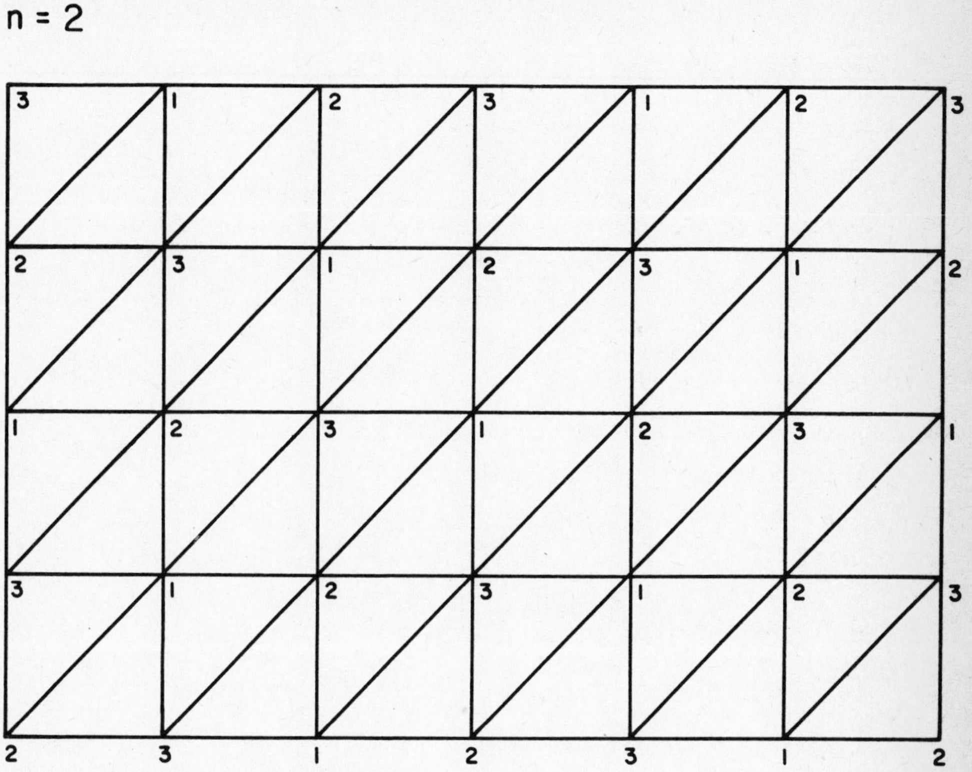

Figure 10.3

Proof: $\underline{\ell}(e^i) = 1$ and $\underline{\ell}(-e) = -n = 1 \bmod n+1$. If x with $e \geqq x \geqq 0$ is integral, then $\underline{\ell}(x) = i$ where $0 \leqq i \leqq n$ according to the number of 1's in x, and similarly for $-e \leqq x \leqq 0$. □

The next lemma states that each n-simplex of **F** has a complete set of labels.

10.10 Lemma: For each (v,π,n) of $i^*\mathbf{F}$, $\underline{\ell}\{v^1, \ldots, v^{n+1}\} = \mu$ and, more particularly, $\underline{\ell}(v^{i+j}) = \underline{\ell}(v^i) + j \bmod n+1$ for i and j in μ.

Proof:

$$\underline{\ell}(v^{i+1}) = \underline{\ell}(v^i + q^i) = \underline{\ell}(v^i) + \underline{\ell}(q^i) = \underline{\ell}(v^i) + 1 \bmod n+1 \, . \qquad \square$$

The converse of the previous lemma is established in the next lemma.

10.11 Lemma: For (v,π,n) in $i_*\mathbf{F}$ let $w^1, \ldots, w^{n+1}$ be the vertices of $(v;\pi;n)$ in an arbitrary order. If $\underline{\ell}(w^{i+1}) = \underline{\ell}(w^i) + 1$ for i in ν, then there is a pair (w,γ) with $(v;\pi;n) = (w;\gamma;n)$, $w^1 = w$, and $w^{i+1} = w^i + q^{\gamma i}$ for i in ν.

Proof: If $\underline{\ell}(w^{i+1}) = \underline{\ell}(w^i) + 1$ and w^{i+1} and w^i are both vertices of $(v;\pi;n)$, then $w^{i+1} - w^i = q^{\gamma i}$ for some γi in μ, see Lemma 10.9. Furthermore, we cannot have $\gamma i = \gamma j$ for $i \neq j$, see Lemma 6.8. □

10.12 Lemma: If $(u;\gamma;n) = (v;\pi;n)$, $\underline{\ell}(u) = k$ and $\underline{\ell}(v) = k+j$, then

$$u^{i+j} = v^{i}$$

for i in μ where indices are regarded modulo $n+1$.

Proof: $u^k = v^h$ if and only if $\underline{\ell}(u^k) = \underline{\ell}(v^h)$ from Lemma 10.10. Also $\underline{\ell}(u^{i+j}) = \underline{\ell}(u)+i+j-1$ and $\underline{\ell}(v^i) = \underline{\ell}(v)+i-1$ modulo $n+1$. □

We shall say that the index (v,π,n) or simplex $(v;\pi;n)$ is coordinated by $\underline{\ell}$ if $\underline{\ell}(v^i) = \underline{\ell}(s^i)$ for i in μ.

$$\underline{\ell}(v^2) = \underline{\ell}(s^2) = 1$$

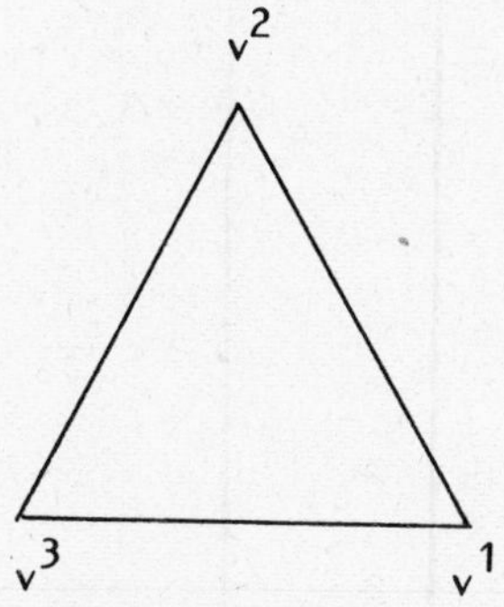

$$\underline{\ell}(v^3) = \underline{\ell}(s^3) = 2 \qquad \underline{\ell}(v^1) = \underline{\ell}(s^1) = 0$$

10.13 Example: $(0,\pi)$ where $\pi = (1, \ldots, n)$ is $\underline{\ell}$ coordinated. Of course, $\underline{\ell}(s^i) = i-1$ for i in μ. □

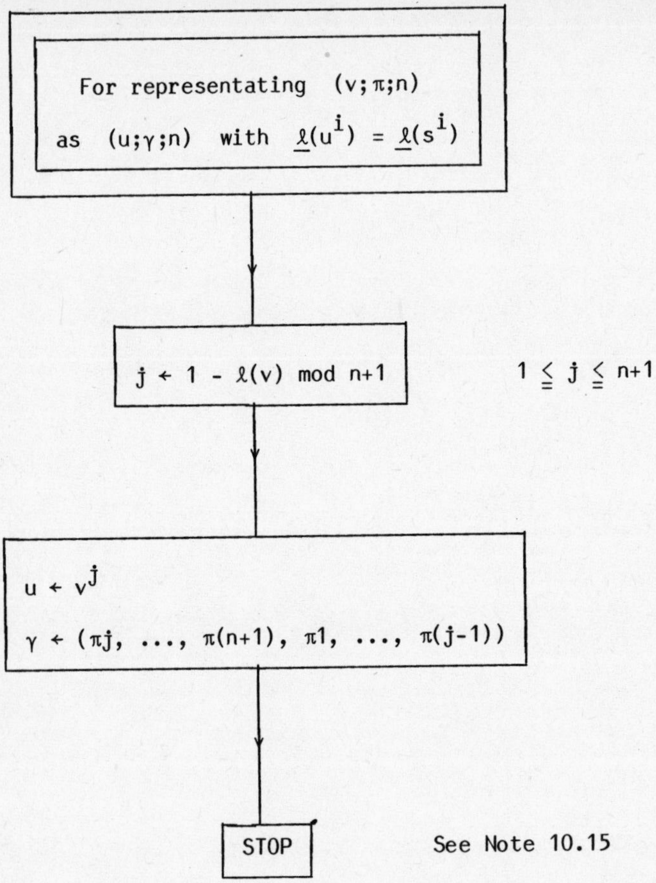
For representating (v;π;n)
as (u;γ;n) with ℓ(u^i) = ℓ(s^i)
j ← 1 - ℓ(v) mod n+1
1 ≦ j ≦ n+1
u ← v^j
γ ← (πj, ..., π(n+1), π1, ..., π(j-1))
STOP
See Note 10.15

Figure 10.4

The next lemma observes that there is always a representation that is coordinated by $\underline{\ell}$.

10.14 Lemma: Any n-simplex $(v;\pi;n)$ of **F** can be represented as $(u;\gamma;n)$ which is coordinated by $\underline{\ell}$.

Proof: Select j with $\underline{\ell}(v^j) = 0$. Set $u = v^j$ and $\gamma = (\pi j, \ldots, \pi(n+1), \pi 1, \ldots, \pi(j-1))$. Clearly $(v;\pi;n) = (u;\gamma;n)$ and $\underline{\ell}(u^i) = \underline{\ell}(s^i)$ for i in μ. □

The chart of Figure 10.4 encapsulates this process and will be employed in replacement rules later. Observe that if $\underline{\ell}(v) = j$, then $\underline{\ell}(v^{1-j}) = 0$ where the superscript is regarded modulo $n+1$.

10.15 Note (for Figure 10.4): $(v;\pi;n) = (u;\gamma;n)$ and $\underline{\ell}(u^i) = \underline{\ell}(s^i)$ for i in μ. □

We shall sometimes be required to perform replacements wherein the representations should remain, before and after, coordianted with $\underline{\ell}$. Such replacement rules are given in the the chart of Figure 10.5. Given $(v;\pi;n)$ and t the chart computes an adjacent simplex $(\hat{v};\hat{\pi};n)$ that does not have v^t as a vertex; if $(v;\pi;n)$ is $\underline{\ell}$ coordinated, then $(\hat{v};\hat{\pi};n)$ will be also.

10.16 Note (for Figure 10.5): The new vertex is $\hat{v}^t$. The correspondence between vertices of $(v;\pi;n)$ and $(\hat{v};\hat{\pi};n)$

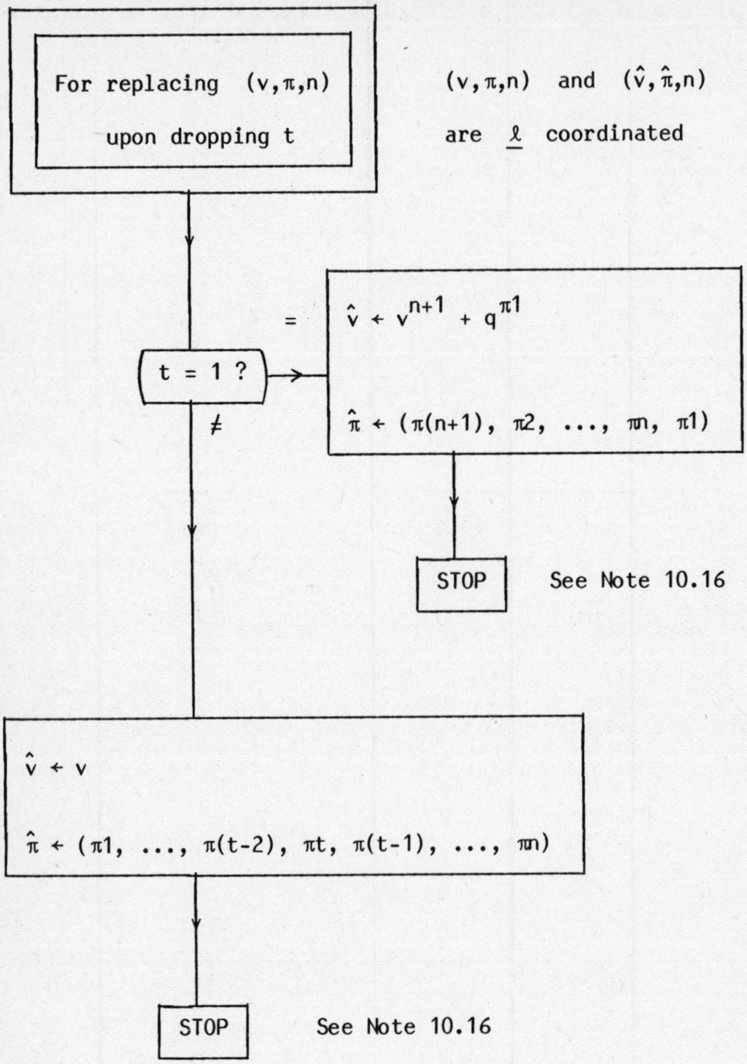
For replacing (v,π,n)
upon dropping t
(v,π,n) and (v̂,π̂,n)
are ℓ coordinated
=
v̂ ← v^{n+1} + q^{π1}
π̂ ← (π(n+1), π2, ..., πn, π1)
t = 1 ?
≠
STOP
See Note 10.16
v̂ ← v
π̂ ← (π1, ..., π(t-2), πt, π(t-1), ..., πn)
STOP
See Note 10.16

Figure 10.5

$$\begin{array}{ccccccc}
1 & 2 & & \boxed{t} & & n & n+1 \\
| & | & \cdots & & \cdots & | & | \\
1 & 2 & & \boxed{t} & & n & n+1
\end{array}$$

where the boxes indicate the dropped and adjoined vertices. □

10.17 Bibliographical Notes: That the vertices of **F** can be labeled with 1, ..., n+1 in such a way that each n-simplex receives a complete set of labels was first observed and proved in Eaves and Saigal [1972]. Todd [1976b] found the function $\underline{\ell}$ which simplified matters considerably. Van der Laan and Talman [1979] were the first to utilize the fact that each n-simplex of **F** had n+1 representations. □

11. TRIANGULATION $F|Q_\alpha$

As **F** is a refinement of **Q**, the natural restriction $F|Q_\alpha$ of **F** to a cone Q_α of **Q** triangulates Q_α, see Lemma 6.22 and Figure 11.1. Our purpose is to develop representation and replacement of rules for $F^k|Q_\alpha \triangleq (F|Q_\alpha)^k$ where $k = \#\alpha$ and $\alpha \subset \mu$.

Recall $\mu = \{1, \ldots, n+1\}$, $q = (e^1, e^2, \ldots, e^n, -e)$, and $Q_\alpha = \{q^\alpha y : y \geq 0\}$.

Define $i(F|Q_\alpha)^k$ to be the collection of all tuples (v,π,k) in i_*F with

(a) $(\pi|k)\mu = \alpha$

(b) $v \in Q_\alpha$.

In view of (a) and (b) we see that the simplex $(v;\pi;k)$ is contained in Q_α.

We proceed to show that the collections of k-simplexes $(F|Q_\alpha)^k$ and $\{(v;\pi;k) : (v,\pi,k) \in i(F|Q_\alpha)^k\}$ are equal. Of course, we know that if $(v;\pi;k) = (u;\gamma;k)$ where (v,π,k) and (u,γ,k) are in $i(F|Q_\alpha)^k$ then $(v, \pi|k, k) = (u, \gamma|k, k)$. See Lemma 6.3.

11.1 Lemma: $(F|Q_\alpha)^k$ is the collection of all $(v;\pi;k)$ with (v,π,k) in $i(F|Q_\alpha)^k$.

Proof: If $v \in Q_\alpha \cap Z^n$ and $(\pi|k)\mu = \alpha$, then $(v;\pi;k)$ is contained in Q_α and $(v;\pi;k)$ is a k-simplex. It only remains to show that the collection of $(v;\pi;k)$ covers Q_α. Let x be a point in Q_α. As **F** is a refinement of **Q**, Q_α is covered by $F^k|Q_\alpha$. Some

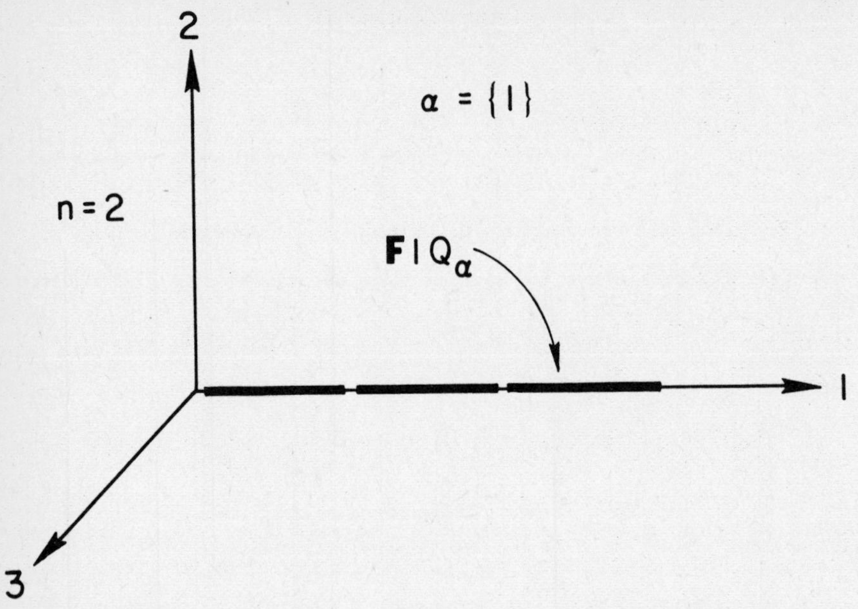

Figure 11.1

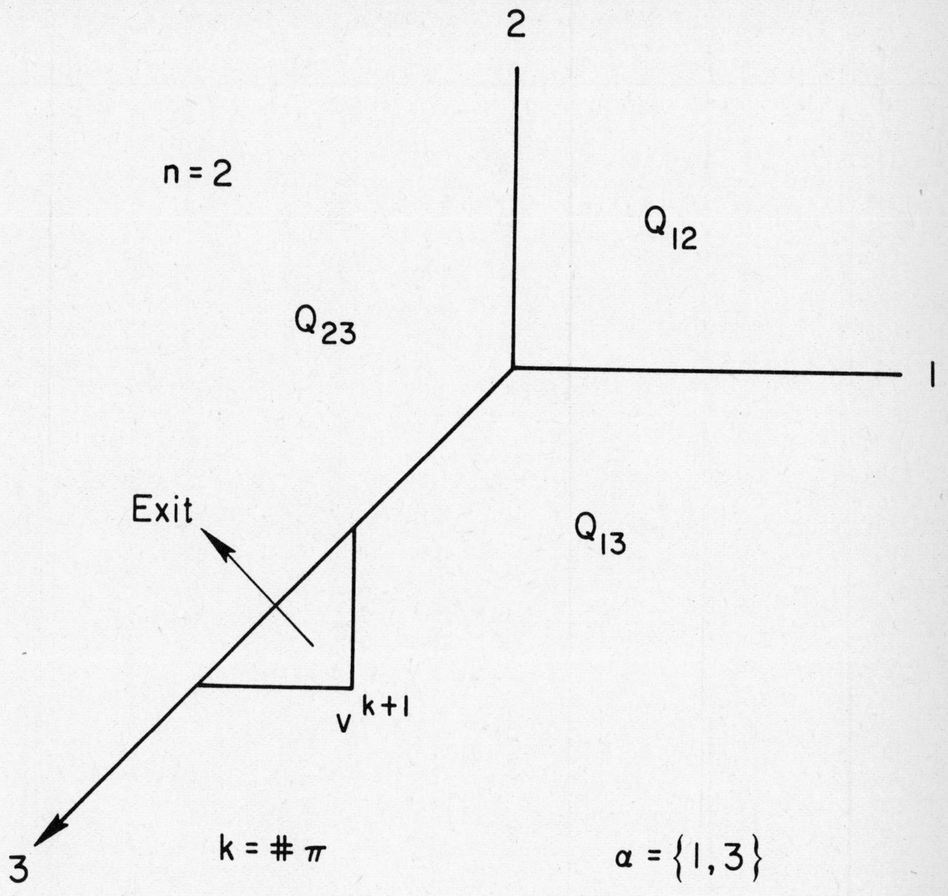

Figure 11.2

k-simplex τ in $\mathbf{F}^k|Q_\alpha$ contains x and some σ in $\mathbf{F}^n$ has τ as a face. The simplex σ is in some Q_β with $\beta \supseteq \alpha$, so represents σ as $(u;\gamma;n)$ where $(\gamma|n)\mu = \beta$. We then have

$$u^1 \leq_\beta u^2 \leq_\beta \cdots \leq_\beta u^{n+1} .$$

So once the sequence u^i leaves Q_α it never returns. It follows that $\tau = (u;\gamma;k)$ and $(u,\gamma,k) \in i(\mathbf{F}|Q_\alpha)^k$. □

Therefore, we see that $i(\mathbf{F}|Q_\alpha)^k$ is a representation set for $(\mathbf{F}|Q_\alpha)^k = \mathbf{F}^k|Q_\alpha$.

11.2 Lemma: For $\alpha \subset \mu$ and $\beta \subset \mu$ we have $(\mathbf{F}|Q_\alpha) \cap (\mathbf{F}|Q_\beta) = \mathbf{F}|Q_{\alpha\cap\beta}$.

Proof: $\sigma \in \mathbf{F}$, $\sigma \subset Q_\alpha$ and $\sigma \subset Q_\beta$, if and only if $\sigma \in \mathbf{F}$ and $\sigma \subset Q_\alpha \cap Q_\beta$. Further, $Q_{\alpha\cap\beta} = Q_\alpha \cap Q_\beta$, see Lemma 5.3. □

Another way of stating the lemma above is

$$(\mathbf{F}|Q_\alpha)|Q_\beta = \mathbf{F}|Q_{\alpha\cap\beta} .$$

11.3 Exercise: For τ in $\mathbf{F}$ let α be the smallest subset of μ with τ in Q_α and let $k = \#\alpha$. Prove τ is in $(\mathbf{F}|Q_\alpha)^k$ if and only if $\dim \tau \geq k$. □

The replacement rules of $(\mathbf{F}|Q_\alpha)^k$ are very nearly those of $\mathbf{F}$, but, of course, a replacement may not be possible, which is to say, an exit may occur, see Figure 11.2.

For the replacement rules of $(\mathbf{F}|Q_\alpha)^k$ let us adjoin some redundant information to each of the representatives. Namely, redefine $i(\mathbf{F}|Q_\alpha)^k$ to be the collection of all (v,π,k,η) such that

(a) $(v,\pi,k) \in i_*\mathbf{F}$

(b) $(\pi|k)\mu = \alpha$

(c) $v \in Q_\alpha$

(d) $v = q\eta,\ 0 \nless \eta \geqq 0$

Given v, of course, η is uniquely determined and could be calculated, however, η is tested in replacement rules, and it appears to be more efficient to carry it along and update it rather than recompute it.

We desire to replace $(v;\pi;k;\eta)$ on dropping v^t where $1 \leqq t \leqq k+1$. Once again we have three cases corresponding to $t = 1$, $2 \leqq t \leqq k$, and $t = k+1$. As we shall see the exit cannot occur in the first and second cases. When an exit occurs it is because all vertices $(v;\pi;k;\eta)$ except v^t lie in a facet of Q_α; identification of this facet is included in our exposition of the replacement rules.

Case $t = 1$: Let $\hat{v} = v^2$, $\hat{\eta} = \eta + e^{\pi 1}$, and $\hat{\pi} = (\pi 2, \ldots, \pi k, \pi 1, \pi(k+1), \ldots, \pi(n+1))$. Clearly $(\hat{v};\hat{\pi};\hat{k};\hat{\eta})$ is in $i(\mathbf{F}|Q_\alpha)^k$. Also $(v;\pi;k;\eta)$ and $(\hat{v};\hat{\pi};k;\hat{\eta})$ both contain the vertices $v^2 = \hat{v}^1$, $v^3 = \hat{v}^2, \ldots, v^{k+1} = \hat{v}^k$ but the latter does not contain v^1 thus $(\hat{v};\hat{\pi};k;\hat{\eta})$ is the replacement. □

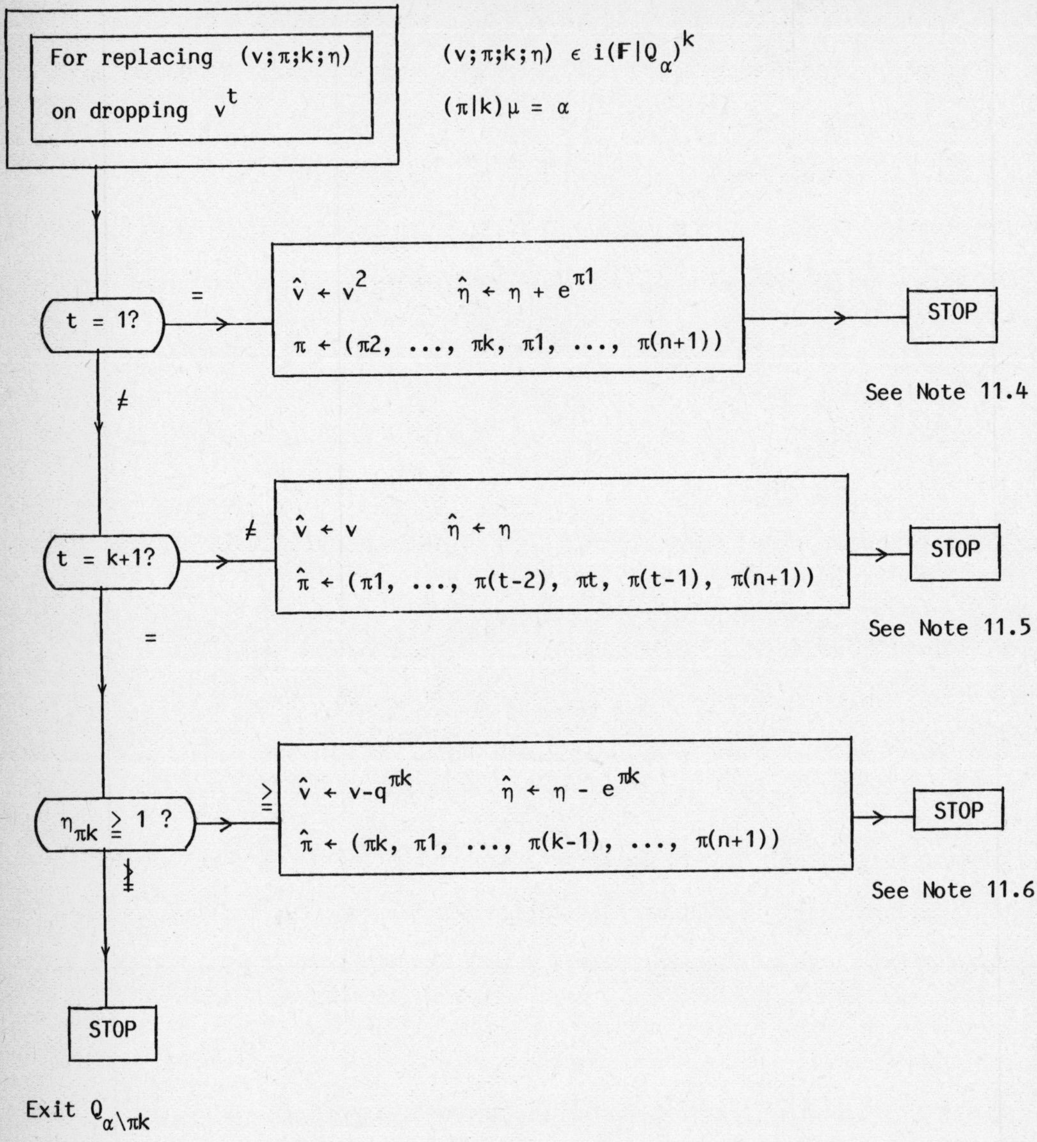
For replacing (v;π;k;η)
on dropping v^t
(v;π;k;η) ∈ i(F|Q_α)^k
(π|k)μ = α
t = 1?
=
v̂ ← v^2   η̂ ← η + e^π1
π ← (π2, ..., πk, π1, ..., π(n+1))
STOP
See Note 11.4
≠
t = k+1?
≠
v̂ ← v   η̂ ← η
π̂ ← (π1, ..., π(t-2), πt, π(t-1), π(n+1))
STOP
See Note 11.5
=
η_πk ≧ 1 ?
≧
v̂ ← v-q^πk   η̂ ← η - e^πk
π̂ ← (πk, π1, ..., π(k-1), ..., π(n+1))
STOP
See Note 11.6
≱
STOP
Exit Q_α\πk

Figure 11.3

Case 2 $(2 \leqq t \leqq k)$**:** Let $\hat{v} = v$, $\hat{\eta} = \eta$, and $\hat{\pi} = \pi 1, \ldots, \pi t, \pi(t-1), \ldots, \pi(n+1))$. Clearly $(\hat{v},\hat{\pi},k,\hat{\eta})$ is in $i(F|Q_\alpha)^k$. The simplexes $(v;\pi;k;\eta)$ and $(\hat{v};\hat{\pi};k;\hat{\eta})$ both contain $v^1 = \hat{v}^1, \ldots, v^{t-1} = \hat{v}^{t-1}$, $v^{t+1} = \hat{v}^{t+1}, \ldots, v^{k+1} = \hat{v}^{k+1}$ and thus $(\hat{v};\hat{\pi};k;\hat{\eta})$ is the replacement. □

Case $t = k+1$**:** If $\eta_{\pi k} \geqq 1$ then $\hat{v} = v - q^{\pi k}$ is in Q_α. Therefore letting $\hat{\eta} = \eta - e^{\pi k}$ and $\hat{\pi} = (\pi k, \pi 1, \ldots, \pi(k-1), \pi(k+1), \ldots, \pi(n+1))$ the index $(\hat{v},\hat{\pi},k,\hat{\eta})$ is in $i(F|Q_\alpha)^k$, $(v;\pi;k;\eta)$ and $(\hat{v};\hat{\pi};k;\hat{\eta})$ are adjacent. $(\hat{v};\hat{\pi};k;\hat{\eta})$ does not contain v^{k+1}, and we have $(\hat{v};\hat{\pi};k;\hat{\eta})$ as the replacement of $(v;\pi;k;\eta)$ on dropping v^{k+1}. If $\eta_{\pi k} = 0$ then $v^1, \ldots, v^k$ all lie in $Q_{\alpha\backslash\pi k}$ and the replacement attempt exits through facet $Q_{\alpha\backslash\pi k}$, see Lemma 3.10. □

11.4 Note (for Figure 11.3): The adjoined vertex is $\hat{v}^{k+1}$. The correspondence between vertices of $(v;\pi;k;\eta)$ and $(\hat{v};\hat{\pi};k;\hat{\eta})$ is:

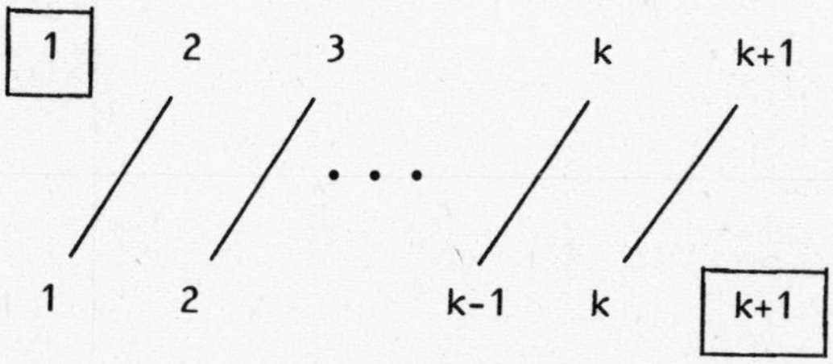

□

11.5 Note (for Figure 11.3): The adjoined vertex is $\hat{v}^t$. The correspondence of vertices is

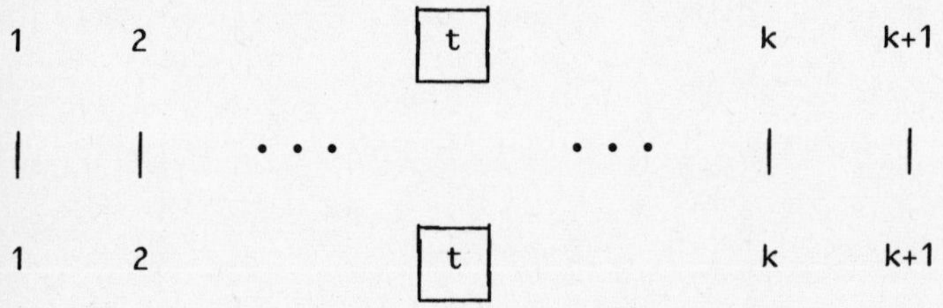

□

11.6 Note (for Figure 11.3): The adjoined vertex is $\hat{v}$. The correspondence of vertices is

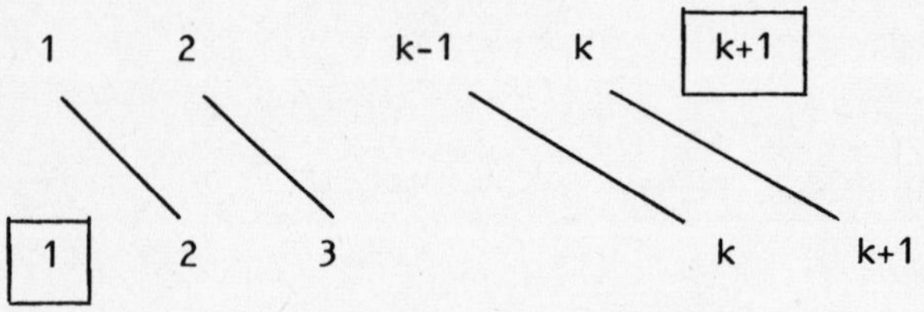

□

11.7 Remark: Let σ be a simplicial cone with generators $u^1, \ldots, u^n$. A class of subdivisions of σ is easily obtained as follows. Let $\alpha_1, \ldots, \alpha_n$ be any positive numbers and define $f : Q_\nu \to \sigma$ by

$$f(x) \equiv \sum_{i=1}^{n} u^i \alpha_i x_i$$

where $Q_\nu \triangleq \{x \in \mathbb{C}^n : x \geqq 0\} \equiv \mathbb{C}^n_+$. For any subdivision $\mathbf{G}$ of Q_ν, for example, $\mathbf{F}|Q_\nu$, the triangulation obtained by restricting the

Freudenthal triangulation to $\mathbb{C}^n_+$, then $f(\mathbf{G})$ subdivides σ. Of course, $f(\mathbf{G})$ and $\mathbf{G}$ are isomorphic. □

11.8 Bibliographical Notes: Van der Laan and Talman [1979] were the first to use $\mathbb{F}|Q_\alpha$. □

12. JUXTAPOSITIONING WITH $\underline{\ell}$

We examine one of the many ways in which subdivisions can be combined in order to form larger ones; this method will be referred to as juxtapositioning.

Let S be the standard n-simplex, $U = [a,b]$ a subset of $\mathbb{C}$, and $\mathbf{X}$ an $(n+1)$-subdivision of $S \times U$.

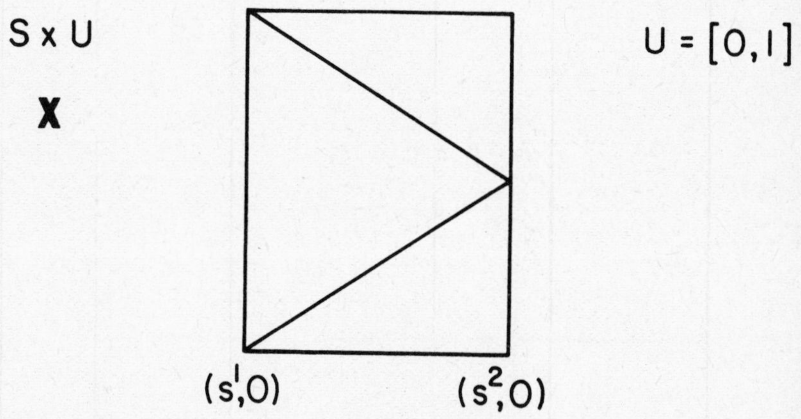

By appropriate placement, called juxtapositioning, of copies of the subdivision $\mathbf{X}$ of $S \times U$ on the n-simplexes of the Freudenthal triangulation of $\mathbb{C}^n$ we obtain an $(n+1)$-subdivision $\mathbf{J}$ of $J \equiv \mathbb{C}^n \times U$, see Figure 12.1. If $\mathbf{X}$ is a triangulation, then $\mathbf{J}$ will be a triangulation also.

For each simplex σ of $\mathbf{F}$ let us define $\underline{\ell}(\sigma)$ to be the collection of numbers $\underline{\ell}(v)$ (see Section 10) where v is a vertex of σ. For each simplex σ of $\mathbf{F}$ we define a linear map $T_\sigma : S_{\underline{\ell}(\sigma)} \to \sigma$ from the face $S_{\underline{\ell}(\sigma)}$ of S to σ by setting

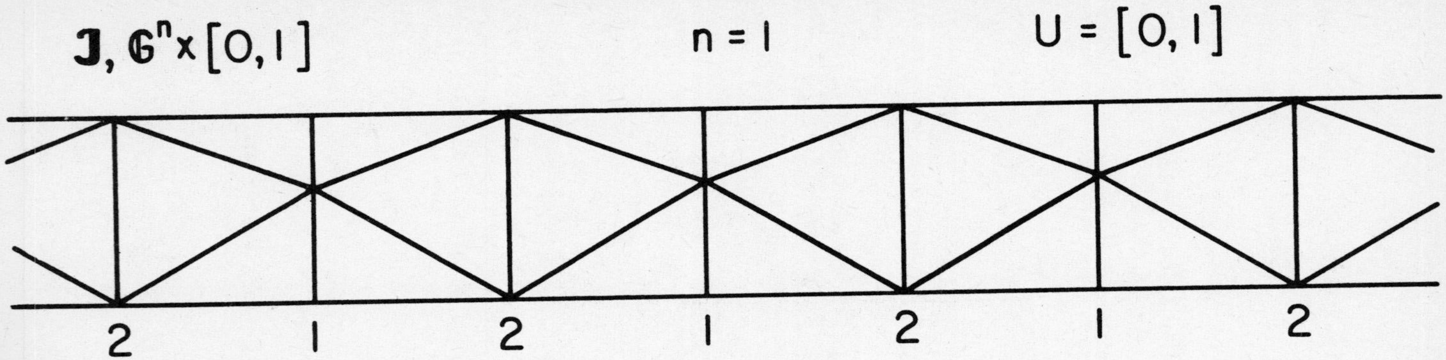

Figure 12.1

$$T_\sigma(s^i) = w^i ,$$

where s^i is a vertex of S and w^i is a vertex of σ with $\underline{\ell}(w^i) = \underline{\ell}(s^i)$. Clearly T_σ is one-to-one and onto. If $\sigma = (v;\pi;n)$ is a simplex in **F** and $\underline{\ell}(v^i) = \underline{\ell}(s^i)$ for i in μ then $T_\sigma(s^i) = v^i$ for i in μ.

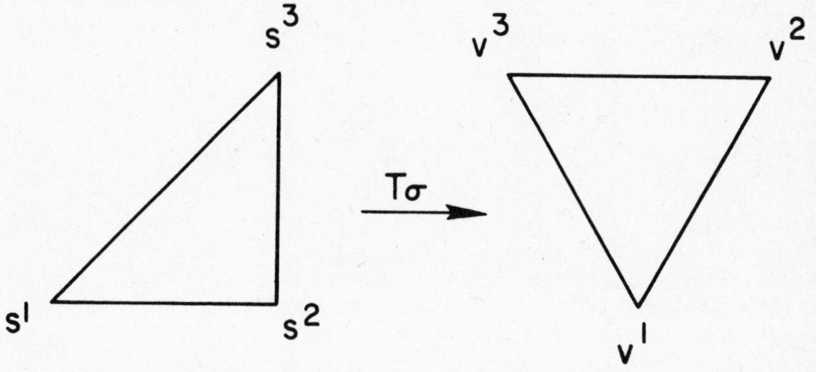

$$s^i \to v^i \qquad \underline{\ell}(v^i) = \underline{\ell}(s^i) = i-1 .$$

12.1 Lemma: If (v,π,n) in i_***F** is $\underline{\ell}$ coordinated then

$$T_\sigma(x) = v + q^{\pi|n}x$$

where $\sigma = (v;\pi;n)$.

Proof: See Lemma 6.2. □

Lemma 4.1 indicates the efficient way of evaluating $q^{\pi/n}x$. Let us extend the map T_σ to a linear map on $S_\alpha \times U$ by $T_\sigma(x,t) = (T_\sigma(x),t)$, that is, T_σ is the identity for the $(n+1)^{st}$ coordinate.

Let $\mathbf{X}$ be an unknown subdivision of $S \times U$ and we proceed to define the subdivision $\mathbf{J}$ of $J \equiv \mathbb{C}^n \times U$. $\mathbf{J}$ is defined to be the collection of all cells

$$T_\sigma(\rho)$$

where σ is in $\mathbf{F}^n$ and ρ is in $\mathbf{X}$, see Figure 12.1.

12.2 Theorem: $\mathbf{J}$ subdivides $\mathbb{C}^n \times U$.

Proof: As $T_\sigma(\mathbf{X}) \subseteq \mathbf{J}$ covers $\sigma \times U$ for σ in $\mathbf{F}^n$ it is clear that $\mathbf{J}$ covers $\mathbb{C}^n \times U$. Now consider $T_\sigma(\mathbf{X})$ and $T_\tau(\mathbf{X})$ for σ and τ in $\mathbf{F}$. We have

$$T_\sigma(\mathbf{X})|(\tau \times U) = T_{\sigma\cap\tau}(\mathbf{X}|S_{\underline{\ell}(\sigma\cap\tau)}) = T_\tau(\mathbf{X})|(\sigma \times U)$$

which is a subdivision of $\sigma \cap \tau \times U$, see Figure 12.2. Thus from Lemma 3.8 the cells of $T_\sigma(\mathbf{X})$ and $T_\tau(\mathbf{X})$ meet in a common face, and consequently, cells of $\mathbf{J}$ meet in a common face.

The n-cells η of $\mathbf{J}$ are images of T_σ of an n-cell ξ of $\mathbf{X}$ for some σ in $\mathbf{F}^n$. If ξ is contained in $\partial S \times U$ then η lies in two (n+1)-cells of $\mathbf{J}$, otherwise η is contained in as many (n+1)-cells as ξ, namely one or two. □

12.3 Exercise: Extend the results heretofore of this section to the situation where U is a subset of $\mathbb{C}^k$. □

Our next step is to set $U = [0,1]$, to specify further the nature of the subdivision $\mathbf{X}$ of $S \times U = S \times [0,1]$, and then to draw the attendant conclusions about $\mathbf{J}$.

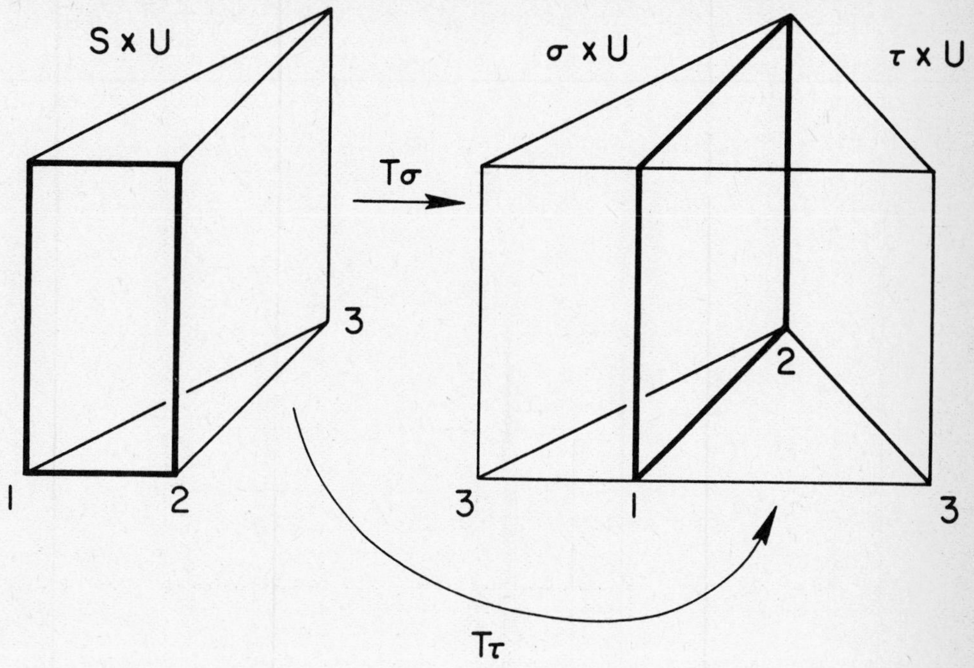

Figure 12.2

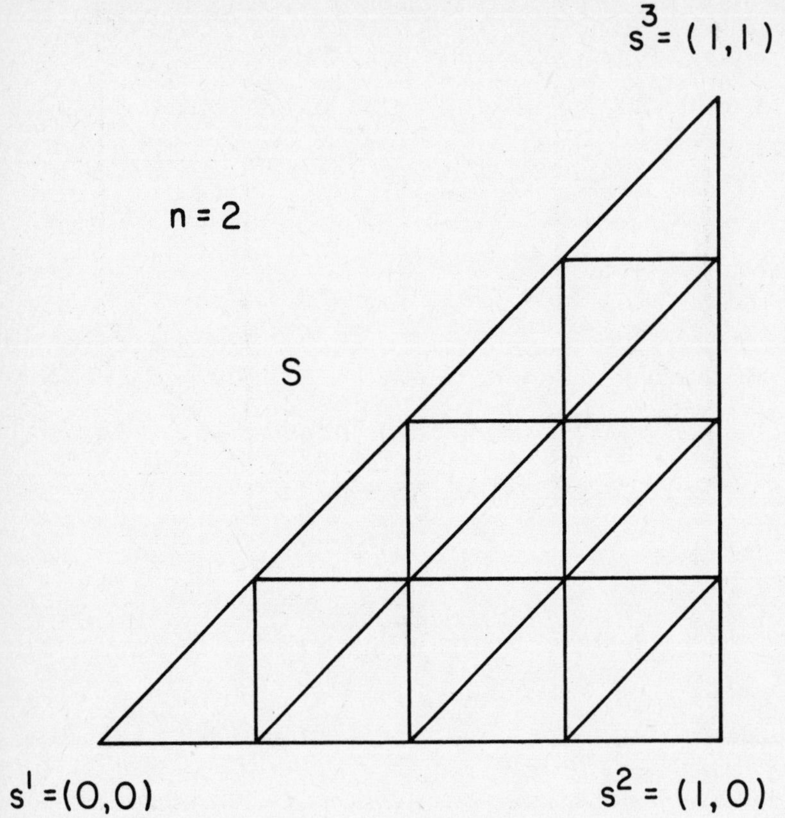

Figure 12.3

Select some $r = 1, 2, \ldots$ called the grid refinement factor and consider the triangulation $r^{-1}\mathbf{F}$ where $\mathbf{F}$ is the Freudenthal triangulation of $\mathbb{C}^n$. As S the standard simplex is in $\mathbf{F}$ the natural restriction of $r^{-1}\mathbf{F}$ to S forms a subdivision of S, see Figure 12.3.

Let us assume that $\mathbf{X}$ is a subdivision of $S \times [0,1]$ with the following two properties.

1) $S \times 0$ is an element of $\mathbf{X}$

2) the restriction of $\mathbf{X}$ to $S \times 1$ is $(r^{-1}\mathbf{F}|S) \times 1$.

Figure 12.4 illustrates $\mathbf{X}$ on $S \times 0$ and $S \times 1$ without indicating anything about the nature of $\mathbf{X}$ in $S \times (0,1)$; the latter matter is labored upon in the construction of the triangulation $\mathbf{V}[r,p]$ of $S \times [0,1]$ in Section 16.

Under these two assumptions on $\mathbf{X}$ the question is, what is the restriction of $\mathbf{F}$ to $\mathbb{C}^n \times 1$? The answer is $r^{-1}\mathbf{F} \times 1$! See Figures 12.5 and 12.6.

With this observation in hand we could repeat the juxtapositioning on $\mathbb{C}^n \times 1$ to obtain a triangulation of $\mathbb{C}^n \times [0,2]$ etc., and indeed this is exactly what is done in Section 17 to obtain the variable rate refining homotopy triangulation $\mathbf{S}$. For the moment, however, let us be content with analyzing the triangulation $\mathbf{J}$ on $\mathbb{C}^n \times [0,1]$. The following theorem establishes that $\mathbf{J}$ restricted to $\mathbb{C}^n \times 1$ is $r^{-1}\mathbf{F} \times 1$, that is, $\mathbf{J}|(\mathbb{C}^n \times 1) = r^{-1}\mathbf{F}$. Let σ be an n-simplex of $\mathbf{F}$.

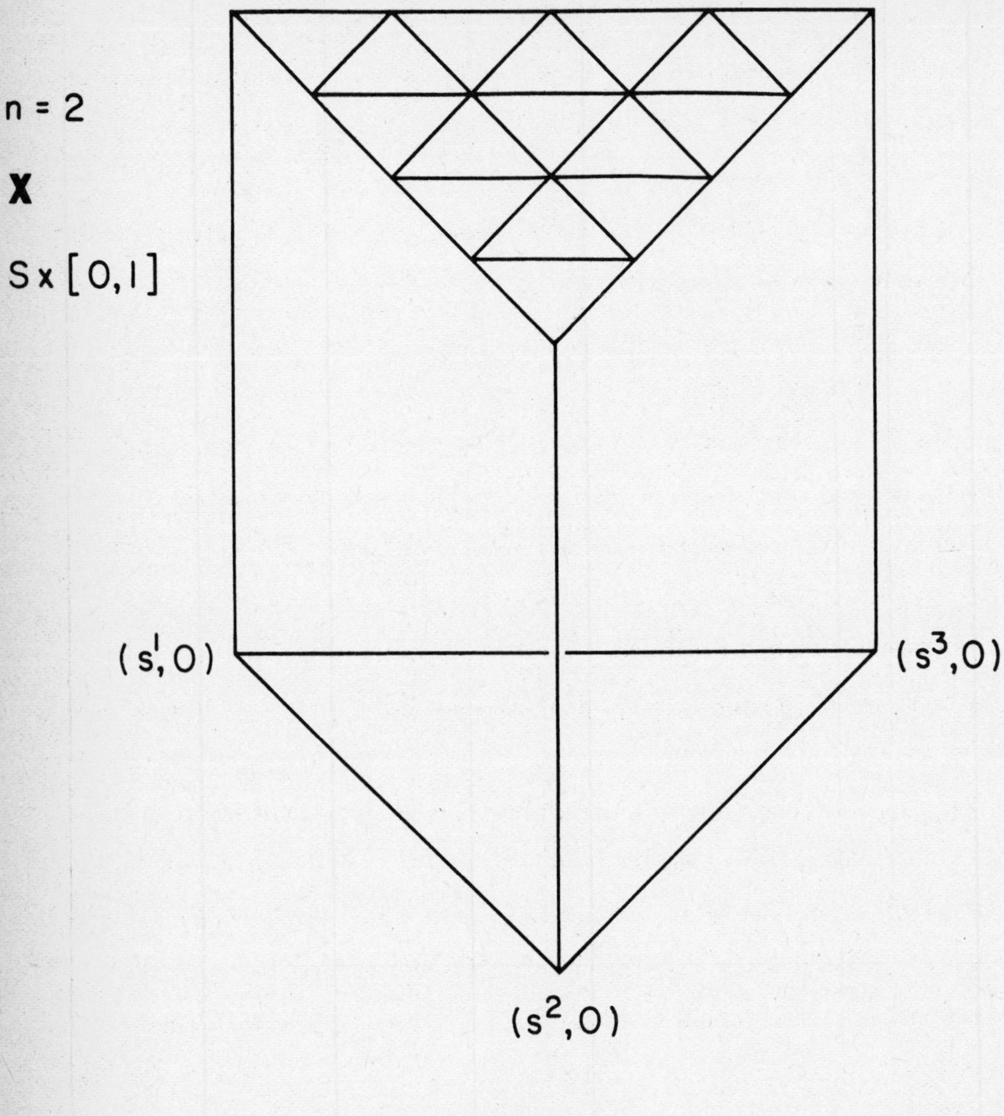

Figure 12.4

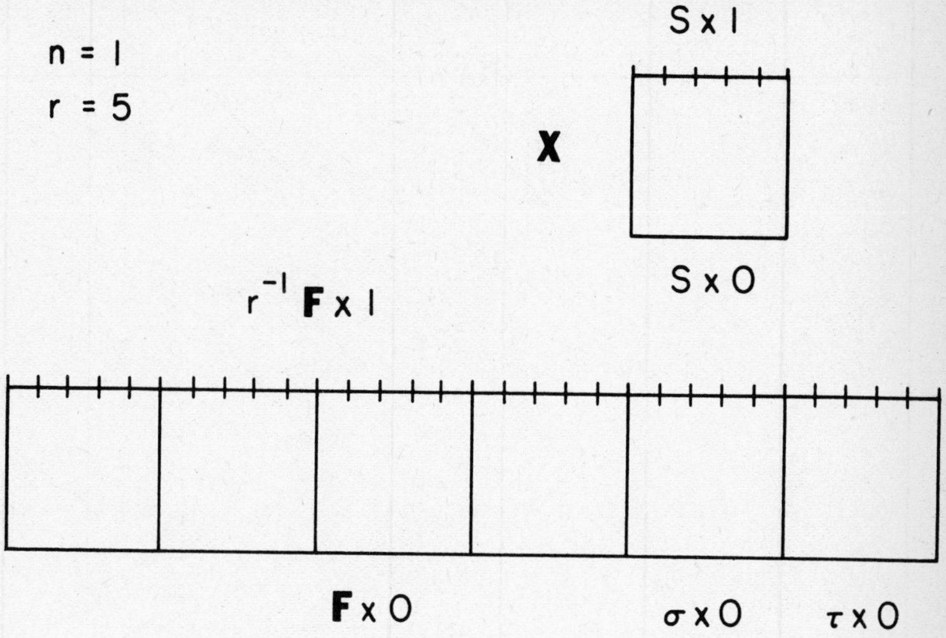

Figure 12.5

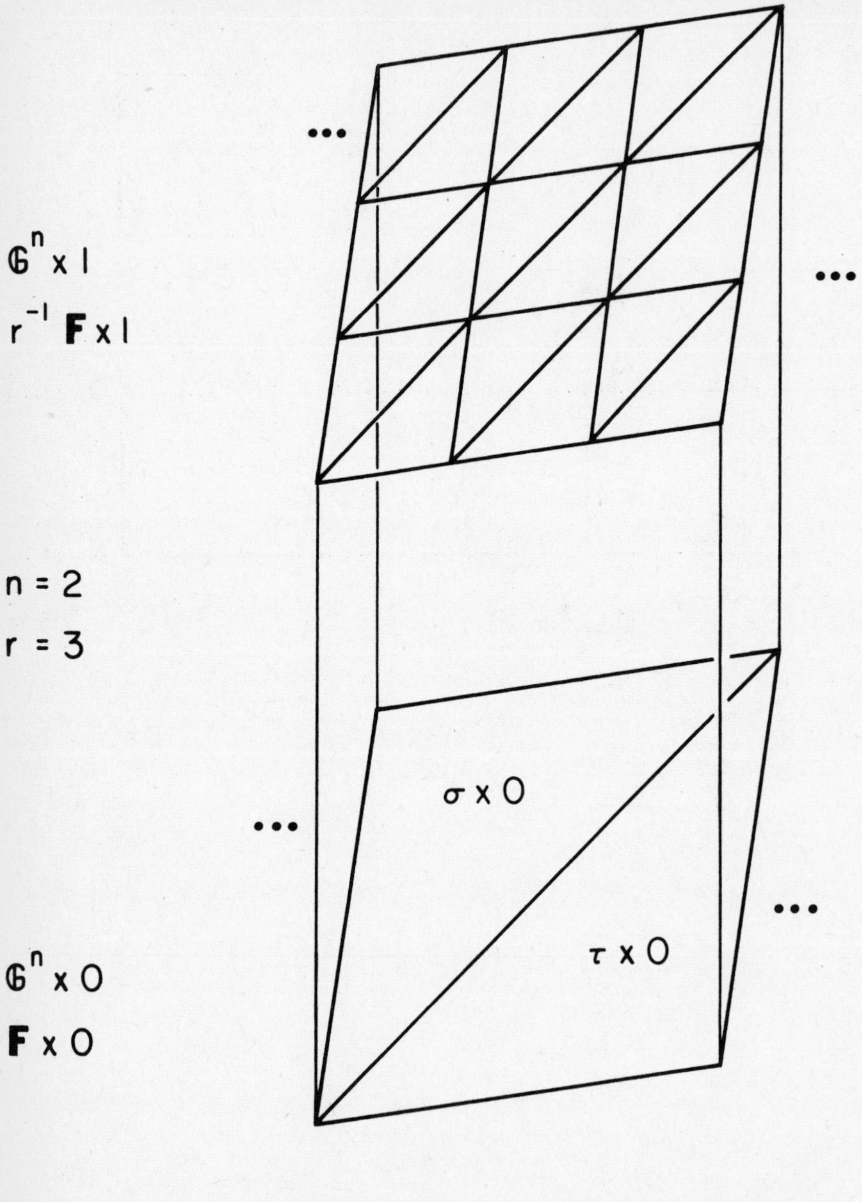

$\mathbb{G}^n \times I$
$r^{-1}\,\mathbf{F} \times I$
$n = 2$
$r = 3$
$\sigma \times 0$
$\tau \times 0$
$\mathbb{G}^n \times 0$
$\mathbf{F} \times 0$

Figure 12.6

12.4 Lemma: $T_\sigma((r^{-1}F)|S) = (r^{-1}F)|\sigma$.

Proof: Extend T_σ to all of $\mathbb{C}^n$ linearly; it is sufficient to show that $T_\sigma(r^{-1}\mathbf{F}) = r^{-1}\mathbf{F}$. Let $\sigma = (v;\pi;n)$ with $\underline{\ell}(v^i) = \underline{\ell}(s^i)$ for $i \in \mu$. Then $T_\sigma(s^i) = v^i$. Thus from Lemma 10.5 and 10.6 we have $T_\sigma(\mathbf{F}) = \mathbf{F}$ and $T_\sigma(r^{-1}\mathbf{F}) = r^{-1}\mathbf{F}$. □

Now assuming $U = [0,1]$ and $\mathbf{X}$ triangulates $S \times [0,1]$ we discuss, as far as is possible, the representation and replacement rules of $\mathbf{J}$.

For the representation rules define $i\mathbf{J}^{n+1}$ to be the collection of all tuples (v,π,n,ρ) where

(a) $(v;\pi;n)$ is an n-simplex of $\mathbf{F}$ with $\underline{\ell}(v^i) = \underline{\ell}(s^i)$ for i in μ

(b) ρ is an $(n+1)$-simplex in $\mathbf{X}$.

For (v,π,n,ρ) in $i\mathbf{J}^{n+1}$ we let it index the $(n+1)$-simplex

$$T_{(v;\pi;n)}(\rho) \ .$$

Clearly $\mathbf{J}^{n+1}$ is the collection of $T_{(v;\pi;n)}(\rho)$ with (v,π,n,ρ) in $i\mathbf{J}^{n+1}$.

Having specified the representation set and rule we consider the replacement rules. Let $(w^1, \ldots, w^{n+2})$ be the vertices of ρ, then $T_{(v;\pi;n)}(w^1), \ldots, T_{(v;\pi;n)}(w^{n+2})$ are the vertices of $T_{(v;w)}(\rho)$. Given (v,π,n,ρ) and t we seek the replacement $(\hat{v},\hat{\pi},n,\hat{\rho})$ in $i\mathbf{J}^{n+1}$, thus, the simplexes $T_{(v;\pi;n)}(\rho)$ and $T_{(\hat{v};\hat{\pi};n)}(\hat{\rho})$ meet in the facet of $T_{(v;\pi;n)}(\rho)$ not containing the vertex $T_{(v;\pi;n)}(w^t)$. There are four

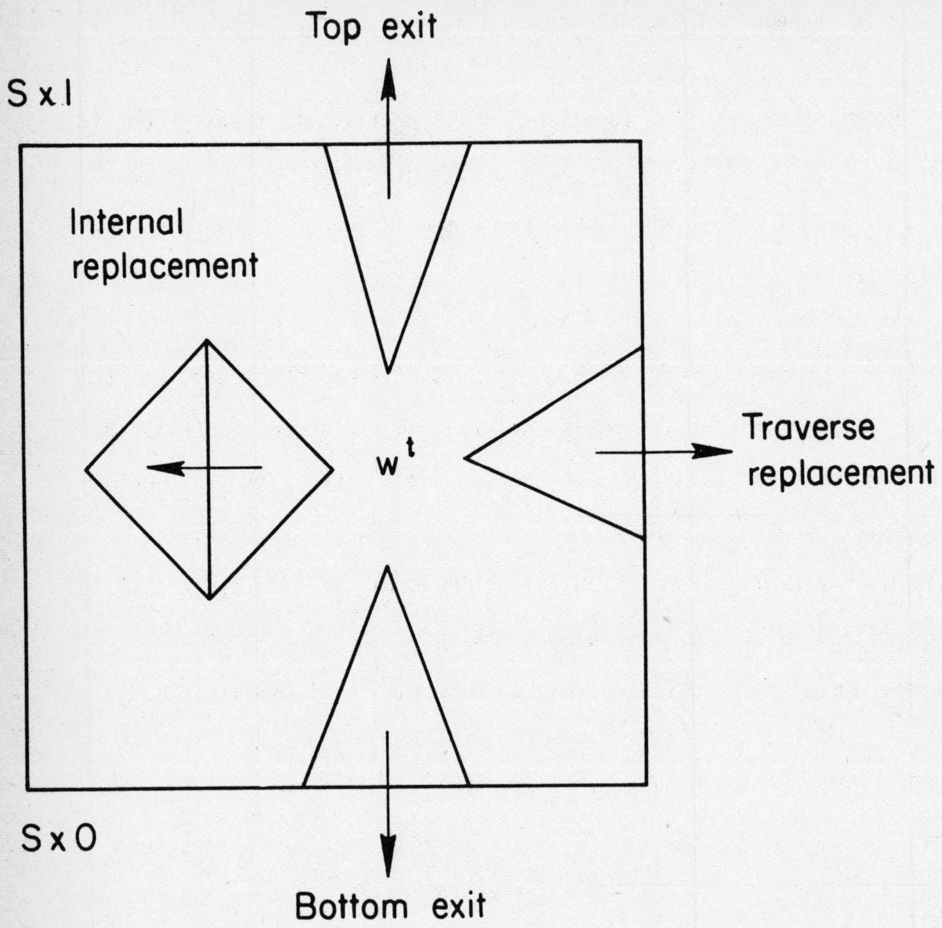

Top exit
S x 1
Internal replacement
w^t
Traverse replacement
S x 0
Bottom exit

Figure 12.7

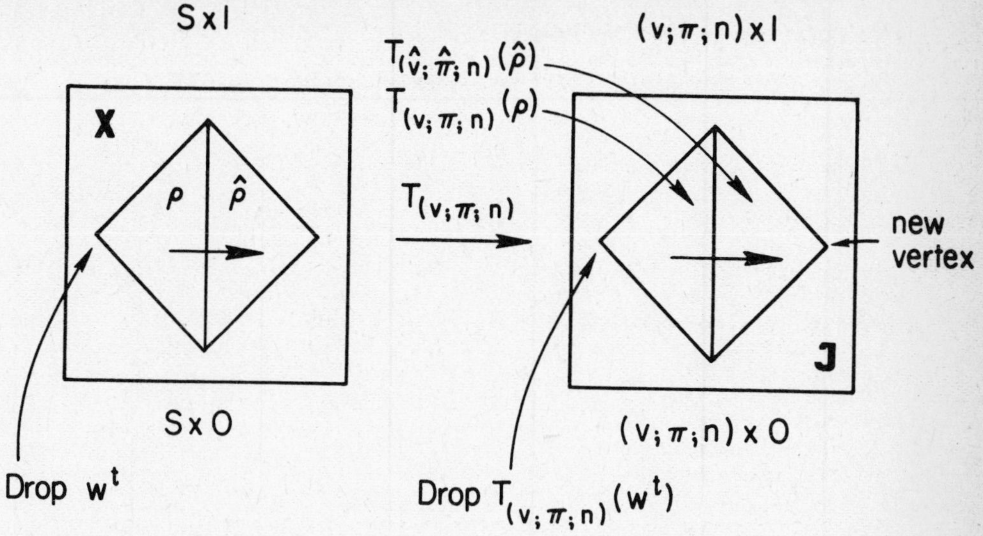
S x I
X
ρ
ρ̂
S x O
Drop w^t
$T_{(v;\pi;n)}$
$T_{(\hat{v};\hat{\pi};n)}(\hat{\rho})$
$T_{(v;\pi;n)}(\rho)$
$(v;\pi;n) \times I$
new vertex
J
$(v;\pi;n) \times O$
Drop $T_{(v;\pi;n)}(w^t)$

Figure 12.8

cases to be considered which we shall refer to as internal replacement, traverse replacement, top exit, and bottom exit, see Figure 12.7.

Internal Replacement: If there is a replacement $\hat{\rho}$ for ρ on dropping w^t in **X** let $(\hat{v},\hat{\pi},n) = (v,\pi,n)$ then the replacement for (v,π,n,ρ) on dropping t is $(\hat{v},\hat{\pi},n,\hat{\rho})$, see Figure 12.8. We refer to this replacement as internal because both $T_{(v,\pi,n)}(\rho)$ and $T_{(\hat{v};\hat{\pi};n)}(\hat{\rho})$ lie in $T_{(v;\pi;n)}(S \times [0,1])$. □

Traverse Replacement: Suppose vertices w^i of ρ for i in $\{1,\ldots, n+2\}\backslash t$ all lie in the facet $S_{\mu\backslash k} \times [0,1]$ of $S \times [0,1]$. In this case, there is no $(n+1)$-simplex in **X** that replaces ρ on dropping w^t. Thus let $\hat{\rho} = \rho$ and let $(\hat{v};\hat{\pi};n)$ be the n-simplex in **F** coordinated by $\underline{\ell}$ obtained by replacing $(v;\pi;n)$ in **F** on dropping v^k, see Figure 12.9. As $T_{(v;\pi;n)}$ and $T_{(\hat{v};\hat{\pi};n)}$ agree on $S_{\mu\backslash k} \times [0,1]$, $T_{(\hat{v};\hat{\pi};n)}(w^i) = T_{(v;\pi;n)}(w^i)$ for i in $\{1, \ldots, n+2\}\backslash t$. Thus $T_{(v;\pi;n)}(\rho)$ and $T_{(\hat{v};\hat{\pi};n)}(\rho)$ are adjacent; note $T_{(v;\pi;n)}(w^t)$ is not in $T_{(\hat{v};\hat{\pi};n)}(S \times [0,1])$. Thus $(\hat{v},\hat{\pi},n,\hat{\rho})$ replaces (v,π,n,ρ) on dropping t.

This replacement is called a traverse replacement for the obvious reasons, to wit, the movement is sideways. □

Top Exit: Suppose vertices w^i for i in $\{1, \ldots, n+2\}\backslash t$ all lie in the facet $S \times 1$ of $S \times [0,1]$. Then, obviously there is no simplex in **J** adjacent to $T_{(v;\pi;n)}(\rho)$ and dropping $T_{(v;\pi;n)}(w^t)$, see Figure 12.10. □

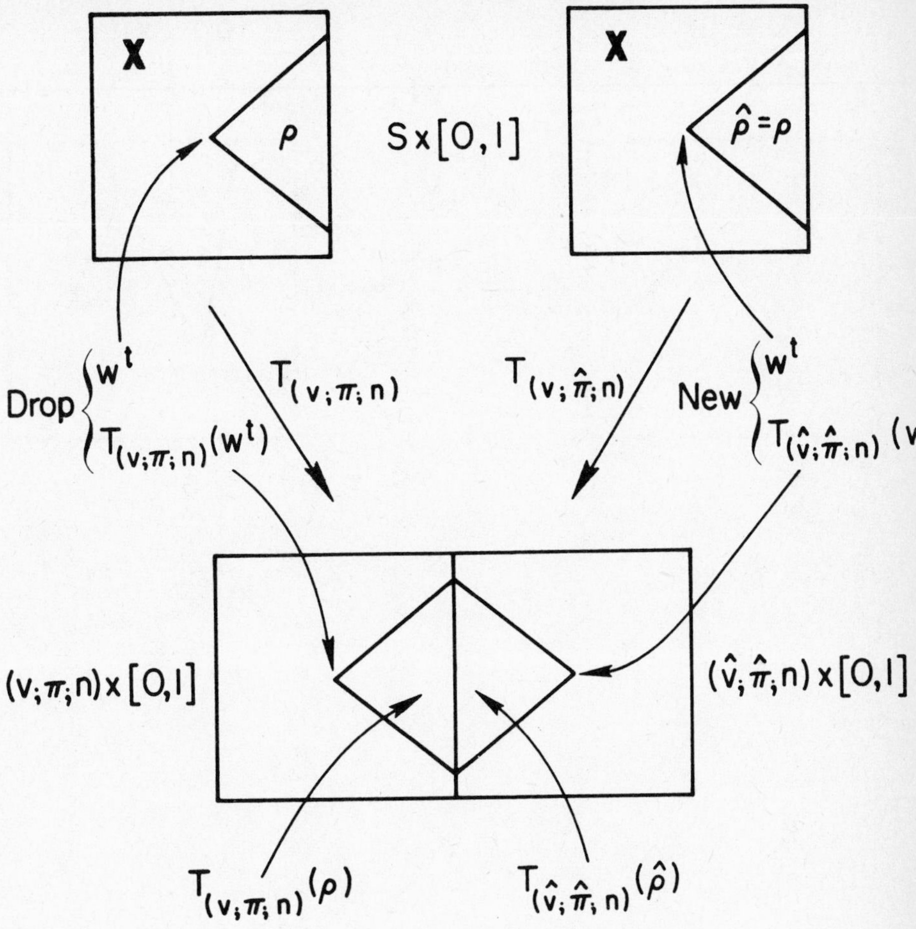
X
ρ
S x [0, 1]
X
ρ̂ = ρ
Drop
w^t
T_(v; π; n)(w^t)
T_(v; π; n)
T_(v; π̂; n)
New
w^t
T_(v̂; π̂; n) (w^t)
(v; π; n) x [0, 1]
(v̂; π̂; n) x [0, 1]
T_(v; π; n)(ρ)
T_(v̂; π̂; n)(ρ̂)

Figure 12.9

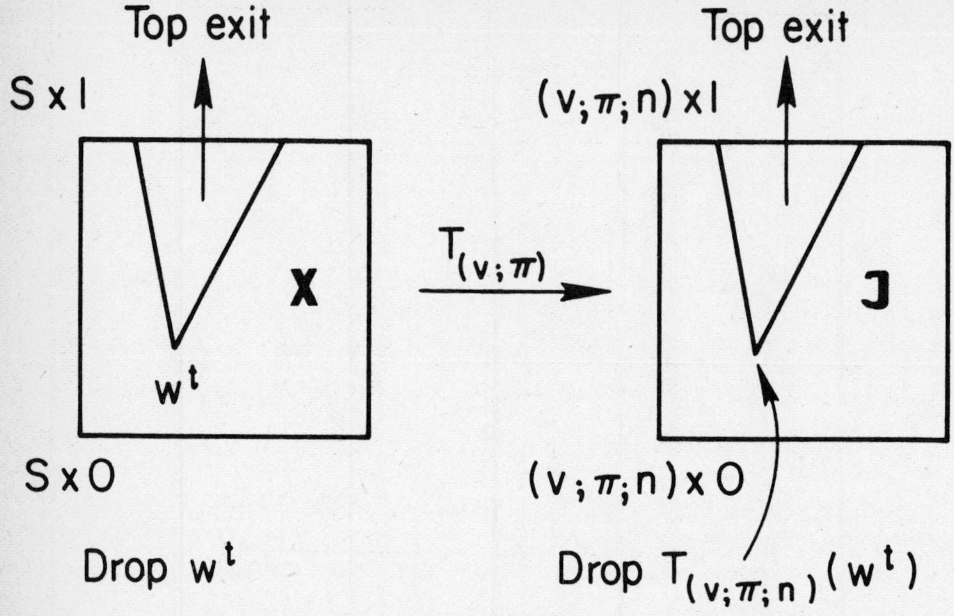
Top exit
S x I
X
w^t
S x O
Drop w^t
T(v;π)
Top exit
(v;π;n) x I
J
(v;π;n) x O
Drop T(v;π;n)(w^t)

Figure 12.10

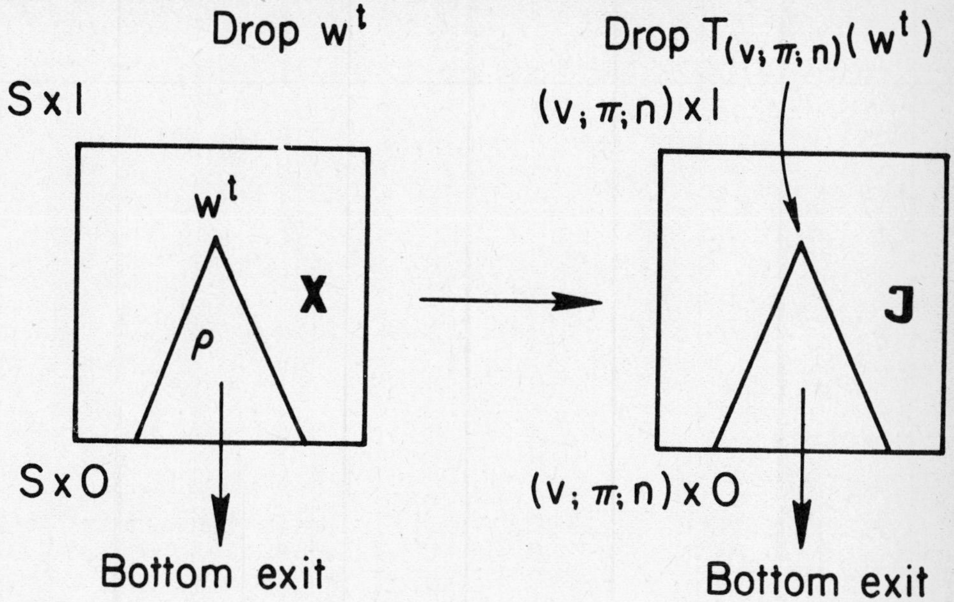
Drop w^t
S x I
w^t
X
ρ
S x O
Bottom exit
Drop T(v; π; n)(w^t)
(v; π; n) x I
J
(v; π; n) x O
Bottom exit

Figure 12.11

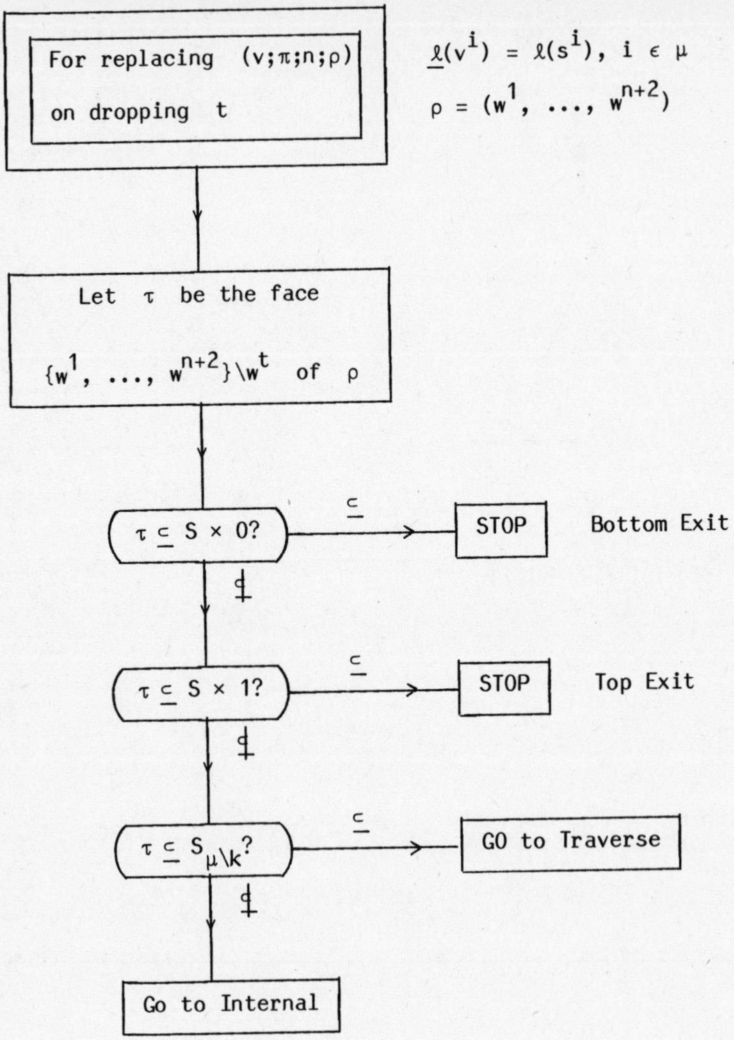
For replacing (v;π;n;ρ)
on dropping t
ℓ(v^i) = ℓ(s^i), i ∈ μ
ρ = (w^1, ..., w^{n+2})
Let τ be the face
{w^1, ..., w^{n+2}}\w^t of ρ
τ ⊆ S × 0?
⊆
STOP
Bottom Exit
⊈
τ ⊆ S × 1?
⊆
STOP
Top Exit
⊈
τ ⊆ S_{μ\k}?
⊆
GO to Traverse
⊈
Go to Internal

Figure 12.12

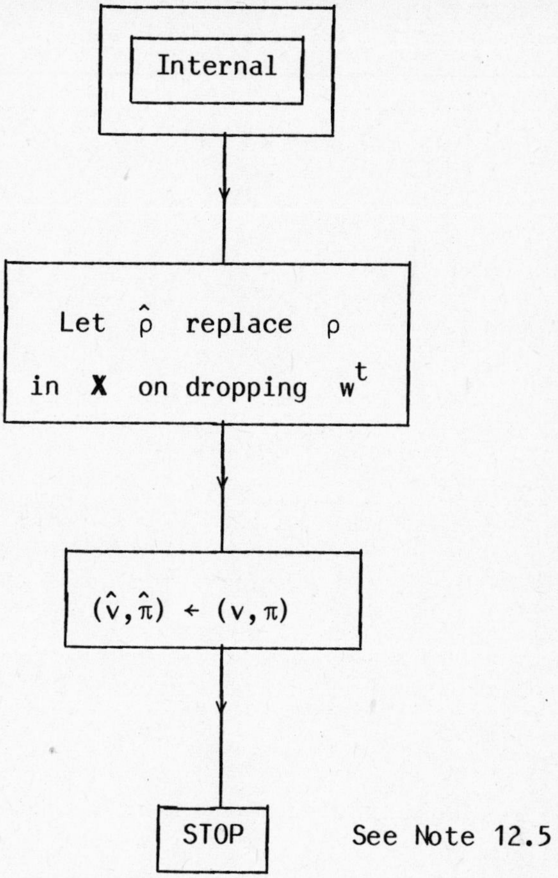
Internal
Let $\hat{\rho}$ replace ρ
in **X** on dropping w^t
$(\hat{v},\hat{\pi}) \leftarrow (v,\pi)$
STOP
See Note 12.5

Figure 12.13

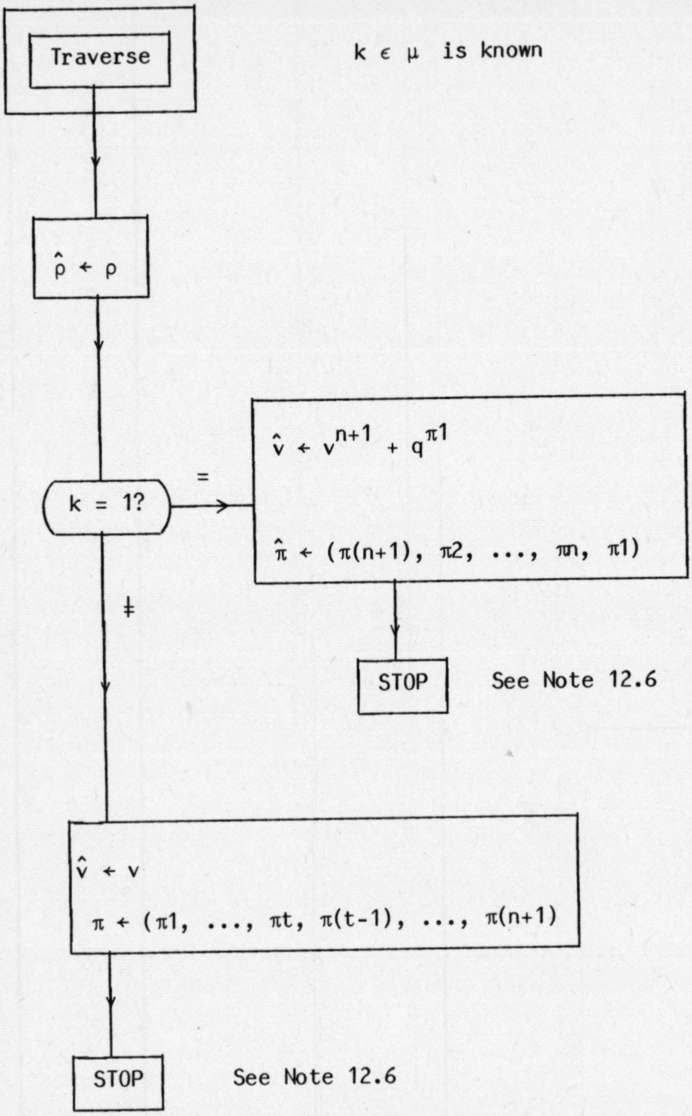
Traverse
k ∈ μ is known
$\hat{\rho} \leftarrow \rho$
k = 1?
=
$\hat{v} \leftarrow v^{n+1} + q^{\pi 1}$
$\hat{\pi} \leftarrow (\pi(n+1), \pi 2, \ldots, \pi n, \pi 1)$
STOP
See Note 12.6
≠
$\hat{v} \leftarrow v$
$\pi \leftarrow (\pi 1, \ldots, \pi t, \pi(t-1), \ldots, \pi(n+1)$
STOP
See Note 12.6

Figure 12.14

Bottom Exit: Suppose vertices w^i for i in $\{1, \ldots, n+2\}\backslash t$ all lie in the facet $S \times 0$ of $S \times [0,1]$. Then obviously there is no simplex in $\mathbf{J}$ adjacent to $T_{(v;\pi;n)}(\rho)$ and not containing $T_{(v;\pi;n)}(w^t)$, see Figure 12.11. □

As the vertices w^i for i in $\{1,\ldots, n+2\}\backslash t$ either do or do not lie in one facet of $S \times [0,1]$ the four cases treated above exhaust all possibilities. Later as our triangulations become more involved the top and bottom exits will become ascent and descent replacements.

Our final act in the section is to chart the replacement rules.

12.5 Note (for Figure 12.13): The adjoined vertex is $T_{(v;\pi;n)}(w^*)$ where w^* is a vertex of $\hat{\rho}$ but not ρ. The correspondence between vertices of $(v;\pi;n;\rho)$ and $(\hat{v};\hat{\pi};n;\hat{\rho})$ and the dropped and adjoined vertex is exactly that correspondence between those of ρ and $\hat{\rho}$. □

12.6 Note (for Figure 12.14): The adjoined vertex is $T_{(\hat{v};\hat{\pi};n)}(w^t)$. The correspondence between vertices of $(v;\pi;n;\rho)$ and $(\hat{v};\hat{\pi};n;\hat{\rho})$ is

$$\begin{array}{ccccccc} 1 & 2 & \cdots & \boxed{t} & \cdots & n+1 & n+2 \\ | & | & \cdots & & \cdots & | & | \\ 1 & 2 & & \boxed{t} & & n+1 & n+2 \end{array}$$ □

12.7 Remark: There is an analogue of juxtapositioning for simplicial cone subdivision $\mathbf{W}$. Let $w^1, \ldots, w^k$ be a distinct list of the generators of the cones of $\mathbf{W}$ and let $\ell : \{w^1, \ldots, w^k\} \to \nu \triangleq \{1,\ldots,n\}$ be a labeling of the generators. If each n-cell σ of $\mathbf{W}$ has a full set of labels that is ℓ (generators of σ) = ν, then a subdivision $\mathbf{X}$ of $\mathbb{C}^n_+ \times U$ can be juxtaposed to obtain a subdivision $\mathbf{J}$ of $J \triangleq W \times U$ where W is the manifold of $\mathbf{W}$. Namely for an n-cell σ of $\mathbf{W}$ with generators $u^1 = w^{i_1}, \ldots, u^n = w^{i_n}$ subdivide $\sigma \times J$ by $T_\sigma(\mathbf{X})$ where $T_\sigma : \mathbb{C}^n_+ \times J \to \sigma \times J$ is defined by

$$T_\sigma(x,y) \equiv \Big(\sum_{i=1}^{n} u^i x_{\ell(u^i)}, y \Big) .$$

If the procedure is executed for each n-cell σ of $\mathbf{W}$ then $\mathbf{J} = \cup_\sigma T_\sigma(\mathbf{X})$ subdivides $J \triangleq W \times U$. A ready example of $\mathbf{X}$ and J is $\mathbf{X} = \mathbf{F}|\mathbb{C}^n_+$ and $J = \{0\}$. □

12.8 Bibliographical Note: Juxtapositioning was first used in Eaves and Saigal [1972]. □

13. SUBDIVISION $\mathbf{P}$ OF $\mathbb{C}^n \times (-\infty,1]$

Beginning with the subdivision $\mathbf{Q}$ of $\mathbb{C}^n$ of Section 5 we will construct a subdivision $\mathbf{P}$ of $\mathbb{C}^n \times (-\infty,1]$ where $(-\infty,1] \triangleq \{x \in \mathbb{C} : x \leqq 1\}$. The subdivision $\mathbf{P}$ is an encoarsement of the subdivision $\mathbf{V}$ of $V = \mathrm{cvx}((S \times 0) \cup (G^n \times 1))$ in Section 15. As a preview of $\mathbf{P}$ consider Figure 13.1; each element $P_{\alpha\beta}$ of $\mathbf{P}$ is a simplicial cone translated by $(0,1)$.

Define $\mathbf{iP}$ to be the collection of all pairs (α,β) where $\alpha \subset \mu$, $\beta \subseteq \mu$, and $\alpha \cap \beta = \phi$. Note that α is permitted to be any subset of μ excluding μ itself, whereas β can be any subset whatever of μ so long as β and α do not intersect.

For each (α,β) in $\mathbf{iP}$ define $P_{\alpha\beta}$ to be the set of all points of form

$$\binom{0}{1} + \binom{q^\alpha}{0}x + \binom{s^\beta}{-e}y$$

in $\mathbb{C}^n \times (-\infty,1]$ where $x \geqq 0$ and $y \geq 0$ are $\#\alpha$ and $\#\beta$-vectors, respectively. Recall $s = (s^1, \ldots, s^{n+1})$ and $q = (e^1, \ldots, e^n, -e)$. Thus $P_{\alpha\beta}$ is seen to be a cone translated by $(0,1)$. We define $\mathbf{P}$ to be the collection of all $P_{\alpha\beta}$ with (α,β) in $\mathbf{iP}$ and the empty set, see Figures 13.2 and 13.3.

To show that $\mathbf{P}$ is a subdivision of $\mathbb{C}^n \times (-\infty,1]$ we proceed as with $\mathbf{W}$, $\mathbf{O}$, and $\mathbf{Q}$ and consider the system LCP(1)

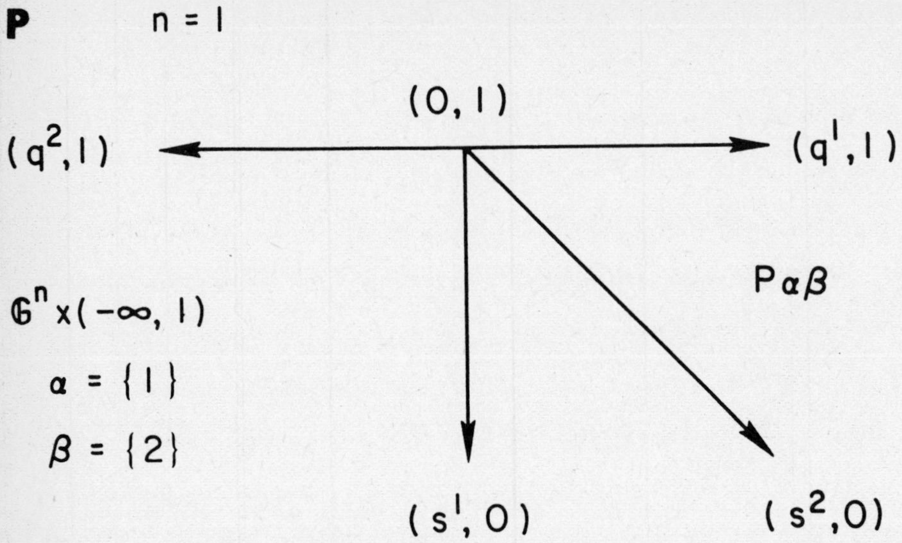

Figure 13.1

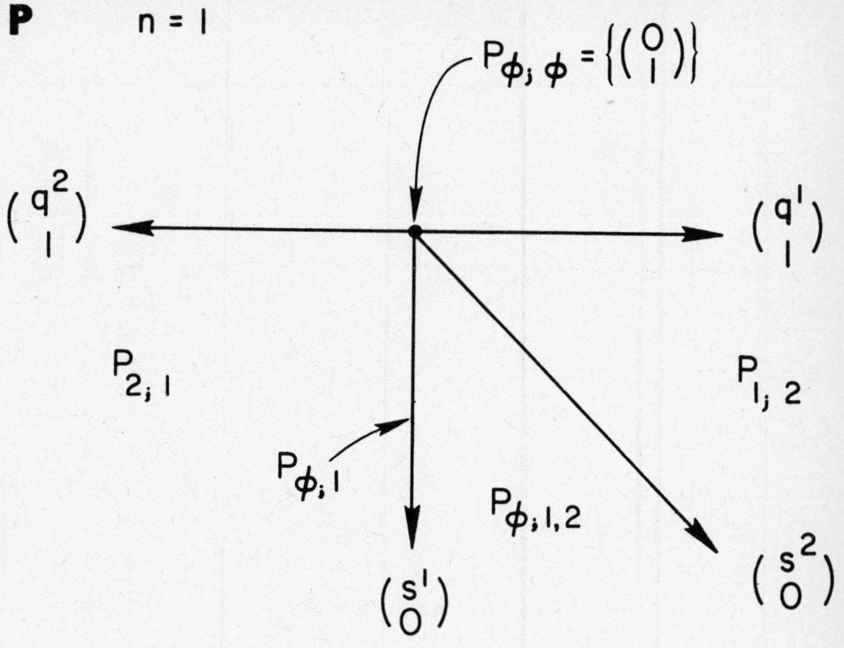
P
n = 1
$P_{\phi;\phi} = \{\binom{0}{1}\}$
$\binom{q^2}{1}$
$\binom{q^1}{1}$
$P_{2;1}$
$P_{1;2}$
$P_{\phi;1}$
$P_{\phi;1,2}$
$\binom{s^1}{0}$
$\binom{s^2}{0}$

Figure 13.2

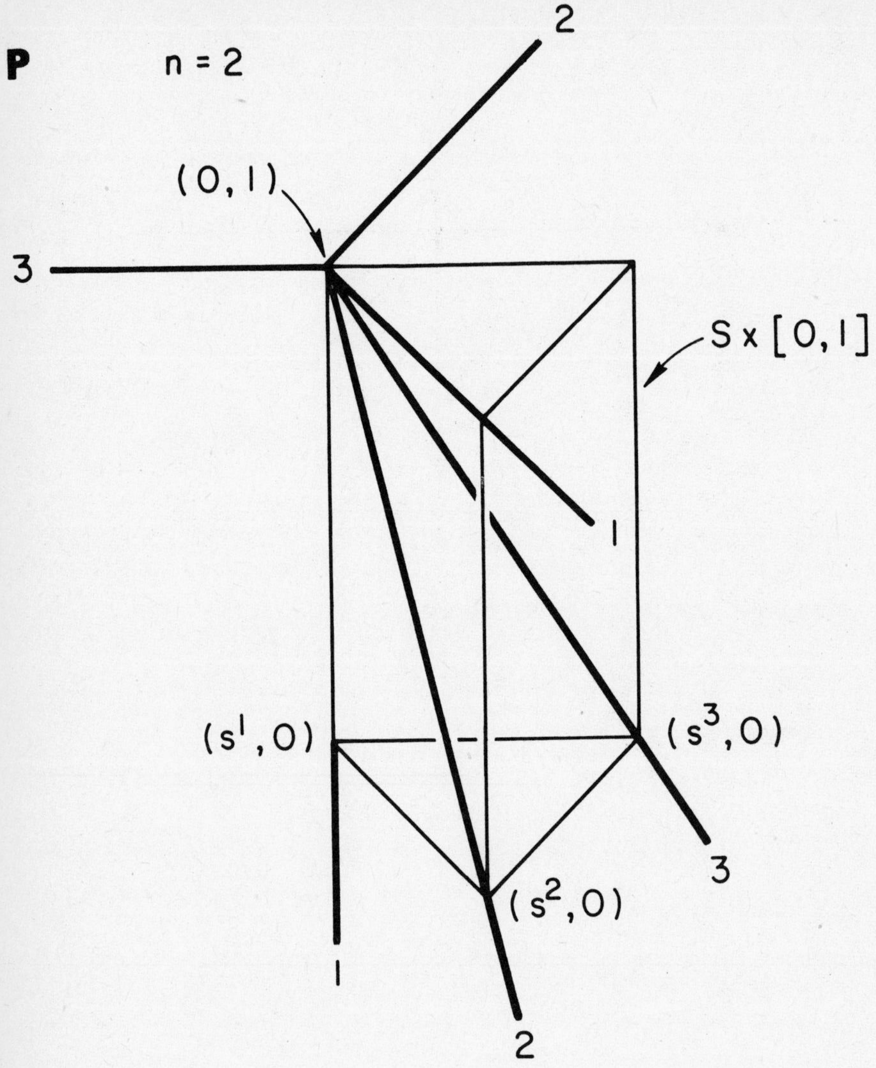

Figure 13.3

$$\binom{q}{0}x + \binom{s}{-e}y = \binom{w}{\theta}$$

$$0 \nless x \geqq 0 \qquad y \geqq 0 \qquad x \cdot y = 0$$

where $x = (x_1, \ldots, x_{n+1})$ and $y = (y_1, \ldots, y_{n+1})$, see Section 2. We shall show that this system, that is, LCP(1) has a unique solution for each (w,θ) for w in $\mathbb{C}^n$ and $\theta \leqq 0$, and as a consequence we will be able to conclude that **P** is a subdivision whose cells are simplicial cones translated by $(0,1)$ and whose manifold is $\mathbb{C}^n \times (-\infty,1]$, see Lemma 3.11.

The following lemma concerning the $(n+1) \times (n+1)$ matrix

$$\begin{pmatrix} J & e \\ -e & 0 \end{pmatrix}$$

where

$$J \triangleq \begin{bmatrix} 1 & & 0 \\ \vdots & \ddots & \\ 1 & \cdots & 1 \end{bmatrix} = (s^2, \ldots, s^{n+1})^T$$

will be employed in the proof that the complementary system has a unique solution.

13.1 Lemma: For each nonzero y in $\mathbb{C}^n$, there is an i in μ with $(y,1)_i > 0$ and $\rho(y,1) > 0$ where ρ is the ith row of

$$\begin{pmatrix} J & e \\ -e & 0 \end{pmatrix}.$$

Proof: If $ey < 0$ then $i = n+1$. If $ey \geqq 0$ let $i = 1, \ldots, n$ be the smallest i such that $y_i > 0$ and $J_i y \geqq 0$. □

13.2 Lemma: LCP(1) has a unique solution for all (w,θ) in $\mathbb{C}^n \times (-\infty,0]$.

Proof: For $\theta = 0$ there is, obviously, a unique solution (x,y) and for the solution $y = 0$, see Section 5. To show LCP(1) has a unique solution for all w in $\mathbb{C}^n$ with $\theta < 0$ is equivalent to showing that the next system has a unique solution for all w in $\mathbb{C}^n$,

$$\binom{q}{0}x + \binom{s}{e}y = \binom{w}{1}$$

$$x \geqq 0 \qquad y \geqq 0 \qquad x \cdot y = 0 \ .$$

This system is referred to as LCP(2). Beginning with LCP(2) exchange the last two columns of the matrices and premultiply the system by the inverse of the new first matrix to get LCP(3).

$$\begin{bmatrix} I & 0 \\ 0 & 1 \end{bmatrix}x - \begin{bmatrix} J & e \\ -e & 0 \end{bmatrix}y = \begin{bmatrix} v \\ 1 \end{bmatrix}$$

$$x \geqq 0 \qquad y \geqq 0 \qquad x \cdot y = 0$$

where

$$J = \begin{bmatrix} 1 & & & 0 \\ \cdot & \cdot & & \\ \cdot & & \cdot & \\ \cdot & & & \cdot \\ 1 & \cdot & \cdot & \cdot & 1 \end{bmatrix}$$

is $n \times n$. LCP(2) has a unique solution for all w in $\mathbb{C}^n$ if and only LCP(3) has a unique solution for all v in $\mathbb{C}^n$. LCP(3) has a solution there is no secondary ray for $d = (e,0)$, namely, if the next system LCP(4 has no solution, see Theorem 2.6.

$$\begin{bmatrix} I & 0 \\ 0 & 1 \end{bmatrix} x - \begin{bmatrix} J & e \\ -e & 0 \end{bmatrix} y - \begin{bmatrix} e \\ 0 \end{bmatrix} z = \begin{bmatrix} v \\ 1 \end{bmatrix}$$

$$\begin{bmatrix} I & 0 \\ 0 & 1 \end{bmatrix} \bar{x} - \begin{bmatrix} J & e \\ -e & 0 \end{bmatrix} \bar{y} - \begin{bmatrix} e \\ 0 \end{bmatrix} \bar{z} = \begin{bmatrix} 0 \\ 0 \end{bmatrix}$$

$$x, y, z, \bar{x}, \bar{y}, \bar{z} \geqq 0 \qquad (x+\bar{x}) \cdot (y+\bar{y}) = 0$$

$$z > 0 \qquad (\bar{x}, \bar{y}, \bar{z}) \neq 0 \qquad (y, \bar{y}) \neq 0 .$$

Suppose for the moment that LCP(4) has a solution. Then $\bar{x}_{n+1} = 0$ and $\bar{y}_\nu = 0$ where $\nu = \{1, \ldots, n\}$. Thus $\bar{x}_\nu > 0$, $y_\nu = 0$, $x_{n+1} > 0$, $(y_{n+1}, \bar{y}_{n+1}) = 0$, and $(y, \bar{y}) = 0$ which contradicts $(y, \bar{y}) \neq 0$. Therefore LCP(3) has a solution for all ν in $\mathbb{C}^n$. It remains to show that LCP(3) has a unique solution for all ν.

Let (x,y) and $(\bar{x}, \bar{y})$ be two solutions. If $y = \bar{y}$ then $x = \bar{x}$, so let us assume $y \neq \bar{y}$. We have

$$\left[\begin{array}{c|c} J & e \\ \hline -e & 0 \end{array}\right] (y-\bar{y}) = (x-\bar{x}) .$$

If $y_{n+1} = \bar{y}_{n+1}$ then let i be the smallest index with $y_i \neq \bar{y}_i$. Then

$$y_i - \bar{y}_i = x_i - \bar{x}_i$$

$$0 < (y-\bar{y}_i)^2 = -x_i\bar{y}_i - \bar{x}_i y_i \leq 0$$

which is a contradiction. So let us assume that $y_{n+1} > \bar{y}_{n+1}$. But examining LCP(3) we see that $y_\nu = \bar{y}_\nu$ implies $y_{n+1} = \bar{y}_{n+1}$ so we have $y_\nu \neq \bar{y}_\nu$. Now apply Lemma 13.1 to obtain an i with $(y_i-\bar{y}_i)\cdot\rho(y-\bar{y}) > 0$ where ρ is the ith row of

$$\left[\begin{array}{c|c} J^T & e \\ \hline -e & 0 \end{array}\right].$$

But we also have

$$(y_i-\bar{y}_i)\ \rho(y-\bar{y}) = -x_i\bar{y}_i - \bar{x}_i y_i \leq 0$$

which is a contradiction. Thus $y = \bar{y}$ and LCP(3) has a unique solution. □

Thus we know that LCP(1) has a unique solution for all (w,θ) in $\mathbb{C}^n \times (-\infty,0]$. As an immediate consequence of Lemma 13.2 is:

13.3 Theorem: **P** is a finite subdivision of $\mathbb{C}^n \times (-\infty,1]$ whose cells $P_{\alpha\beta}$ are simplicial cones translated by $(0,1)$. Furthermore

(i) dim $P_{\alpha\beta}$ has dimension $\#\alpha + \#\beta$

(ii) $P_{\alpha\beta} \cap P_{\gamma\delta} = P_{\alpha\cap\gamma,\beta\cap\delta}$

(iii) $P_{\alpha\beta} \supseteq P_{\gamma\delta}$, if and only if $\alpha \supseteq \gamma$ and $\beta \supseteq \delta$

(iv) $P_{\alpha\beta} = P_{\gamma\delta}$, if and only if $\alpha = \gamma$ and $\beta = \delta$

(v) ρ is a proper face of $P_{\alpha\beta}$, if and only if $\rho = P_{\gamma\delta}$ for some $\gamma \subseteq \alpha$, $\delta \subseteq \beta$, and $(\alpha,\beta) \neq (\gamma,\delta)$. □

Thus **iP** is a representation set of **P**$\backslash\phi$. Clearly $P_{\alpha\beta} \cap (\mathbb{C}^n \times 1)$ is $(Q_\alpha,1)$ where Q_α is an element of **Q**. Consequently the natural restriction of **P** to $\mathbb{C}^n \times 1$ is $(\mathbf{Q},1)$.

For the next lemma let $\alpha \subset \mu$, $\bar{\alpha} \subset \mu$, $\beta = \mu\backslash\alpha$, and $\bar{\beta} = \mu\backslash\bar{\alpha}$.

13.4 Lemma: If $a+b = \bar{a}+\bar{b}$ where $(a,b) \in Q_\alpha \times S_\beta$ and $(\tilde{a},\tilde{b}) \in Q_{\hat{\alpha}} \times S_{\tilde{\beta}}$ then $a = \tilde{a}$ and $b = \tilde{b}$.

Proof: The conclusion follows immediately from Theorem 13.2. □

The next lemma indicates how the cells of **P** intersect with $S \times 0$; let $\alpha \subset \mu$ and $\beta \subseteq \mu\backslash\alpha$.

13.5 Lemma: $P_{\alpha\beta} \cap (S \times 0) = S_\beta \times 0$.

Proof: Let (w,θ) be a point in $S \times 0$ and $P_{\alpha\beta}$; thus

$$\binom{w}{0} = \binom{0}{1} + \binom{q^\alpha}{0}x + \binom{s^\beta}{-e}y$$

for some $x \geqq 0$ and $y \geqq 0$. Then $w = a+b$ with $a \in Q_\alpha$ and $b \in S_\beta$. From Lemma 13.4 with $\bar{\beta} = \mu$, $a = 0$ and $w = b \in S_\beta$. □

Let $i\mathbf{P}^{n+1} = \{(\alpha,\beta) : \alpha \subset \mu,\ \beta = \mu\backslash\alpha\}$ be the representation set for the $2^{n+1}-1$ $(n+1)$-cells $P_{\alpha\beta}$ of **P**. Each such $P_{\alpha\beta}$ has $n+1$ facets $P_{\alpha\backslash t,\beta\backslash t}$ for t in μ. Given (α,β) in $i\mathbf{P}^{n+1}$ and t in μ let us find the replacement $(\hat{\alpha},\hat{\beta})$ so that $P_{\alpha\beta}$ and $P_{\hat{\alpha}\hat{\beta}}$ meet in the n-cell $P_{\alpha\backslash t,\beta\backslash t}$.

From Lemma 13.3 we see that

$$\left.\begin{aligned}\hat{\alpha} &= \alpha\backslash t\\ \hat{\beta} &= \beta \cup \{t\}\end{aligned}\right\} \quad \text{if } t \in \alpha$$

and

$$\left.\begin{aligned}\hat{\alpha} &= \alpha \cup \{t\}\\ \hat{\beta} &= \beta\backslash t\end{aligned}\right\} \quad \text{if } t \in \beta$$

unless $\hat{\alpha} = \mu$, and then we have a top exit, see Figure 13.4.

13.6 Exercise: Define $h : \mathbb{C}^n \times (-\infty,1] \to \mathbb{C}^n \times (-\infty,1]$ by $h(x,\theta) = (x+\theta p,\ \theta)$. Argue that $h(\mathbf{P})$ is a subdivision of $\mathbb{C}^n \times (-\infty,1]$. In a later exercise $h(\mathbf{P})$ is called upon. □

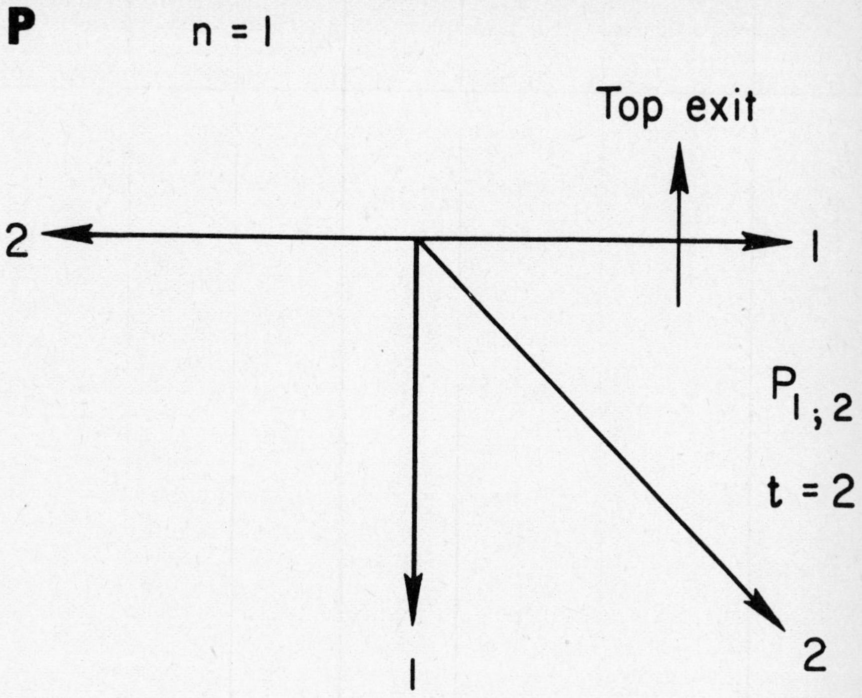

Figure 13.4

13.7 Bibliographical Notes: In explaining the variable dimension algorithm of van der Laan and Talman [1979], Todd [1978b] constructed a subdivision like **P** and used the linear complementary problem to prove it, as we have done here, see Note 15.22. A Proof is available which avoids the LCP theory, see the subdivision theorem in Broadie and Eaves [1983]. □

14. CONING TRANSVERSE AFFINELY DISJOINT SUBDIVISIONS

Herein we develop coning which is another of the many ways of combining two subdivisions **K** and **L** to form a larger one **M**. As a preview consider Figure 14.1 where one of the two manifolds is a point. We shall employ coning in the construction of the triangulation **V** in the next section.

Let σ and τ be two sets in $\mathbb{C}^n$. We say that σ and τ are transverse if tng σ and tng τ meet only at the origin. We say that σ and τ are affinely disjoint if aff σ and aff τ are disjoint. Let **K** and **L** be k and ℓ-subdivisions whose manifolds K and L are transverse and affinely disjoint.

The coning operations for combining **K** and **L** to get the subdivision **M** with carrier cvx(K $\cup$ L) is simply to take all convex combinations cvx($\sigma \cup \tau$) of cells σ in **K** and τ in **L** as the cells for the new subdivision **M** which, as we shall see, is an (ℓ+k+1) subdivision, see Figures 14.2 and 14.3.

Clearly the convex hull of two cells may not be a cell, however, the convex hull of two bounded cells is a cell. We begin with a lemma that indicates the dimension of the convex hull of two cells. Let lim σ indicate the smallest subspace containing σ.

14.1 Lemma: Let σ and τ be two sets. Then the dimension of cvx($\sigma \cup \tau$) is

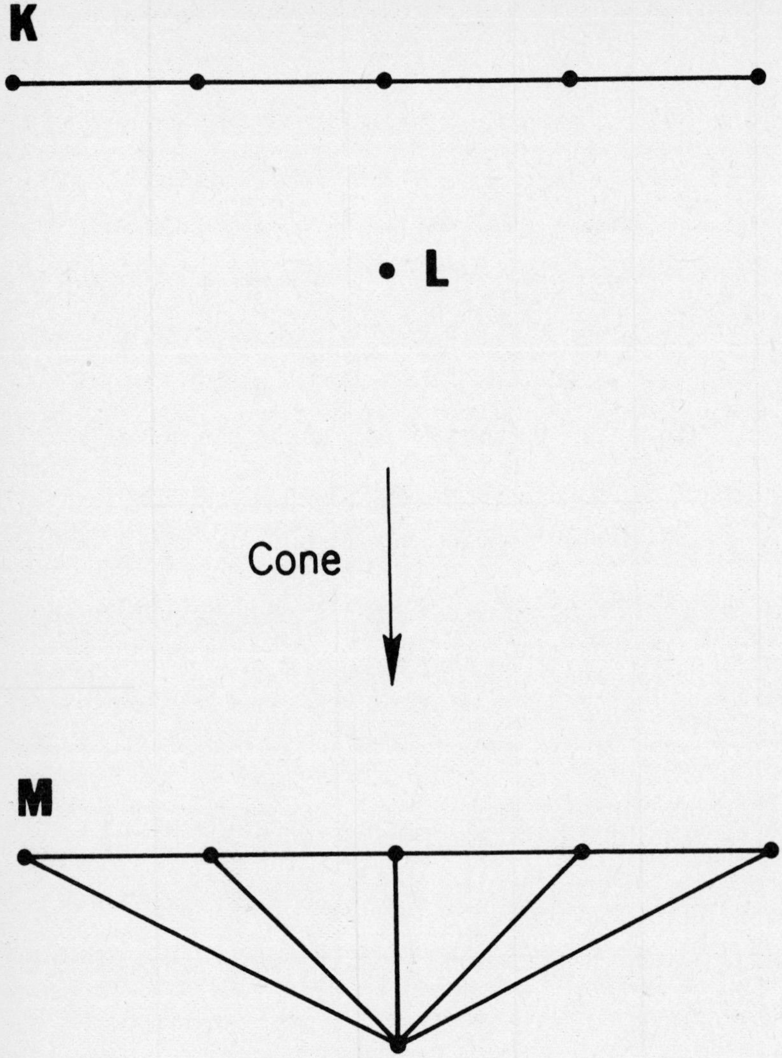

Figure 14.1

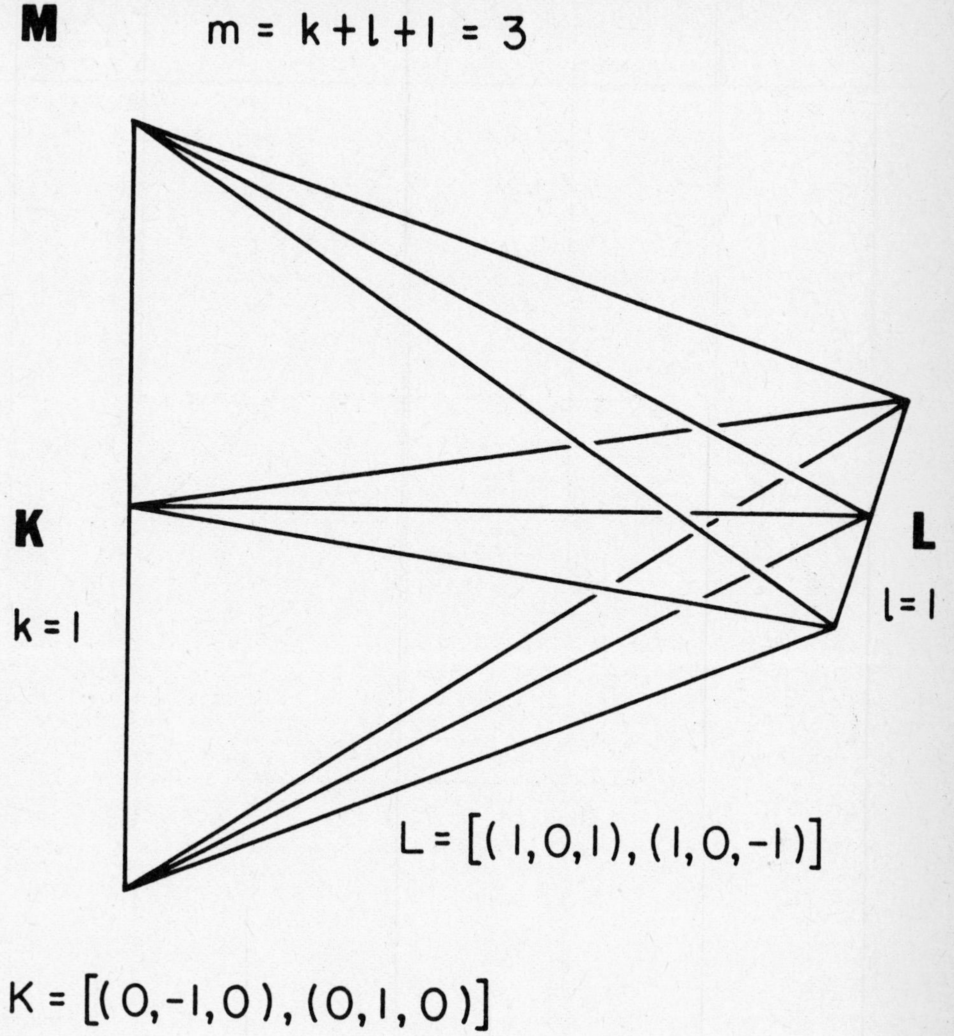
M
m = k+l+1 = 3
K
k = 1
L
l = 1
L = [(1,0,1), (1,0,-1)]
K = [(0,-1,0), (0,1,0)]

Figure 14.2

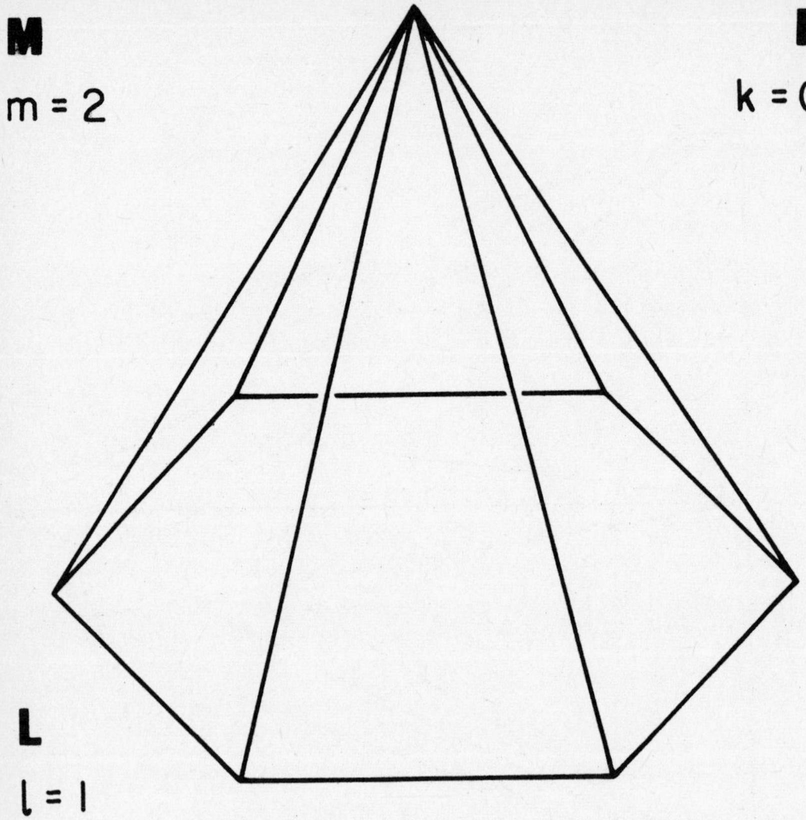

Figure 14.3

$$\zeta + \dim \sigma + \dim \tau - \dim(\text{tng}\ \sigma \cap \text{tng}\ \tau)$$

where $\zeta = 1$ or 0 as $\text{aff}\ \sigma \cap \text{aff}\ \tau$ is empty or not.

Proof: For x in $\text{aff}\ \sigma$,

$$\begin{aligned}
\dim \text{cvx}(\sigma \cap \tau) &= \dim \text{cvx}((\sigma \cup \tau) - x) \\
&= \dim \text{cvx}((\sigma - x) \cup (\tau - x)) \\
&= \dim \text{lin}((\sigma - x) \cup (\tau - x) \\
&= \dim \text{lin}(\text{lin}(\sigma - x) \cup \text{lin}(\tau - x)) \\
&= \dim \text{lin}(\sigma - x) + \dim \text{lin}(\tau - x) - \dim(\text{lin}(\sigma - x) \cap \text{lin}(\tau - x)) \\
&= \dim \sigma + \dim \text{lin}(\tau - x) - \dim(\text{tng}\ \sigma \cap \text{lin}(\tau - x)) \ .
\end{aligned}$$

If x can be chosen in $\text{aff}\ \sigma \cap \text{aff}\ \tau$, then $\dim \text{cvx}(\sigma \cap \tau) = \dim \sigma + \dim \tau - \dim(\text{tng}\ \sigma \cap \text{tng}\ \tau)$. If $\text{aff}\ \sigma \cap \text{aff}\ \tau = \phi$ then $0 \not\in \text{aff}(\tau - x)$, $\dim \text{lin}(\tau - x) = 1 + \dim \text{tng}\ \tau = 1 + \dim \tau$ and $\text{tng}\ \sigma \cap \text{lin}(\tau - x) = \text{tng}\ \sigma \cap \text{tng}\ \tau$. To see the last equality, consider a point in $\text{tng}\ \sigma$ and $\text{lin}(\tau - x)$; namely

$$p = \sum x_i \lambda_i = \sum (y_i - x)\ \theta_i$$

$$\sum \lambda_i = 0$$

where $x_i \in \sigma$ and $y_i \in \tau$. If $\sum \theta_i = 0$ then $p = \sum y_i \theta_i$ is in $\text{tng}\ \tau$. Otherwise divide p by $\sum \theta_i$ to get $\tilde{p} = y - x$ in $\text{tng}\ \sigma$ where y is in $\text{aff}\ \tau$. But then $\tilde{p} + x = y$ is in $\text{aff}\ \sigma \cap \text{aff}\ \tau$ which is a contradiction. □

From the lemma we see that if σ and τ are transverse and affinely disjoint then $\dim \mathrm{cvx}(\sigma,\tau) = \dim \sigma + \dim \tau + 1$.

The next lemma indicates that $\theta x + (1-\theta)y$ is uniquely represented as such for x in K and y in L if K and L are transverse and affinely disjoint.

14.2 Lemma: Let K and L be transverse and affinely disjoint sets. Assume

$$\theta x^1 + (1-\theta)y^1 = \lambda x^2 + (1-\lambda)y^2$$

where $x^i \in K$ and $y^i \in L$ for $i = 1,2$. Then $\theta = \lambda$. If $\theta \neq 0$, $x^1 = x^2$. If $\theta \neq 1$, $y^1 = y^2$.

Proof: Consider $\theta x^1 - \lambda x^2 = (1-\lambda)y^2 - (1-\theta)y^1$. If $\delta = \theta-\lambda = (1-\lambda) - (1-\theta) \neq 0$, then $\delta^{-1}(\theta x^1-\lambda x^2) = \delta^{-1}((1-\lambda)y^2 - (1-\theta)y^1)$. The last points are in aff K and aff L, respectively, which contradicts aff $K \cap$ aff $L = \phi$. Thus $\delta = 0$ and $\theta = \lambda$, and so $\theta(x^1-x^2) = (1-\theta)(y^2-y^1)$. The last two points are in tng K and tng L, respectively. Thus $0 = \theta(x^1-x^2) = (1-\theta)(y^2-y^1)$. □

The next lemma will inform us how the cells of **M** meet; to see that there is something to prove consider Figure 14.4.

14.3 Lemma: Let K and L be transverse and affinely disjoint sets. Let σ_i be a convex subset of K and τ_i a convex subset of L for $i = 1, 2$,. The intersection of the sets $\mathrm{cvx}(\sigma_1 \cup \tau_1)$ and $\mathrm{cvx}(\sigma_2 \cup \tau_2)$ is the set $\mathrm{cvx}((\sigma_1 \cap \sigma_2) \cup (\tau_1 \cap \tau_2))$.

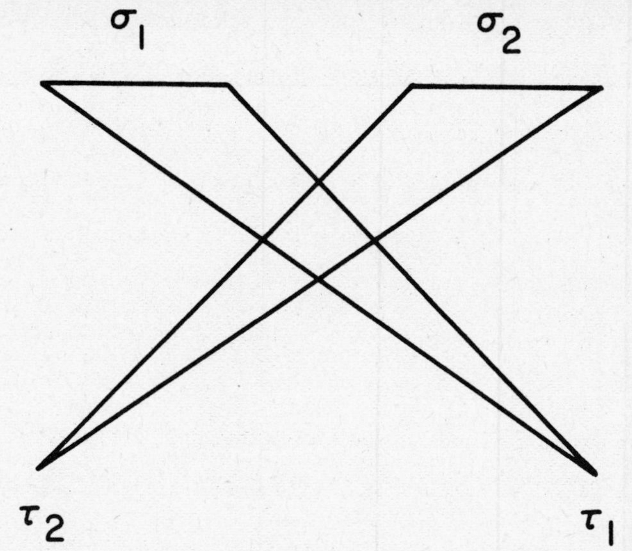

Figure 14.4

Proof: As $\mathrm{cvx}(\sigma_1 \cup \tau_1)$ and $\mathrm{cvx}(\sigma_2 \cup \tau_2)$ both contain $\mathrm{cvx}((\sigma_1 \cup \sigma_2) \cup (\tau_1 \cap \tau_2))$ so does their intersection. Any point u in the intersection can be expressed as $u = \theta_1 x^1 + (1-\theta_1)y^1 = \theta_2 x^2 + (1-\theta_2)y^2$ where $x^i \in \sigma$ and $y^i \in \tau$, and $0 \leqq \theta_i \leqq 1$ for $i = 1, 2$. From Lemma 14.2 $\theta_1 = \theta_2$. If $\theta_1 = 1$ then $u = x^1 = x^2$ is in $\sigma_1 \cap \sigma_2$. If $\theta_1 = 0$ then $u = y^1 = y^2$ is in $\tau_1 \cap \tau_2$. If $0 < \theta_1 < 1$ then $x^1 = x^2$ and $y^1 = y^2$ and u is in $\mathrm{cvx}((\sigma_1 \cap \sigma_2) \cup (\tau_1 \cap \tau_2))$. □

14.4 Lemma: Let σ and τ be transverse and affinely disjoint bounded cells. A set is a face of the cell $\mathrm{cvx}(\sigma \cup \tau)$, if and only if it is of form $\mathrm{cvx}(\zeta \cup \psi)$ where ζ and ψ are faces of σ and τ.

Proof: Let η be a halfspace containing $\mathrm{cvx}(\sigma \cup \tau)$. If $x \in \sigma$, $y \in \tau$, $0 < \theta < 1$, and $\theta x + (1-\theta)y \in \partial\eta$ then x and y are in $\partial\eta$. Thus $x \in \sigma \cap \partial\eta$ and $y \in \tau \cap \partial\eta$. It follows that $\partial\eta \cap \mathrm{cvx}(\sigma \cup \tau)$ $= \mathrm{cvx}((\sigma \cap \partial\eta) \cup (\tau \cap \partial\eta))$. Now let $\bar{\sigma}$ and $\bar{\tau}$ be faces of σ and τ. If $\theta x + (1-\theta)y$ is in $\mathrm{cvx}(\bar{\sigma} \cap \bar{\tau})$ where $x \in \sigma$ and $y \in \tau$, then $x \in \bar{\sigma}$ and $y \in \bar{\tau}$ from Lemma 14.2. Thus $\mathrm{cvx}(\bar{\sigma} \cup \bar{\tau})$ is an extreme set in the cell $\mathrm{cvx}(\sigma \cup \tau)$ and is a face. □

Let **K** and **L** be k and ℓ-subdivisions with bounded cells, $k \geq 0$, and $\ell \geq 0$, and assume that their manifolds K and L are transverse and affinely disjoint. Define **M** to be the collection of all cells of form $\mathrm{cvx}(\sigma \cup \tau)$ where σ and τ are cells of **K** and **L**, respectively.

14.5 Theorem: $\mathbf{M}$ is a $(k+\ell+1)$-subdivision with manifold $M = \text{cvx}(K \cup L)$.

Proof: As all cells of $\mathbf{K}$ and $\mathbf{L}$ are bounded, $\mathbf{M}$ is a collection of bounded cells. From Lemma 14.1 no cell of $\mathbf{M}$ exceeds dimension $k+\ell+1$, from Lemmas 14.3 and 14.4 each cell of $\mathbf{M}$ is a face of a cell of dimension $k+\ell+1$, and from Lemmas 14.3 and 14.4 any two cells of $\mathbf{M}$ meet in a common face. If ρ is a $(k+\ell)$-cell then it is of form $\text{cvx}(\sigma \cup \tau)$ where σ is a k-cell of $\mathbf{K}$ and τ is an $(\ell-1)$-cell or σ is a $(k-1)$-cell in τ is an ℓ-cell. For the first case we argue ρ lies in at most two $(k+\ell+1)$-cells and the second follows by symmetry. The only $(k+\ell+1)$-cells containing $\text{cvx}(\sigma \cup \tau)$ are of form $\text{cvx}(\sigma \cup \bar{\tau})$ where $\bar{\tau}$ is an ℓ-cell of $\mathbf{L}$ containing τ. As there are at most two such $\bar{\tau}$ the theorem is proved. □

Let $\rho = \text{cvx}(\sigma \cup \tau)$ be a $(k+\ell+1)$-cell of $\mathbf{M}$ and $\eta = \text{cnv}(\sigma \cup \zeta)$ a facet of the cell. The replacement for ρ with respect to the facet η is $\hat{\rho} = \text{cvx}(\sigma \cup \hat{\tau})$, if it exists, where $\hat{\tau}$ is the replacement for τ with respect to the facet ζ, if it exists. Similar replacement rules apply for a facet of form $\eta = \text{cvx}(\zeta \cup \tau)$.

14.6 Bibliographical Notes

Special cases of the "transverse-affinely disjoint" result have been employed in Todd [1978], Wright [1981], Bárány [1979], van der Laan and Talman [1981], and Kojima and Yamamoto [1982a]. □

15. TRIANGULATION $\mathbf{V}$ OF $V = \text{cvx}((S \times 0) \cup (\mathbb{C}^n \times 1))$

Let V be the convex hull of $\mathbb{C}^n \times 1$ and $S \times 0$. Our purpose here is to describe, prove, and state the replacement rules for the triangulation $\mathbf{V}$ of V. Although the triangulation $\mathbf{V}$, as such, has been used in the solution of equations, restrictions of $\mathbf{V}$ properly squeezed and sheared form the principle "micro structure" for the variable rate refining triangulation $\mathbf{S}$. As a preview of $\mathbf{V}$ consider Figure 15.1. Also as a preview let us cite three conspicuous features that the triangulation $\mathbf{V}$ will have:

(a) $\mathbf{V}^0 \subseteq (S \times 0) \cup (\mathbb{C}^n \times 1)$

(b) $\mathbf{V}|(\mathbb{C}^n \times 1) = \mathbf{F} \times 1$

(c) $S \times 0 \in \mathbf{V}$

that is, vertices of $\mathbf{V}$ lie in $S \times 0$ or $\mathbb{C}^n \times 1$, the restriction of $\mathbf{V}$ to $\mathbb{C}^n \times 1$ is the Freudenthal triangulation $\mathbf{F}$, and the simplex $S \times 0$ itself is in $\mathbf{V}$.

In describing the triangulation $\mathbf{V}$ it is assumed the reader is familiar with the subdivisions $\mathbf{Q}$ of $\mathbb{C}^n$, $\mathbf{F}$ of $\mathbb{C}^n$, $\mathbf{F}|Q_\alpha$ of Q_α, and $\mathbf{P}$ of $\mathbb{C}^n \times (-\infty,1]$ of foregoing sections. Recall $S_\beta = \{s^\beta y : y \geqq 0, ey = 1\}$, $s = (0, e^1, e^1+e^2, \ldots, e)$, $Q_\alpha = \{q^\alpha x : x \geq 0\}$, $q = (e^1, e^2, \ldots, e^n, -e)$. Some aspect of each $\mathbf{Q}$, $\mathbf{F}$, and $\mathbf{P}$ will play a role in $\mathbf{V}$. For example, $\mathbf{V}$ will be a refinement of the forced restriction $\mathbf{P}\|V$ of $\mathbf{P}$ to V; see Figure 15.2. Note that $\mathbf{P}\|V$ is not a subdivision as its elements are not cells.

Let us begin our formal development of $\mathbf{V}$. Fix (α,β) in $\mathbf{iP}$, see Section 13, that is, $\alpha \subset \mu$ and $\beta \subseteq \mu\backslash\alpha$. Define $V_{\alpha\beta}$ by

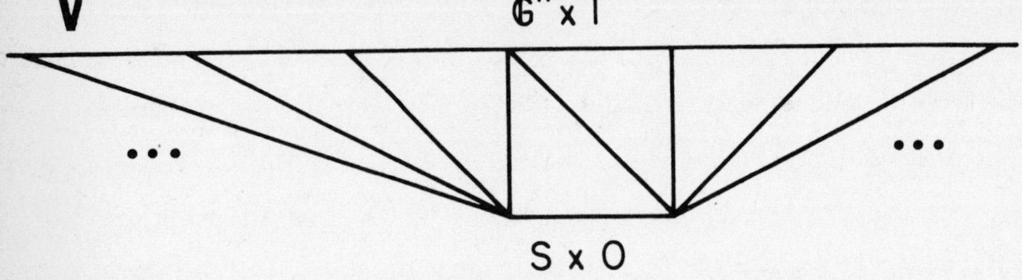

Figure 15.1

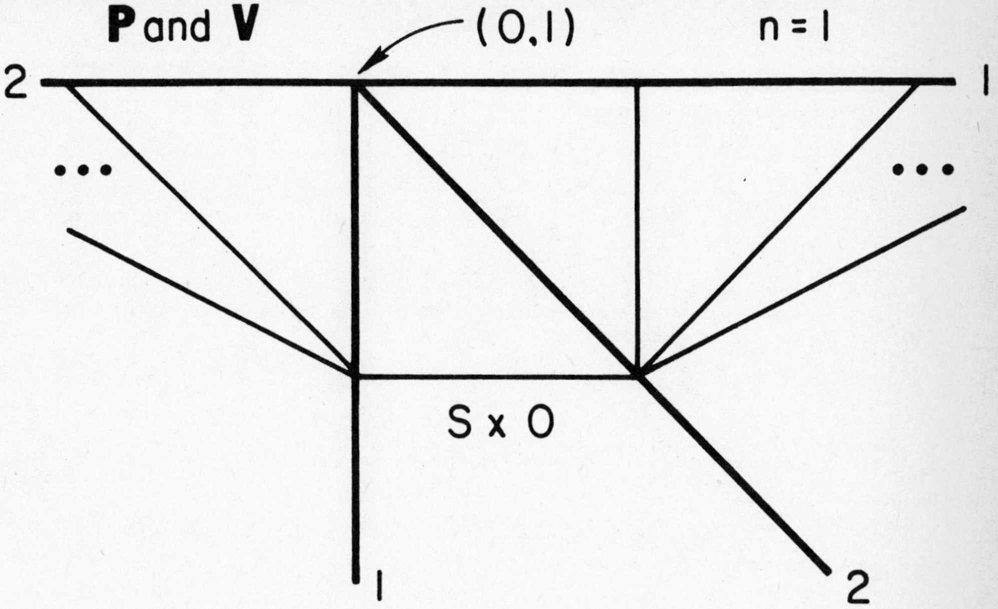

Figure 15.2

$$V_{\alpha\beta} = V \cap P_{\alpha\beta}$$

where $P_{\alpha\beta}$ is by definition the cell

$$\{\binom{0}{1} + \binom{q^{\alpha}}{0}x + \binom{s^{\beta}}{-e}y : x \geqq 0,\ y \geqq 0\}$$

of **P**, see Figure 15.3. The next lemma gives a second characterization of $V_{\alpha\beta}$.

15.1 Lemma: $V_{\alpha\beta}$ is the convex hull of $Q_{\alpha} \times 1$ and $S_{\beta} \times 0$.

Proof: As $V_{\alpha\beta} = V \cap P_{\alpha\beta}$ contains $P_{\alpha\beta} \cap (\mathbb{C}^n \times 1) = Q_{\alpha} \times 1$ and $P_{\alpha\beta} \cap (S \times 0) = S_{\beta} \times 0$, see Lemma 13.5. $V_{\alpha\beta}$ certainly contains their convex combination E. Now let

$$\binom{w}{\theta} = (1-\theta)\binom{t}{0} + \theta\binom{u}{1} = \binom{0}{1} + \binom{q^{\alpha}}{0}x + \binom{s^{\beta}}{-e}y\ ,$$

with $t \in S$, $0 \leqq \theta \leqq 1$, $x \geqq 0$, and $y \geqq 0$ be a point in $V \cap P_{\alpha\beta}$. Clearly $\theta = 1\text{-}ey$. If $\theta = 0$, then $(w,\theta) = (t,0)$ and $(q^{\alpha}x + s^{\beta}y,\ 0) = (a+b,0)$ where $a \in Q_{\alpha}$ and $b \in S_{\beta}$. From Lemma 13.4, setting $\bar{\beta} = \mu$, $a = 0$ and $(w,\theta) \in S_{\beta} \times 0 \subset E$. For $\theta = 1$ then $(w,\theta) = (u,1) = (q^{\alpha}x,1)$ and $(w,\theta) \in Q_{\alpha} \times 1 \subset E$. For $0 < \theta < 1$ we have $(w,\theta) =$

$$\theta\begin{bmatrix} \theta^{-1}q^{\alpha}x \\ 1 \end{bmatrix} + (1-\theta)\begin{bmatrix} (1-\theta)^{-1}\ s^{\beta}y \\ 0 \end{bmatrix}$$

or (w,θ) is in E. □

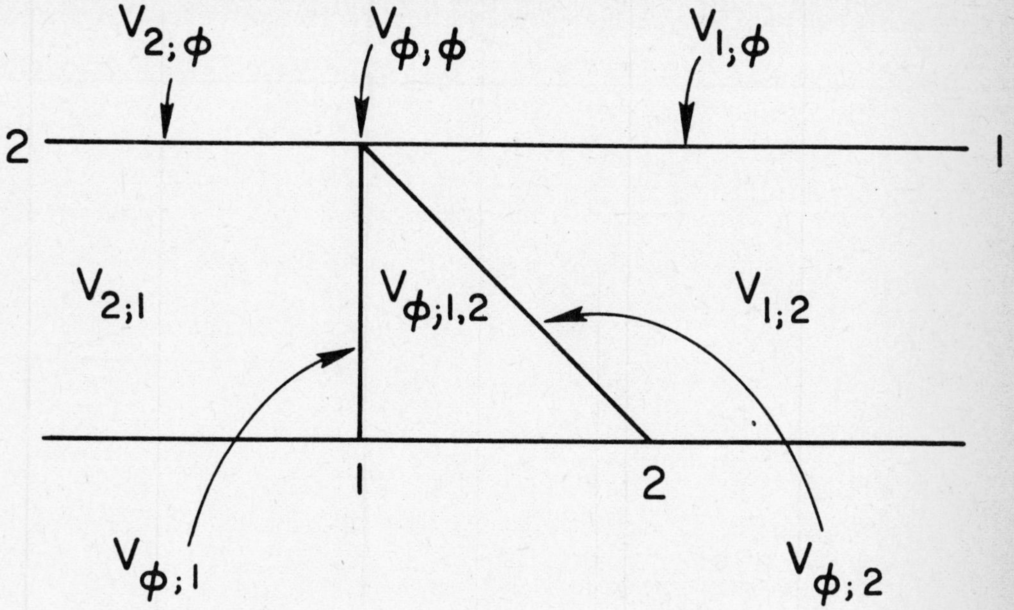

Figure 15.3

Define $\mathbf{V}_{\alpha\beta}$ to be the collection of all cells of form

$$\text{cvx}((\sigma \times 1) \cup (\tau \times 0)) ,$$

where σ is a simplex in $\mathbf{F}|Q_\alpha$ and τ is a face of S_β. Thus each element of $\mathbf{V}_{\alpha\beta}$ is an element of $(\mathbf{F}|Q_\alpha) \times 1$, a face of $S_\beta \times 0$, or the convex hull of a nonempty element of $(\mathbf{F}|Q_\alpha) \times 1$ and a nonempty face of $S_\beta \times 0$.

We proceed to show that $\mathbf{V}_{\alpha\beta}$ triangulates $V_{\alpha\beta} \triangleq V \cap P_{\alpha\beta}$, see Figure 15.4.

We shall employ the coning construction of Section 14 to analyze $\mathbf{V}_{\alpha\beta}$. In this regard we require the next lemma.

15.2 Lemma: Q_α and S_β are transverse for all $\alpha \subset \mu$ and $\beta \subseteq \mu\backslash\alpha$.

Proof: We can assume $\beta = \mu\backslash\alpha$. Consider a point in both tangent spaces

$$\sum_{i\in\alpha} (s^{i+1} - s^i)x_i = \sum_{i\in\beta} (s^i - s^h)x_i$$

where $h \in \beta$ and superscripts are regarded modulo $n+1$. If $\alpha = \phi$ or $\beta = \phi$, obviously, the tangent spaces intersect in $\{0\}$; thus suppose $\alpha \neq \phi$ and $\beta \neq \phi$. Rearrange the expression above to obtain

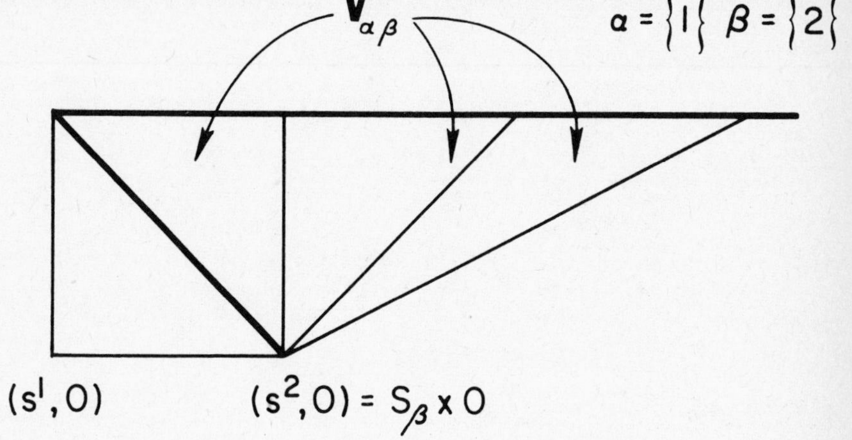

Figure 15.4

$$\sum_{i\in\mu} s^i y_i = 0$$

and we will have $\sum_{i\in\mu} y_i = 0$. As the s^i are in general position, $y_i = 0$. Some s^i for i in α can occur in our first equation only once, since $\beta \neq \phi$. For the i we have $x_i = y_i = 0$. We then have $x_{i+1} = y_{i+1} = 0$, then $x_{i+2} = y_{i+2} = 0$ etc., where the subscripts are regarded modulo $n+1$. □

15.3 Lemma: $\mathbf{V}_{\alpha\beta}$ is a $(\#\alpha + \#\beta)$-subdivision of $V_{\alpha\beta}$.

Proof: $V_{\alpha\beta}$ is the convex hull of $Q_\alpha \times 1$ and $S_\beta \times 0$ which are transverse and affinely disjoint, see Lemma 15.1 and 15.2. $(\mathbf{F}|Q_\alpha) \times 1$ triangulates $Q_\alpha \times 1$, and $\mathbf{T}$ the set of all faces of $S_\beta \times 0$ triangulates $S_\beta \times 0$. $(V_{\alpha\beta}, \mathbf{V}_{\alpha\beta})$ is obtained by coning $(Q_\alpha \times 1, (\mathbf{F}|Q_\alpha) \times 1)$ and $(S_\beta \times 0, \mathbf{T})$, see Theorem 14.5. □

Observe that if $\alpha \neq \phi$ and $\beta \neq \phi$ the $(\#\alpha + \#\beta)$-simplexes in $\mathbf{V}_{\alpha\beta}$ are those of form

$$\mathrm{cvx}((\sigma \times 1) \cup (S_\beta \times 0))$$

where $\dim \sigma = \#\alpha$. If $\beta = \phi$ of course $\mathbf{V}_{\alpha\beta} = \mathbf{F}|Q_\alpha$.

For the next four lemmas let $\alpha \subset \mu$, $\bar{\alpha} \subset \mu$, $\beta \subseteq \mu\backslash\alpha$, $\bar{\beta} \subseteq \mu\backslash\bar{\alpha}$, and $\gamma = \alpha \cap \bar{\alpha}$ and $\delta = \beta \cap \bar{\beta}$.

15.4 Lemma: $V_{\alpha\beta} \cap V_{\bar{\alpha}\bar{\beta}} = V_{\gamma\delta}$.

Proof: $(V \cap P_{\alpha\beta}) \cap (V \cap P_{\bar{\alpha}\bar{\beta}}) = V \cap (P_{\alpha\beta} \cap P_{\bar{\alpha}\bar{\beta}}) = V \cap P_{\gamma\delta}$, see Theorem 13.3. □

15.5 Lemma: $\mathbf{V}_{\alpha\beta} \| V_{\bar{\alpha}\bar{\beta}} = \mathbf{V}_{\alpha\beta} | V_{\bar{\alpha}\bar{\beta}} = \mathbf{V}_{\gamma\delta}$.

Proof: As $V_{\alpha\beta} \cap V_{\bar{\alpha}\bar{\beta}} = V_{\gamma\delta}$, $\mathbf{V}_{\alpha\beta}$ contains $\mathbf{V}_{\gamma\delta}$, and $\mathbf{V}_{\gamma\delta}$ triangulates $V_{\gamma\delta}$, the result follows. □

15.6 Lemma: $\mathbf{V}_{\alpha\beta} \cap \mathbf{V}_{\bar{\alpha}\bar{\beta}} = \mathbf{V}_{\gamma\delta}$.

Proof: $\mathbf{V}_{\alpha\beta} | V_{\bar{\alpha}\bar{\beta}} = \mathbf{V}_{\gamma\delta} = \mathbf{V}_{\bar{\alpha}\bar{\beta}} | V_{\alpha\beta}$. □

Now define $\mathbf{V}$ to be the union of all $\mathbf{V}_{\alpha\beta}$ for $(\alpha, \beta) \in \mathbf{iP}$ that is, all $\alpha \subset \mu$ and $\beta \subseteq \mu \backslash \alpha$. We proceed to show that $\mathbf{V}$ triangulates V.

15.7 Lemma: $\mathbf{V}$ is an $(n+1)$-subdivision triangulating V.

Proof: The sets $V_{\alpha\beta} = V \cap P_{\alpha\beta}$ in $\mathbf{P} \| V$ with $\#\alpha + \#\beta = n+1$ cover V, see Theorem 13.1. The cells of $\mathbf{V}_{\alpha\beta}$ cover $V_{\alpha\beta}$, see Lemma 15.3. As V is $n+1$ dimensional in $\mathbb{C}^{n+1}$ it only remains to show that cells of $\mathbf{V}$ meet in a common face. From Lemma 15.5 $\mathbf{V}_{\alpha\beta} | V_{\gamma\delta} = \mathbf{V}_{\gamma\delta} = \mathbf{V}_{\bar{\alpha}\bar{\beta}} | V_{\gamma\delta}$ and the result follows, see Lemma 3.9. □

15.8 Lemma: $\mathbf{V}|(\mathbb{C}^n \times 1) = \mathbf{F} \times 1$.

Proof: $\mathbf{V}_{\alpha\beta}|(\mathbb{C}^n \times 1) = (\mathbf{F}|Q_\alpha) \times 1$. But $\mathbf{Q}$ subdivides $\mathbb{C}^n$ and $\mathbf{F}$ refines $\mathbf{Q}$, see Lemma 6.22. Therefore $\cup_{(\alpha,\beta)}(V_{\alpha\beta}|\mathbb{C}^n \times 1)$ $= \cup_{(\alpha,\beta)}(F|Q_\alpha) \times 1$ where (α,β) ranges over $\mathbf{iP}$ yields the result. □

15.9 Lemma: If σ and τ are in $\mathbf{V}$, then $\sigma \cap \tau$ is the convex hull of $\sigma \cap \tau \cap (\sigma^n \times 1)$ and $\sigma \cap \tau \cap (S \times 0)$.

Proof: Let $\sigma \in \mathbf{V}_{\alpha\beta}$ and $\tau \in \mathbf{V}_{\bar{\alpha}\bar{\beta}}$. $\sigma \cap \tau = \sigma \cap \tau \cap V_{\gamma\delta}$ from Lemma 15.4. From Lemma 15.5 $\sigma \cap V_{\gamma\delta}$ is of form $\text{cvx}((\rho \times 1) \cup (S_\phi \times 0))$ where $\rho \in Q_\gamma$ and $\phi \subseteq \delta$; similarly $\tau \cap V_{\gamma\delta} = \text{cvx}((\bar{\rho} \times 1) \cup (S_{\bar{\phi}} \times 0))$ where $\bar{\rho} \in Q_\gamma$ and $\bar{\phi} \subseteq \delta$. But from Lemma 14.3, the intersection of $\text{cvx}((\rho \times 1) \cup (S_\phi \times 0))$ and $\text{cvx}((\bar{\rho} \times 1) \cup (S_{\bar{\phi}} \times 0))$ is $\text{cvx}((\rho \cap \bar{\rho} \times 1)$ $\cup (S_{\phi\cap\bar{\phi}} \times 0)) \subseteq \text{cvx}((\sigma \cap \tau \cap (\mathbb{C}^n \times 1)) \cup (\sigma \cap \tau \cap (S \times 0)))$. □

$\mathbf{V}$ is not locally finite as each proper face S_β of S is contained in infinitely many cells of $\mathbf{V}$. However, for any (x,t) in V with $t > 0$ there is a neighborhood meeting at most a finite number of cells of $\mathbf{V}$. It follows that each n-simplex of $\mathbf{V}$ of form $\text{cvx}((\tau \times 1) \cup (S_\beta \times 0))$ with $\tau \neq \phi$ and $\beta \neq \phi$ lies in exactly two $(n+1)$-simplexes of $\mathbf{V}$. This point, as others, is argued again in the replacement rules in that the replacement can be made across such a facet.

We now move to consider the representation and replacement rules of $\mathbf{V}$. Define $\mathbf{iV}^{n+1}$ to be the collection of 4-tuples (v,π,k,η) where

(a) $v \in Z^n$

(b) π permutes $\mu \triangleq \{1, \ldots, n+1\}$

(c) $k = 0, 1, \ldots, n$

(d) $v = q\eta$, $0 \nleq \eta \geqq 0$

(e) $\eta_{\pi i} = 0$, $i = k+1, \ldots, n+1$

(f) $\pi(k+1) < \pi(k+2) < \cdots < \pi(n+1)$.

We shall give some meaning to (a)-(f) as quickly as possible. Recall $\pi|k$ is the ordering $(\pi 1, \ldots, \pi k)$ and that $(\pi|k)\mu = \{\pi 1, \ldots, \pi k\}$. Note that (d) and (e) are equivalent to $v \in Q_{(\pi|k)\mu}$. If $(v,\pi,k,\eta) \in i\mathbf{V}^{n+1}$ then $(v,\pi,k) \in i_*\mathbf{F}$ and we continue to use (v,π,k) to index the k-simplex $(v;\pi;k)$ in $\mathbf{F}$, see Section 6.

Give the quadruple (v,π,k,η) in $i\mathbf{V}^{n+1}$ we define $(v;\pi;k;\eta)$ to be the convex hull of $(v;\pi;k) \times 1$ in $Q_\alpha \times 1$ and $S_\beta \times 0$ where $\alpha = (\pi|k)\mu$ and $\beta = \mu\backslash\alpha$, see Figure 15.5.

That is, $(v;\pi;k;\eta)$ is the convex hull of the $n+2$ points

$$w^1 = (v_1, 1), \ \ldots, \ w^{k+1} = (v^{k+1}, 1)$$

$$w^{k+2} = (s^{\pi(k+1)}, 0) \ , \ \ldots, \ w^{n+2} = (s^{\pi(n+1)}, 0)$$

where $v^1 = v$, $v^{i+1} = v^i + q^{\pi i}$ for $i = 1, \ldots, k$, and $s^{\pi i}$ is the $(\pi i)^{th}$ vertex of S.

If $(v,\pi,k,\eta) \in i\mathbf{V}^{n+1}$ and $\pi|k$ is known, then π is completely specified since it is required that $\pi(k+1) < \pi(k+2) < \cdots < \pi(n+1)$ and $\{\pi(k+1), \ldots, \pi(n+1)\} = \mu\backslash(\pi|k)\mu$. The next lemma shows that $i\mathbf{V}^{n+1}$ is an index set for $\mathbf{V}^{n+1}$.

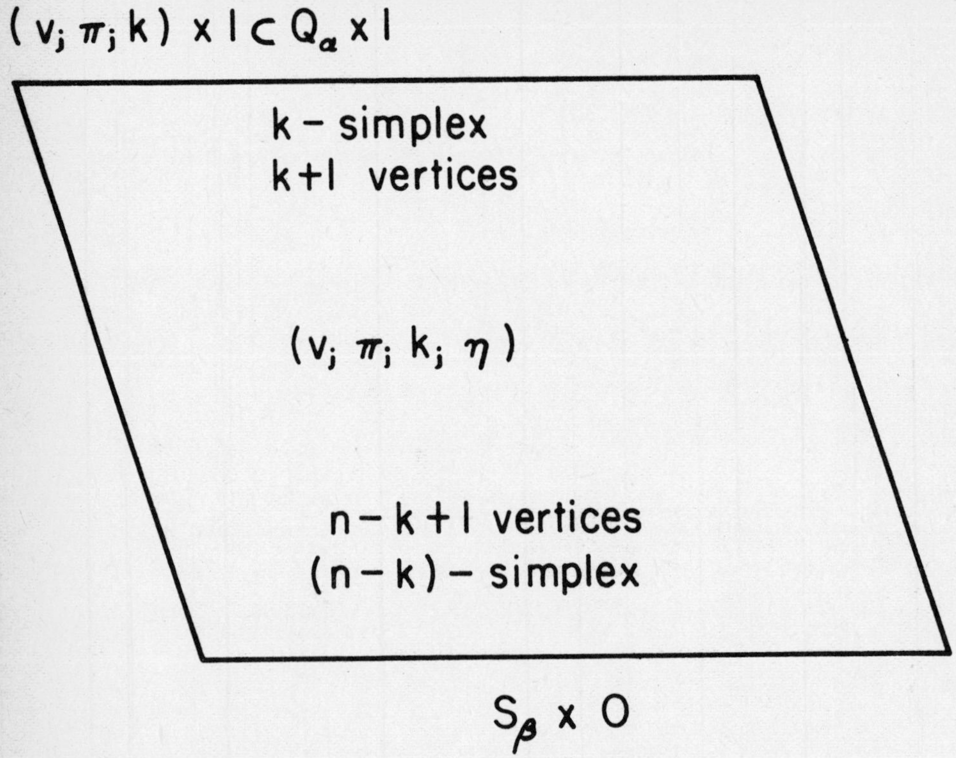
(v; π; k) x I ⊂ Q_α x I
k - simplex
k+I vertices
(v; π; k; η)
n - k + I vertices
(n - k) - simplex
S_β x 0

Figure 15.5

15.10 Lemma: $\mathbf{V}^{n+1} = \{(v;\pi;k;\eta) : (v,\pi,k,\eta) \in i\mathbf{V}^{n+1}\}$.

Proof: Let $\alpha = (\pi|k)\mu$ and $\beta = \mu\backslash\alpha$. Then $(v;\pi;k;\eta)$ $= \mathrm{cvx}((v;\pi;k) \times 1) \cup (S_\beta \times 0)$ which is an $(n+1)$-simplex of $\mathbf{V}_{\alpha\beta}$, see Lemma 14.1. For $\alpha \subset \mu$ and $\beta = \mu\backslash\alpha$ let τ be an $\#\alpha$-simplex of $\mathbf{F}_\alpha$. Represent τ as $(v;\pi;k)$ with $v \in Q_\alpha$ and $\pi(k+1) < \cdots < \pi(n+1)$, see Lemma 11.1. Let $v = q\eta$ with $0 \nleq \eta \geqq 0$ and we have $\mathrm{cvx}((\tau \times 1) \cup (S_\beta \times 0)) = (v;\pi;k;\eta)$. □

Let us consider several examples. As a distinguished case consider $(v,\pi,k,\eta) = (0,\pi,0,0)$, then $(v;\pi;k;\eta)$ is the convex hull of $(0,1)$ and $S \times 0$, see Figures 15.6 and 15.7.

The next lemma established uniqueness of the representation.

15.11 Lemma: If $(v;\pi;k;\eta) = (\bar{v};\bar{\pi};\bar{k};\bar{\eta})$ then (v,π,k,η) $= (\bar{v},\bar{\pi},\bar{k},\bar{\eta})$.

Proof: $(v;\pi;k;\eta) \cap (\mathbb{C}^n \times 1) = (\bar{v};\bar{\pi};\bar{k};\bar{\eta}) \cap (\mathbb{C}^n \times 1) = (v;\pi;k) \times 1$ $= (\bar{v},\bar{\pi},\bar{k}) \times 1$ and we conclude $k = \bar{k}$. As $(v;\pi;k) \subset Q_{(\pi|k)\mu}$ and $(\bar{v};\bar{\pi};k) \subset Q_{(\bar{\pi}|k)\mu}$ we see that $Q_{(\pi|k)\mu}$ and $Q_{(\bar{\pi}|k)\mu}$ share a relative interior point, thus $Q_{(\pi|k)\mu} = Q_{(\bar{\pi}|k)\mu}$ and $(\pi|k)\mu =$ $(\bar{\pi}|k)\mu$. From Lemma 6.3 we get $v = \bar{v}$ and $\pi|k = \bar{\pi}|k$, and consequently, $\eta = \bar{\eta}$. □

As usual, to prove validity of a replacement $(\hat{v},\hat{\pi},\hat{k},\hat{\eta})$ for (v,π,k,η) on dropping t in $\mathbf{V}$ one needs only to check that

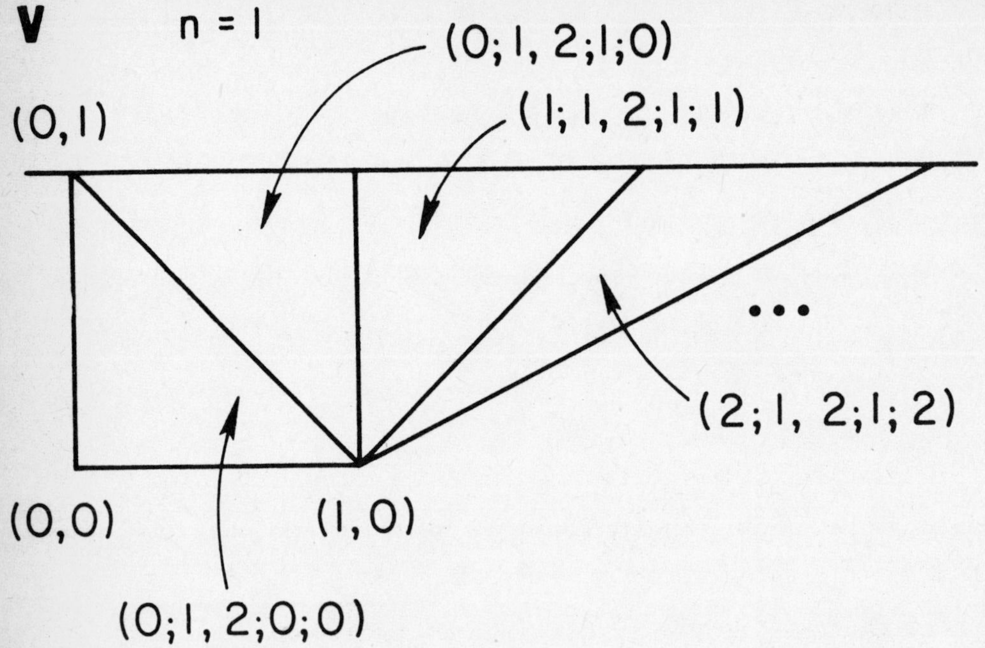

Figure 15.6

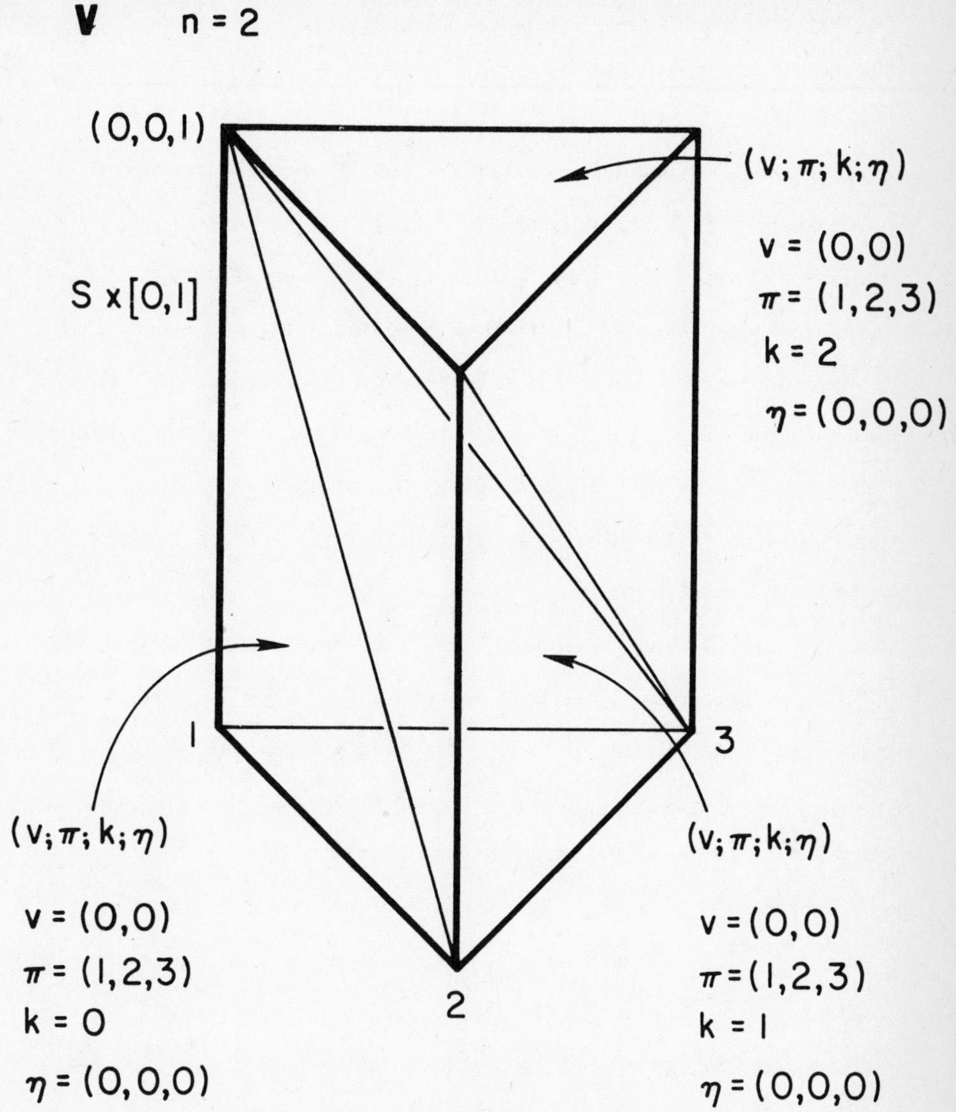
V n = 2
(0,0,1)
S x [0,1]
(v; π; k; η)
v = (0,0)
π = (1,2,3)
k = 2
η = (0,0,0)
1
3
2
(v; π; k; η)
v = (0,0)
π = (1,2,3)
k = 0
η = (0,0,0)
(v; π; k; η)
v = (0,0)
π = (1,2,3)
k = 1
η = (0,0,0)

Figure 15.7

(a) $(\hat{v},\hat{\pi},\hat{k},\hat{\eta}) \in i\mathbf{V}^{n+1}$

(b) $(\hat{v},\hat{\pi},\hat{k},\hat{\eta}) \neq (v,\pi,k,\eta)$

(c) $(\hat{v};\hat{\pi};\hat{k};\hat{\eta})$ contains w^i for $i \neq t$.

Consider the $(n+1)$-simplex $(v;\pi;k;\eta)$ of $\mathbf{V}$ and assume vertex w^t is to be dropped. If no replacement $(\hat{v},\hat{\pi},\hat{k},\hat{\eta})$ exists it is because all vertices w^i with $i \neq t$ lie either in $S \times 0$ or $\mathbb{C}^n \times 1$ and we shall refer to these cases as a bottom exit or top exit, see Figure 15.8.

We proceed to give a narrative description of the replacement rules for $\mathbf{V}$ and hasten to point out that the narration parallels the rules as specified in the charts of Figures 15.9 through 15.11. As there are a number of cases the reader might find it fruitful to vaselate between the narration and the charts.

Given (v,π,k,η) and the index t of the vertex to be dropped the first step is to determine if w^t is in $S \times 0$ or $\mathbb{C}^n \times 1$. This is resolved by comparing t and k. For $t \leqq k+1$, w^t lies in $\mathbb{C}^n \times 1$ and for $t \geqq k+2$, w^t lies in $S \times 0$, see the definition of $(v;\pi;k;\eta)$.

The top and bottom exits are readily detected. If $t = 1$ and $k = 0$ then all vertices of $(v;\pi;k;\eta)$ are in $S \times 0$ except w^t, and we have a bottom exit. If $t = n+2$ and $k = n$, then all vertices of $(v;\pi;k;\eta)$ are in $\mathbb{C}^n \times 1$ except w^t which is in $S \times 0$, and we have a top exit. Clearly these are the only conditions which lead to a bottom or top exit. If there is a top exit, $(v;\pi;k;\eta) \cap (\mathbb{C}^n \times 1)$ is the n-simplex $(v;\pi;k)$, and if there is a bottom exit, $(v;\pi;k;\eta) \cap (S \times 0) = S \times 0$.

Aside from the top and bottom exits there are four remaining cases:

Case A: $1 = t \leqq k$

Case B: $2 \leqq t = k+1$

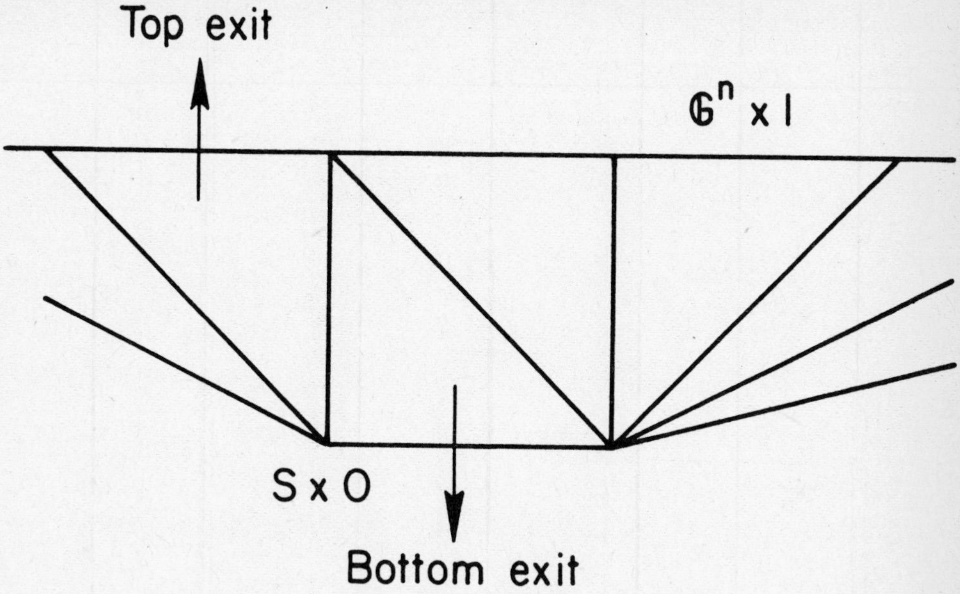

Figure 15.8

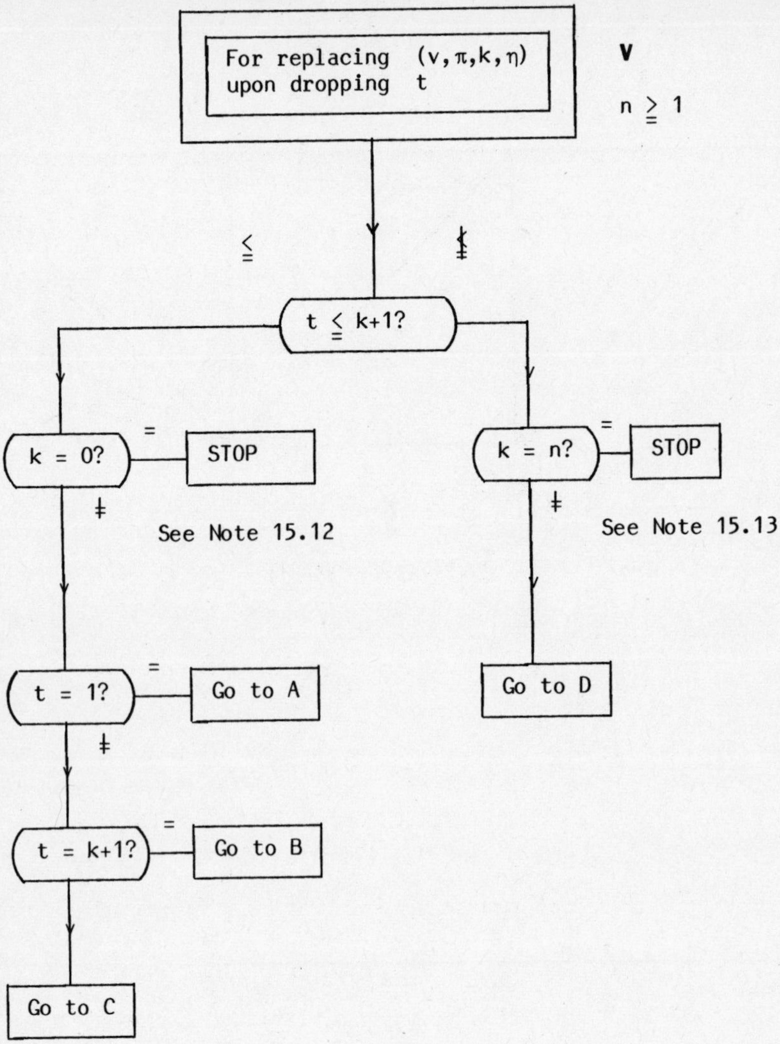
For replacing (v,π,k,η)
upon dropping t
V
n ≧ 1
≦
≰
t ≦ k+1?
=
k = 0?
STOP
≠
See Note 15.12
=
k = n?
STOP
≠
See Note 15.13
=
t = 1?
Go to A
≠
=
t = k+1?
Go to B
Go to C
Go to D

Figure 15.9

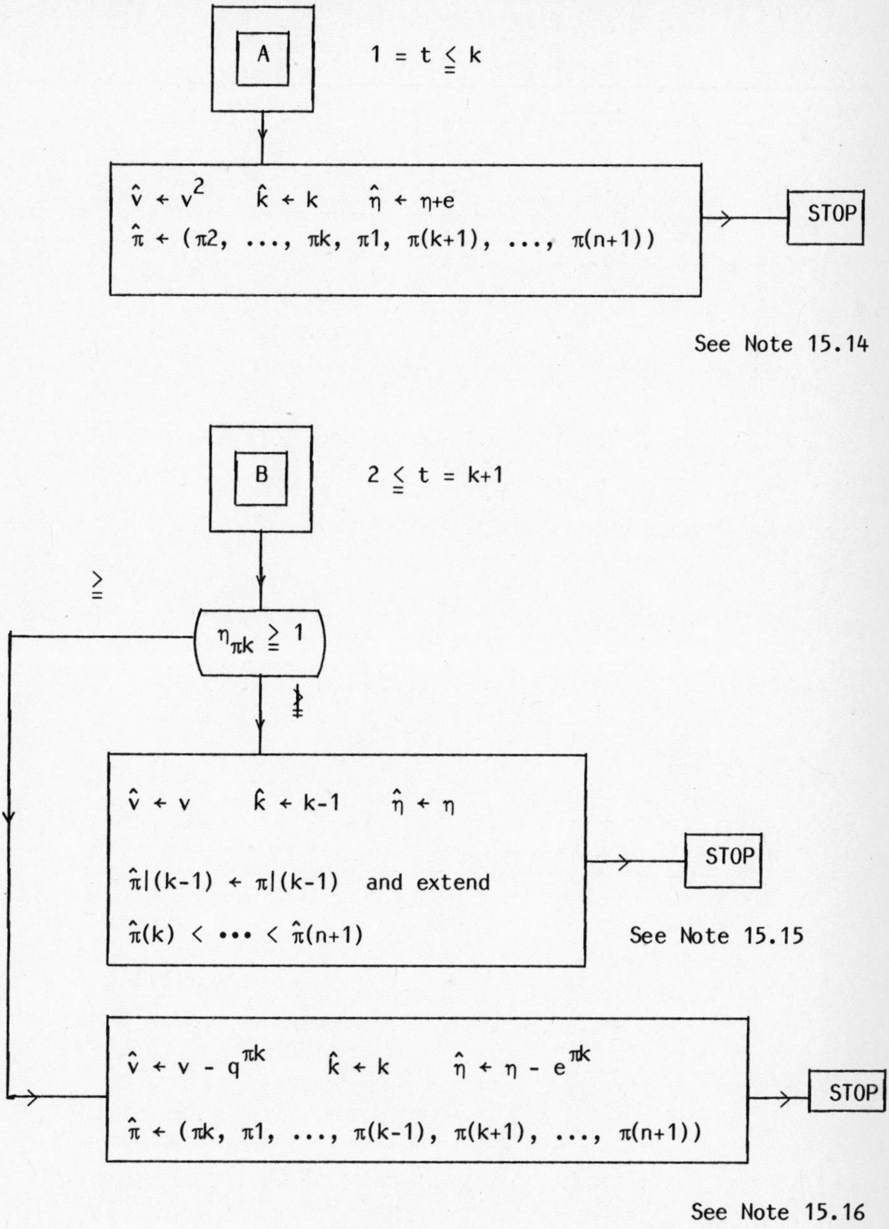
A
1 = t ≦ k
v̂ ← v² k̂ ← k η̂ ← η+e
π̂ ← (π2, ..., πk, π1, π(k+1), ..., π(n+1))
STOP
See Note 15.14
B
2 ≦ t = k+1
≧
η_πk ≧ 1
≱
v̂ ← v k̂ ← k-1 η̂ ← η
π̂|(k-1) ← π|(k-1) and extend
π̂(k) < ••• < π̂(n+1)
STOP
See Note 15.15
v̂ ← v - q^πk k̂ ← k η̂ ← η - e^πk
π̂ ← (πk, π1, ..., π(k-1), π(k+1), ..., π(n+1))
STOP
See Note 15.16

Figure 15.10

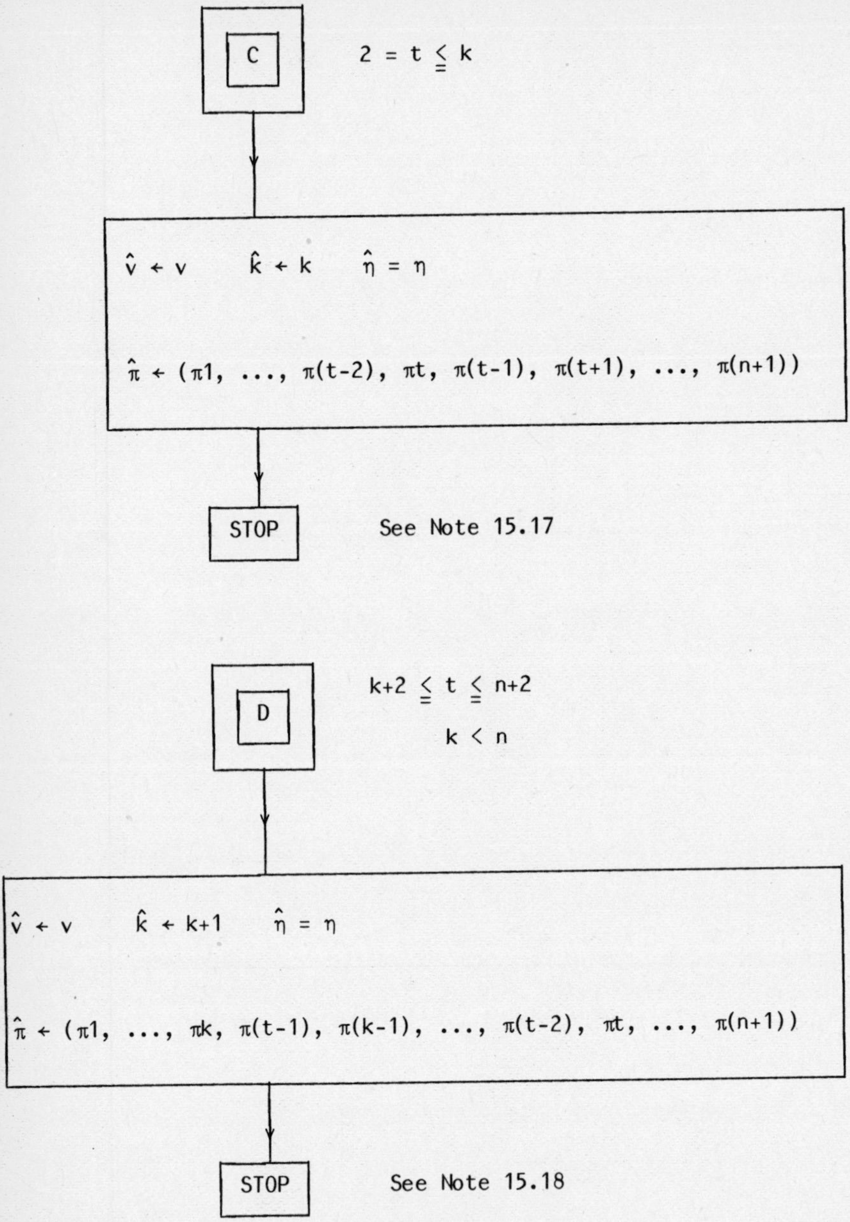
C
2 = t ≦ k
v̂ ← v   k̂ ← k   η̂ = η
π̂ ← (π1, ..., π(t-2), πt, π(t-1), π(t+1), ..., π(n+1))
STOP
See Note 15.17
D
k+2 ≦ t ≦ n+2
k < n
v̂ ← v   k̂ ← k+1   η̂ = η
π̂ ← (π1, ..., πk, π(t-1), π(k-1), ..., π(t-2), πt, ..., π(n+1))
STOP
See Note 15.18

Figure 15.11

Case C: $2 \leqq t \leqq k$

Case D: $k+2 \leqq t \leqq n+2,\ k < n$.

Case A: $(1 = t \leqq k)$: w^1 is dropped and $k \geqq 1$. Let $\hat{v} = v_2$, $\hat{\pi} = (\pi 2, \ldots, \pi k, \pi 1, \pi(k+1), \ldots, \pi(n+1))$, $\hat{k} = k$, and $\hat{\eta} = \eta + e^{\pi 1}$. We have $\hat{v} = q\hat{\eta}$ and $\hat{\eta}_\beta = 0$ where $\beta = \mu\backslash(\hat{\pi}|k)\mu$. Since $(\pi|k)\mu = (\hat{\pi}|k)\mu$, $(\hat{v};\hat{\pi};k)$ and $(v;\pi;k)$ have k vertices in common, and since, $\mu\backslash(\pi|k)\mu = \mu\backslash(\hat{\pi}|k)\mu$ the $(n+1)$-simplex $(v;\pi;k;\eta)$ and $(\hat{v};\hat{\pi};\hat{k};\hat{\eta})$ have $k + (n-k+1) = n+1$ vertices in common. It follows that $(\hat{v};\hat{\pi};\hat{k};\hat{\eta})$ is the replacement of $(v;\pi;k;\eta)$ upon dropping w^t. □

Case B: $(2 \leqq t = k+1)$: $w^{k+1} = (v^{k+1},1)$ is dropped and $k \geqq 1$. First suppose that $\eta_{\pi k} \geqq 1$. In this case $\hat{v} = v - q^{\pi k}$ remains in $Q_{(\pi|k)\mu}$. Note $\hat{v} = q\hat{\eta}$ where $\hat{\eta} = \eta - e^{\pi k}$ and $\hat{\eta}_\beta = 0$ where $\beta = \mu\backslash(\pi|k)\mu$. Let $\hat{\pi} = (\pi k, \pi 1, \ldots, \pi(k-1), \pi(k+1), \ldots, \pi(n+1))$ then clearly $(v;\pi;k)$ and $(\hat{v};\hat{\pi};\hat{k})$ with $k = \hat{k}$ are adjacent and $(\hat{v};\hat{\pi};k)$ does not contain v^{k+1}. Since $(\pi|k)\mu = (\hat{\pi}|k)\mu$ we have $\mu\backslash(\pi|k)\mu = \mu\backslash(\hat{\pi}|k)\mu$. Thus $(\hat{v};\hat{\pi};\hat{k};\hat{\eta})$ and $(v;\pi;k;\eta)$ share $k + (n-k+1) = n+1$ vertices and the former does not contain $(v^{k+1}, 1)$. Consequently $(\hat{v};\hat{\pi};\hat{k};\hat{\eta})$ is the replacement of $(v;\pi;k;\eta)$ upon dropping w^{k+1}. Next let us suppose $\eta_{\pi k} = 0$ or that $\hat{v} = v$ is in Q_α where $\alpha = \pi|(k-1)$. Let $\hat{k} = k-1$ and define $\hat{\pi} = (\pi 1, \ldots, \pi(k-1), \ldots)$ where the last $n-k+2$ elements are rearranged in increasing order. Thus $(\hat{v};\hat{\pi};k-1)$ is in Q_α.

As $(\hat{v};\hat{\pi};\hat{k})$ is a facet of $(v;\pi;k)$ they share k vertices.

As $\mu\backslash(\hat{\pi}|k-1)\mu \supset \mu\backslash(\pi|k)\mu$, $(v;\pi;k;\eta)$ and $(\hat{v};\hat{\pi};\hat{k};\hat{\eta})$ share $k+n-k+1$ vertices. Under Case B the replacement for $(v;\pi;k;\eta)$ upon dropping w^{k+1} has been exhibited for $\eta_{\pi k} \geqq 1$ and $\eta_{\pi k} = 0$. □

Case C $(2 \leqq t \leqq k)$: $w^t = (v^t,1)$ is dropped. Let $\hat{v} = v$, $\hat{\pi} = (\pi 1, \ldots, \pi(t-2), \pi t, \pi(t-1), \pi(t+1), \ldots, \pi(n+1)$ and $\hat{k} = k$. Of course $\hat{v}$ is in $Q_{(\hat{\pi}|k)\mu}$ and $\hat{v} = q\hat{\eta}$ with $\hat{\eta} = \eta$. As $(\hat{v};\hat{\pi};k)$ and $(v;\pi;k)$ share k vertices, $(\hat{v};\hat{\pi};\hat{k})$ does not contain v^t, and $\mu\backslash(\pi|k)\mu = \mu\backslash(\hat{\pi}|\hat{k})\mu$ then $(n+1)$ simplexes $(v;\pi;k;\eta)$ and $(\hat{v};\hat{\pi};\hat{k};\hat{\eta})$ share $k+n-k+1 = n+1$ vertices, and $(\hat{v};\hat{\pi};\hat{k};\hat{\eta})$ does not contain w^t and is the desired replacement. □

Case D $(k+2 \leqq t \leqq n+2, k < n)$: Vertex $w^t = (s^{\pi(t-1)},0)$ is being dropped. Let $\hat{v} = v$, $\hat{\pi} = (\pi 1, \ldots, \pi k, \pi(t-1), \pi(k+1), \ldots, \pi(t-2)$, $\pi t, \ldots, \pi(n+1)$, and $\hat{k} = k+1$. As v is in $Q_{(\pi|k)\mu}$, $\hat{v}$ is in $Q_{(\hat{\pi}|\hat{k})\mu}$. Let $\hat{\eta} = \eta$, then $\hat{v} = q\hat{\eta}$. Clearly $(v;\pi;k)$ is a facet of $(\hat{v};\hat{\pi};\hat{k})$ and they share all vertices of $(\hat{v};\hat{\pi};\hat{k})$ except $\hat{v}^{k+2} = v^{k+1} + q^{\pi(t-1)}$. As $(\mu\backslash(\hat{\pi}|\hat{k})\mu) \cup \{\phi(t-1)\} = \mu\backslash(\pi|k)\mu$, $(v;\pi;k;\eta)$ and $(\hat{v};\hat{\pi};\hat{k};\hat{\eta})$ share all vertices of $(v;\pi;k;\eta)$ except $w^t = (s^{\pi(t-1)},0)$. Thus $(\hat{v};\hat{\pi};\hat{k};\hat{\eta})$ is the replacement for $(v;\pi;k;\eta)$ upon dropping w^t. □

15.12 Note (for Figure 15.9): A bottom exit has occurred through $S \times 0$. □

15.13 Note (for Figure 15.9): A top exit has occurred through $(v;\pi;n) \times 1$. □

15.14 Note (for Figure 15.10): Replacement. The adjoined vertex is $(\hat{v}^{k+1}, 1)$. The correspondence between vertices of $(v;\pi;k;\eta)$ and $(\hat{v};\hat{\pi};\hat{k};\hat{\eta})$ is

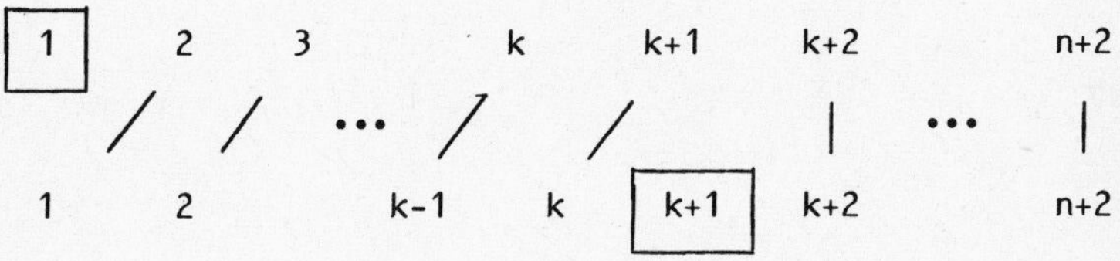

where the squares indicate the dropped and adjoined vertex. □

15.15 Note (for Figure 15.10): Replacement. The adjoined vertex is $(s^{\pi k}, 0)$. If $\pi k < \pi(k+1)$ then $\hat{\pi} <= \pi$ and the correspondence between vertices is

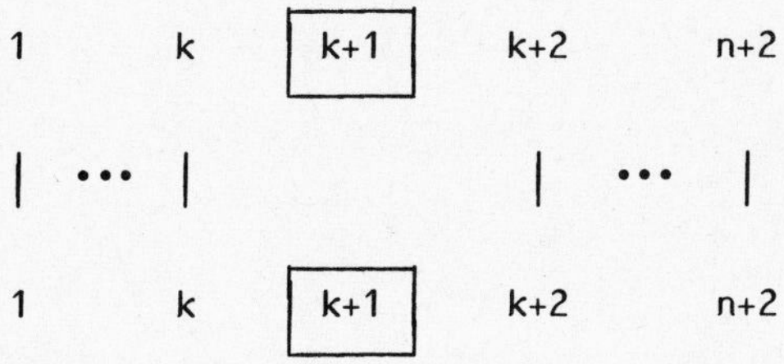

□

If $\pi j < \pi k < \pi(j+1)$ for $j = k+1, \ldots, n$, $\hat{\pi} <= (\pi 1, \ldots, \pi(k-1), \pi(k+1), \ldots, \pi j, \pi k, \pi(j+1), \ldots, \pi(n+1))$ and the correspondence between vertices is

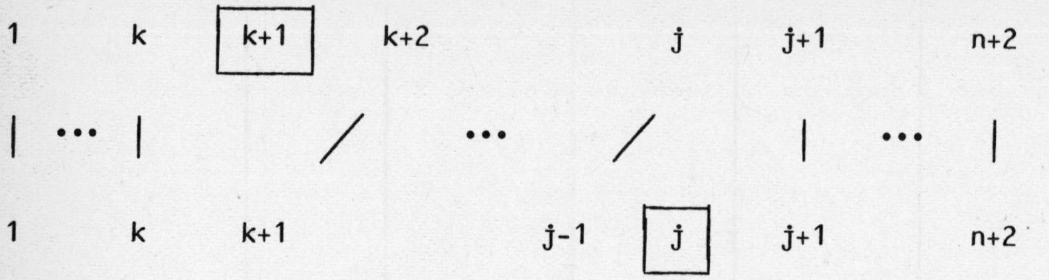

□

If $\pi k > \pi(n+1)$ then $\hat{\pi} = (\pi 1, \ldots, \pi(k-1), \pi(k+1), \ldots, \pi(n+1), \pi k)$ and the correspondence is

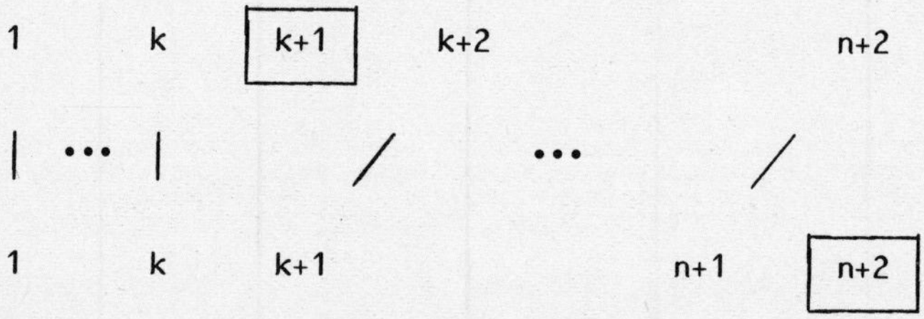

□

15.16 Note (for Figure 15.10): Replacement. The adjoined vertex is $(\hat{v}^1;1)$. The correspondence between vertices is:

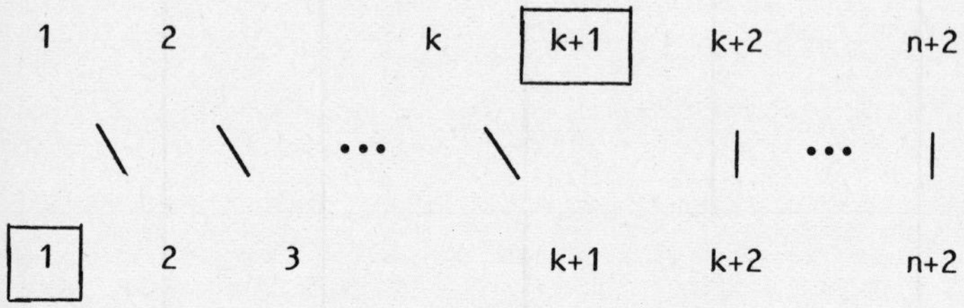

□

15.17 Note (for Figure 15.11): Replacement. The adjoined vertex is $(\hat{v}^t,1)$. The correspondence between vertices is

1	t-1	[t]	t+1	n+2
| ...	|		| ...	|
1	t-1	[t]	t+1	n+2

□

15.18 Note (for Figure 15.11): Replacement. The adjoined vertex is $(\hat{v}^{k+2},1)$. The correspondence between vertices is:

1	k+1	k+2	t-2	[t-1]	t	n+2
| ...	|	\ ...	\		| ...	|
1	k+1	[k+2]	k+3	t-1	t	n+2

□

Now let us squeeze and shear **V**; let r be any positive element of G and p any element of $\mathbb{C}^n$. Define the **V**-PL map $g : V \to V$ by

$$g(x,1) = (rx + p,\ 1)$$

$$g(x,0) = (x,0)$$

where (x,δ) is a vertex of **V**.

From Lemma 9.1 we know that $g(\mathbf{V})$ is a triangulation isomorphic to **V**, see Figure 15.12.

In Lemma 15.19 we will show that $g(V) = V$ hence $g(\mathbf{V})$ triangulation V.

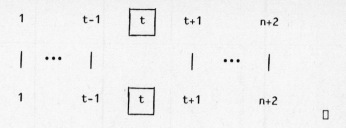

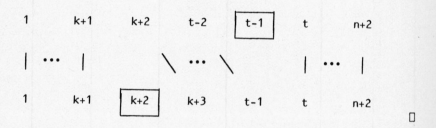

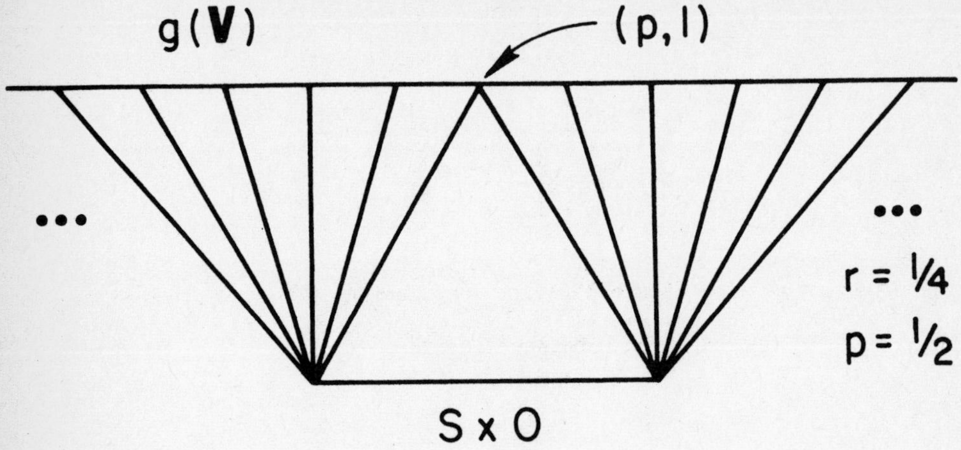

Figure 15.12

Of course the representation set and replacement rules of $g(\mathbf{V})$ are those of $\mathbf{V}$. That is $ig(\mathbf{V})^{n+1} = i\mathbf{V}$ and the replacement rules are <u>identical</u>. For (v,π,k,η) in $ig(\mathbf{V})^{n+1}$ then $(v;\pi;k;\eta)$ in $g(\mathbf{V})$ represents the $(n+1)$-simplex with vertices

$$(rv^1+p,\ 1),\ \ldots,\ (rv^{k+1}+p,\ 1)$$

$$(v^{k+2},\ 0),\ \ldots,\ (v^{n+2},\ 0)$$

where

$$(v^1,\ 1),\ \ldots,\ (v^{k+1},\ 1)$$

$$(v^{k+2},\ 0),\ \ldots,\ (v^{n+2},\ 0)$$

are the vertices of $(v;\pi;k;\eta)$ in $\mathbf{V}$.

15.19 Lemma: $g(V) = V$.

Proof: Let $c = (n,\ n-1,\ \ldots,\ 1)\ (n+1)^{-1}$ be the barycenter of S. Define $V(k,kc)$ to be the convex hull of $(kS-kc) \times 1$ at $S \times 0$ for $k = n+1,\ n+2,\ \ldots$. In Lemma 16.2 it is shown that $\mathbf{V}|V(k,kc)$ triangulates $V(k,kc)$. As the $V(k,kc)$ increase to V it follows from Lemma 9.5 that $g(V)$ is convex and hence $g(V) = V$. □

15.20 Exercise: Show that g is linear on each $V_{\alpha\beta}$ and hence that g is continuous. □

15.21 Exercise: Show that h(**P**) is an encoarsement of g(**V**). See Exercise 13.5 and Figure 15.13. □

15.22 Bibliographical Remarks: If the path followed in solving equations with LP homotopies is projected in a certain manner the projected path wanders in different dimensions, e.g., in $\mathbb{C}^0$, then $\mathbb{C}^1$, $\mathbb{C}^1$, $\mathbb{C}^2$, $\mathbb{C}^3$, $\mathbb{C}^2$, $\mathbb{C}^3$, etc. Some algorithms are more easily viewed in the projected space, see Shapely [1973]. A more recent and general purpose scheme of this sort was described by van der Laan and Talman [1979]; as Todd [1978] explained it, they in essence used a triangulation isomorphic to **V** to implement the restart algorithm of Merrill [1972].

A class of triangulations with a structure like **V** are discussed in van der Laan and Talman [1979], Kojima and Yamamoto [1982a and b], and Yamamoto [1981], and, indeed, certain of these triangulations can be used in the construction of variable rate refining triangulations. □

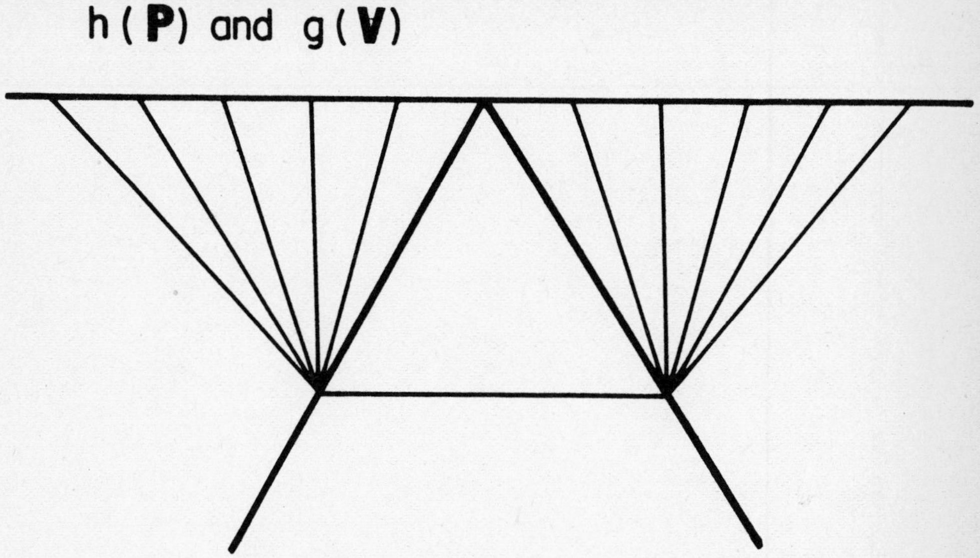

Figure 15.13

16. TRIANGULATION $\mathbf{V}[r,p]$ OF $S \times [0,1]$ BY RESTRICTING, SQUEEZING AND SHEARING $\mathbf{V}$

Given the triangulation $\mathbf{V}$ we analyze certain restrictions which we squeeze and shear to complete the building blocks for the variable rate refining triangulation $\mathbf{S}$.

Let r in Z and p in Z^n be chosen so that $r^{-1}p$ is in S, or equivalently, so that $rS-p$ contains the origin; recall that $r^{-1}p$ is in S if and only if

$$r \geqq p_1 \geqq p_2 \geqq \cdots \geqq p_n \geqq 0 .$$

We refer to r as the refinement factor and $r^{-1}p$ as the focal point. Define $V(r,p)$ to be the convex hull of $S \times 0$ and $(rS-p) \times 1$, see Figure 16.1. Define $\mathbf{V}(r,p)$ to be a natural restriction $\mathbf{V}|V(r,p)$ of $\mathbf{V}$ to $V(r,p)$. We have three tasks before us, first, to show that $\mathbf{V}(r,p)$ triangulates $V(r,p)$, second, to develop the representation and replacement rules for $\mathbf{V}(r,p)$, and finally, to squeeze and shear $\mathbf{V}(r,p)$ to obtain a triangulation $\mathbf{V}[r,p]$ of $S \times [0,1]$.

Recall that $\mathbf{F}|(rS-p)$ and $\mathbf{F}|Q_\alpha$ triangulate $rS-p$ and Q_α, respectively, see Lemmas 6.20 and 6.22. Also recall that $\mathbf{V}|(\mathbb{C}^n \times 1) = \mathbf{F} \times 1$. The next lemma offers the technical step in proving that $\mathbf{V} \| V(r,p) \subseteq \mathbf{V}$, that is, that $\mathbf{V}|V(r,p)$ triangulates $V(r,p)$.

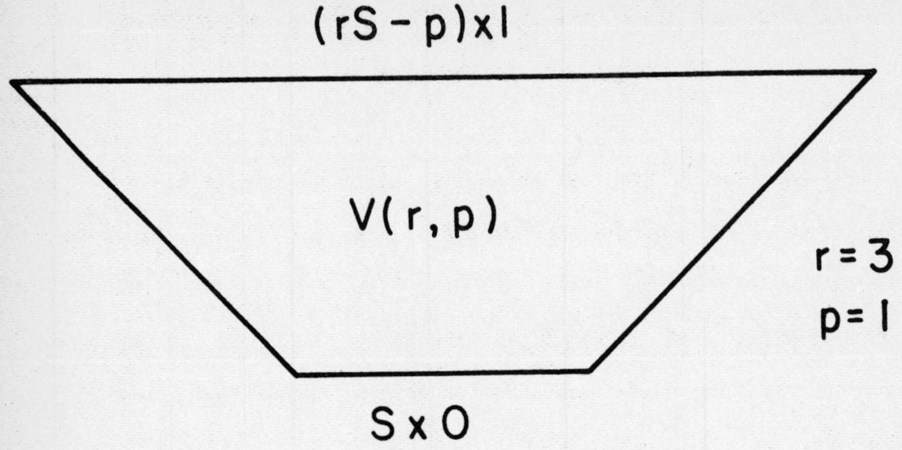

Figure 16.1

16.1 Lemma: Let $a \in Q_\alpha$ and $b \in S_\beta$. If $c \in S$, $\gamma(a+c) + (1-\gamma)b \in S$, and $0 < \gamma \leqq 1$, then $a+c \in S$.

Proof: Let $\gamma(a+c) = (1-\gamma)b = t \in S$. Of course the barycentric coordinates $\Lambda(a,i) = a_{i-1} - a_i$ for i in μ and $\Lambda(t,i) = t_{i-1} - t_i$ are nonnegative for i in μ where $(\cdot)_0 \triangleq 1$ and $(\cdot)_{n+1} \triangleq 0$. We proceed to show that the barycentric coordinates of $a+c$ are nonnegative. Let $a = qx$ and $b = s\lambda$ where $x \geqq 0$, $\lambda \geqq 0$, $x_\beta = 0$, $\lambda_\alpha = 0$, $e\lambda = 1$, and $x \cdot \lambda = 0$. Since $1-c_1 \geqq 0$, if $x_1 - x_{n+1} + c_1 > 1$, then $x_1 > 0$, $\lambda_1 = 0$, $b_1 = 1$, and $t_1 = \gamma(x_1 - x_{n+1} + c_1) + (1-\gamma)(1) > 1$ which contradicts $1-t_1 \geqq 0$. Thus $\Lambda(a+c,1) \geqq 0$. Since $c_1 - c_2 \geqq 0$, if $x_1 - x_{n+1} + c_1 < x_2 - x_{n+1} + c_2$ then $x_2 > 0$, $\lambda_2 = 0$, $b_1 = b_2$ and $t_2 = \gamma(x_2-x_{n+1}+c_2) + (1-\gamma)(b_2) > t_1 = \gamma(x_1-x_{n+1}+c_1) +(1-\gamma)(b_1)$ which contradicts $t_1 \leqq t_2$. Thus $\Lambda(a+c,2) \geqq 0$. Repeating the last argument we have $\Lambda(a+c, j) \geqq 0$ for $j = 1, \ldots, n$. Since $c_n \geqq 0$, if $x_n - x_{n+1} + c_n < 0$ then $x_{n+1} > 0$, $\lambda_{n+1} = 0$, $b_n = 0$ and $t_n = \gamma(x_n - x_{n+1} + c_n) + (1-\gamma)(b_n) < 0$ which contradicts $t_n \geqq 0$. Thus $\Lambda(a+c, n) \geqq 0$ and $a+c$ is in S. □

16.2 Lemma: $\mathbf{V}(r,p)$ is a finite triangulation of $V(r,p)$.

Proof: Let (w,θ) be a point in $V(r,p)$. We need only show that for some τ in $\mathbf{V}$ we have $(w,\theta) \in \tau \subseteq V(r,p)$. Let σ be a simplex of $\mathbf{V}$ containing (w,θ); σ has the form $\mathrm{cvx}((\rho \times 1) \cup (\tau \times 0))$ where $\rho \in \mathbf{F}|Q_\alpha$ and τ is a face of S_β for some $\alpha \subset \mu$ and $\beta = \mu\backslash\alpha$. If $\theta = 0$ then

$(w,\theta) \in \tau \times 0 \subseteq V(r,p)$ and $\tau \times 0 \in \mathbf{V}$. If $\theta = 1$, then $(w,\theta) \in (\rho \cap (rS-p)) \times 1 \subseteq V(r,p)$ and $\rho \cap (rS-p)$ is an element of $\mathbf{F}|Q_\alpha$ as $\mathbf{F}|(rS-p)$ subdivides $rS-p$. Thus let us assume $0 < \theta < 1$. As $(w,\theta) \in V(r,p) \cap \sigma$ it can be expressed as

$$(w,\theta) = (1-\theta)\binom{t^0}{0} + \theta\binom{rt^1-p}{1}$$

$$= \binom{0}{1} + \binom{q^\alpha}{0}x + \binom{s^\beta}{-e}y$$

with $t^0 \in S$, $t^1 \in S$, $x \geqq 0$ and $y \geqq 0$. Thus

$$(w,\theta) = \theta\begin{bmatrix}a\\1\end{bmatrix} + (1-\theta)\begin{bmatrix}b\\0\end{bmatrix}$$

where $a = \theta^{-1}q^\alpha x \in Q_\alpha$ and $b = (1-\theta)^{-1}s^\beta y \in S_\beta$; note $ey = 1-\theta$. Dividing an expression for (w,θ) above by r we have

$$\left(\frac{1-\theta}{r}\right)t^0 + \theta t^1 = \theta r^{-1}a + \theta(r^{-1}p) + \left(\frac{1-\theta}{r}\right)b \quad .$$

Dividing by $\theta + (1-\theta)r^{-1}$ we have

$$t = \gamma(r^{-1}a + r^{-1}p) + (1-\gamma)b$$

with $t \in S$ and $0 < \gamma < 1$. Clearly $r^{-1}a \in Q_\alpha$ and $+r^{-1}p \in S$. Applying Lemma 16.1 we have $r^{-1}a + r^{-1}p \in S$ or $a \in rS-p$. Let ρ' be an element of $\mathbf{F}$ in $(rS-p) \cap Q_\alpha$ containing a. Then (w,θ) is in $\mathrm{cvx}(\rho' \times 1) \cup (\tau \times 0)$ which is in $\mathbf{V}$ and a subset of $V(r,p)$. □

To summarize, $\mathbf{V}(r,p)$ is the collection of all simplexes of form $cvx((\rho \times 1) \cup (\tau \times 0))$ where $\rho \in \mathbf{F}|Q_\alpha \cap (rS-p))$ and $\tau = S_\beta$ for some $\alpha \subset \mu$ and $\beta \subseteq \mu\backslash\alpha$, and further, $\mathbf{V}(r,p)$ triangulates $V(r,p)$. Of course, $\mathbf{V}(r,p)$ is finite.

Note the $\mathbf{V}$ is the union of all $\mathbf{V}(r,p)$ where (r,p) ranges over $Z \times Z^n$ with $r^{-1}p$ in S. In particular, if $(r_i,p^i) \in Z \times Z^n$ and $r_i^{-1}p^i \in S$ for $i = 1, 2, \ldots$, and

$$\bigcup_i (r_iS - p_i) = G^n$$

then

$$\bigcup_i V(r_i,p^i) = V\ , \qquad \bigcup_i \mathbf{V}(r_i,p^i) = \mathbf{V}\ .$$

16.3 Exercise: Show that $V_{\alpha\beta} \cap V(r,p)$ is the convex hull of $(Q_\alpha \cap (rS-p)) \times 1$ and $S_\beta \times 0$, see Section 15. □

We move to consider the representation and replacement rules of $\mathbf{V}(r,p)$. The representation set for $\mathbf{V}(r,p)^{n+1}$ is merely a subset of that for $\mathbf{V}^{n+1}$, and given the replacement rules for $\mathbf{V}$ we need merely to adjoin the check for a side exit to obtain those for $\mathbf{V}(r,p)$, see Figure 16.2. Thus, if upon dropping index t from (v,π,k,η) no replacement $(\hat{v},\hat{\pi},\hat{k},\hat{\eta})$ exists, it is because all vertices w^i of $(v;\pi;k;\eta)$ with $i \neq t$ lie in a facet of $V(r,p)$ and, in particular, in a bottom, side, or top facet of $V(r,p)$, and we refer to these cases as "bottom", "side", and "top" exits, see Lemma 3.10 and Figure 16.2.

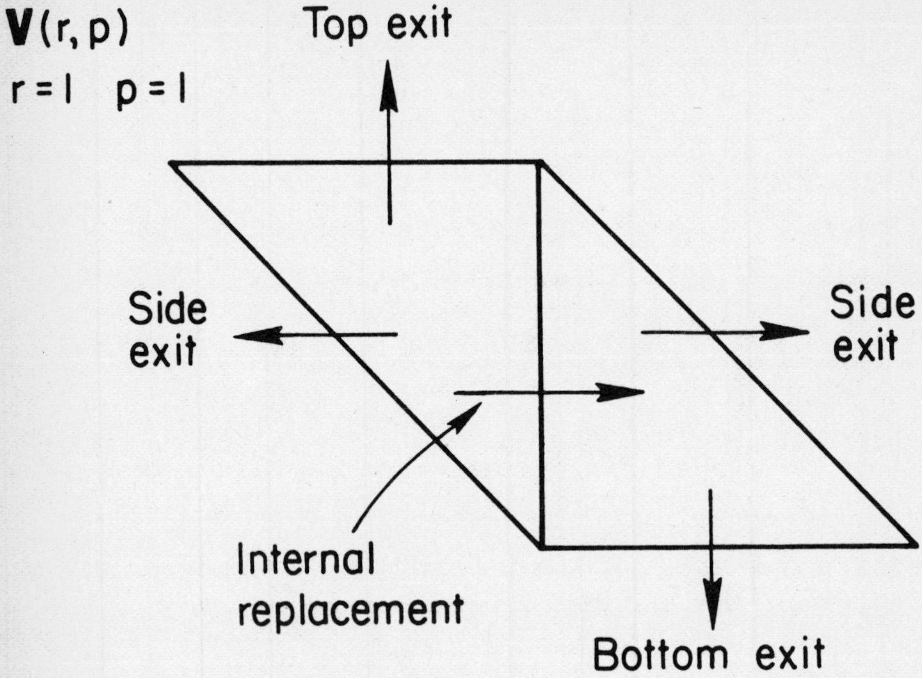
V(r, p)
r = l p = l
Top exit
Side exit
Side exit
Internal replacement
Bottom exit

Figure 16.2

16.4 Lemma: The facets of $V(r,p)$ are $S \times 0$, $(rS-p) \times 1$, and

$$cvx\big(((rS_\beta - p) \times 1) \cup (S_\beta \times 0)\big)$$

for $\beta = \mu \backslash i$ with $i \in \mu$.

Proof: The facets of $S \times [0,1]$ are $S \times 0$, $S \times 1$, and $cvx((S_\beta \times 1) \cup (S_\beta \times 0))$. Now see Lemma 9.2. □

In the next section we will have repeatedly juxtaposed, squeezed and sheared copies of $\mathbf{V}(r,p)$, and the bottom, side, and top exits will become descent, traverse, and ascent replacements; however, for the present we just STOP upon encountering these cases.

The representation set $i\mathbf{V}(r,p)^{n+1}$ for $\mathbf{V}(r,p)$ is a susbset of $i\mathbf{V}^{n+1}$ and the representation maps for $i\mathbf{V}(r,p)^{n+1}$ and $i\mathbf{V}^{n+1}$ are identical. Let $i\mathbf{V}(r,p)^{n+1}$ be the collection of all (v,π,k,η) in $i\mathbf{V}^{n+1}$ such that $(v;\pi;k;\eta)$ is contained in $V(r,p)$, that is, such that $(v;\pi;k)$ is in $rS-p$. In other words, (v,π,k,η) is an element of $i\mathbf{V}(r,p)^{n+1}$ if and only if

(a) $v \in Z^n$

(b) π permutes μ

(c) $k = 0, 1, \ldots, n$

(d) $v = q\eta$, $0 \nless \eta \geqq 0$

(e) $\eta_i = 0$, $i \in \mu \backslash (\pi|k)\mu$

(f) $\pi(k+1) < \pi(k+2) < \cdots < \pi(n+1)$

(g) $(v;\pi;k) \subseteq rS-p$.

Given $(v;\pi;k;\eta)$ in $i\mathbf{V}(r,p)^{n+1}$, its vertices are

$$w^1 = (v^1,1), \ldots, w^{k+1} = (v^{k+1}, 1) ,$$

$$w^{k+2} = (s^{\pi(k+1)}, 0), \ldots, w^{n+2} = (s^{\pi(n+1)}, 0) ,$$

where $v^1, \ldots, v^{k+1}$ are the vertices of $(v;\pi;k)$ and $s^{\pi i}$ is the πi^{th} vertex of S.

Our discussion of the representation set in rules in $\mathbf{V}(r,p)^{n+1}$ is complete and we move to develop the replacement rules.

Given $(v;\pi;k;\eta)$ in $i\mathbf{V}(r,p)^{n+1}$ and a vertex v^t to drop, suppose $(\hat{v};\hat{\pi};\hat{k};\hat{\eta})$ in $i\mathbf{V}^{n+1}$ is the replacement in $\mathbf{V}$. If the adjoined vertex is in $V(r,p)$, then $(\hat{v};\hat{\pi};\hat{k};\hat{\eta})$ is the replacement in $\mathbf{V}(r,p)$ as well. If the adjoined vertex is not in $V(r,p)$, then a side exit has occurred and all vertices of $(v;\pi;k;\eta)$ other than w^t lie in a facet $cvx(rS_\beta\text{-}p \times 1) \cup (S_\beta \times 0)$ of $V(r,p)$ where $\beta = \mu\backslash i$ for some i in μ.

In checking for a side exit we employ barycentric coordinates just as in Section 8. Thus x is in $rS\text{-}p$ if and only if any one of the following holds.

(a) $r^{-1}(x+p) \in S$

(b) $\Lambda(r^{-1}(x+p), i) \geqq 0, \ i \in \mu$

(c) $(x+p)_{i-1} - (x+p)_i \geqq 0, \ i \in \mu$

where $(x+p)_0 \triangleq r$ and $(x+p)_{n+1} \triangleq 0$.

With regard to (c), if x is integral as it will be for replacement in $\mathbf{V}(r,p)$, the check can be conducted without numerical error as all quantities are integral. Naturally, in computations (c) is used.

We proceed to give a narrative description of the replacement rules for **V**(r,p) and, again, hasten to point out that the narration parallels the rules as specified in the charts of Figures 16.3 through 16.7. As there are a number of cases the reader will probably, again, find it fruitful to vaselate between the narration and the charts of **V**(r,p) and **V**.

As in **V** we determine if the vertex w^t to be dropped is in $S \times 0$ or $G^n \times 1$, for $t \leq k+1$, w^t lies in $G^n \times 1$ and for $t \geqq k+2$, w^t lies in $S \times 0$. If $t \leqq k+1$ and $k = 0$ the analysis proceeds as in **V** and we have a bottom exit. If $t \geqq k+2$ and $k = n$ the analysis is as in **V** and we have a top exit. As in **V** there are four cases remaining.

Case A: $1 = t \leqq k$

Case B: $2 \leqq t = k+1$

Case C: $2 \leqq t \leqq k$

Case D: $k+2 \leqq t \leqq n+2$, $k < n$.

Case A $(1 = t \leqq k)$: $w^1 = (v^1,1)$ is dropped and $k \geqq 1$. If $\Lambda(r^{-1}(v^{k+1}+p), \pi 1)$ is positive, then $v^{k+1} + q^{\pi 1}$ is in $rS-p$, and thus $(\hat{v};\hat{\pi};\hat{k})$ is in $rS-p$ where $\hat{v} = v^2$, $\hat{\pi} = (\pi 2, \ldots, \pi k, \pi 1, \pi(k+1), \ldots, \pi(n+1))$ and $\hat{k} = k$. Letting $\hat{\eta} = \eta + e^{\pi 1}$ we have $\hat{v} = q\hat{\eta}$ with $\hat{\eta} \geqq 0$ and $\hat{\eta}_\beta = 0$ where $\beta = \mu\backslash(\hat{\pi}|\hat{k})\mu$. Thus $(\hat{v};\hat{\pi};\hat{k};\hat{\eta})$ is the replacement, see the argument for **V**.

Now let us assume that $\Lambda(r^{-1}(v^{k+1}+p), \pi 1)$ is zero or equivalently, that $v^{k+1} + q^{\pi 1}$ is not in $rS-p$. The following lemma states that all vertices of $(v;\pi;k;\eta)$ but $(v^1,1)$ lie in the side facet $\beta = \mu\backslash\pi 1$ of $V(r,p)$, that is, in $\mathrm{cvx}((rS_\beta - p) \times 1) \cup (S_\beta \times 0))$. Therefore under the assumptions of Case A, we indicated the replacement or showed that it did not exit and our analysis here is complete. □

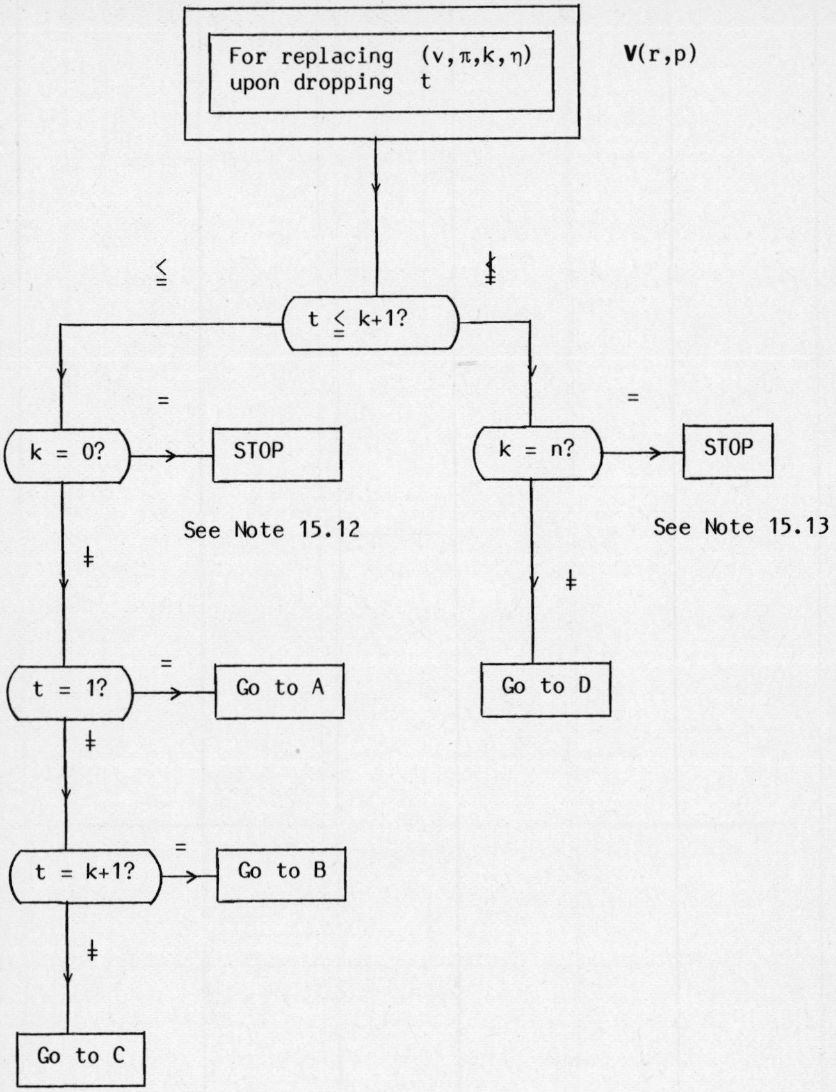
For replacing (v,π,k,η) upon dropping t
V(r,p)
t ≦ k+1?
≦
≰
k = 0?
=
STOP
See Note 15.12
≠
k = n?
=
STOP
See Note 15.13
≠
t = 1?
=
Go to A
≠
Go to D
t = k+1?
=
Go to B
≠
Go to C

Figure 16.3

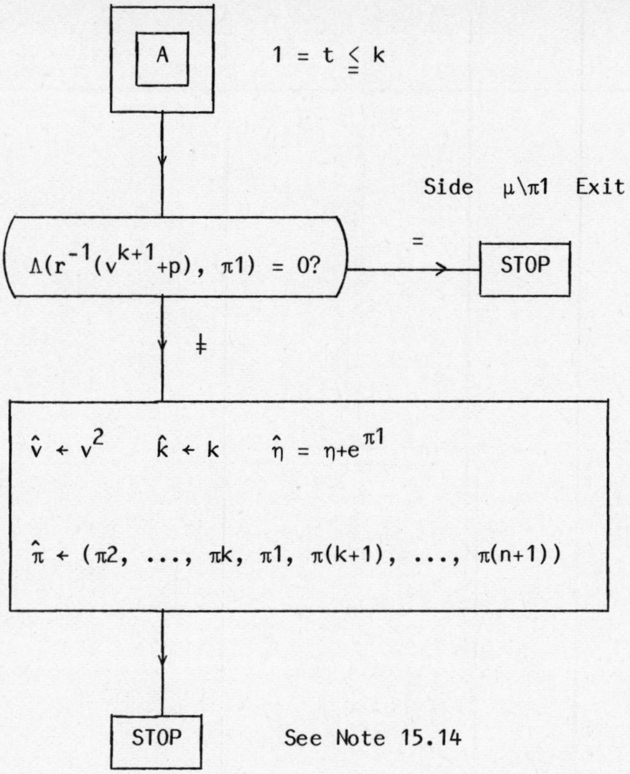
A
1 = t ≦ k
Side μ\π1 Exit
Λ(r^-1(v^k+1+p), π1) = 0?
=
STOP
≠
v̂ ← v^2 k̂ ← k η̂ = η+e^π1
π̂ ← (π2, ..., πk, π1, π(k+1), ..., π(n+1))
STOP
See Note 15.14

Figure 16.4

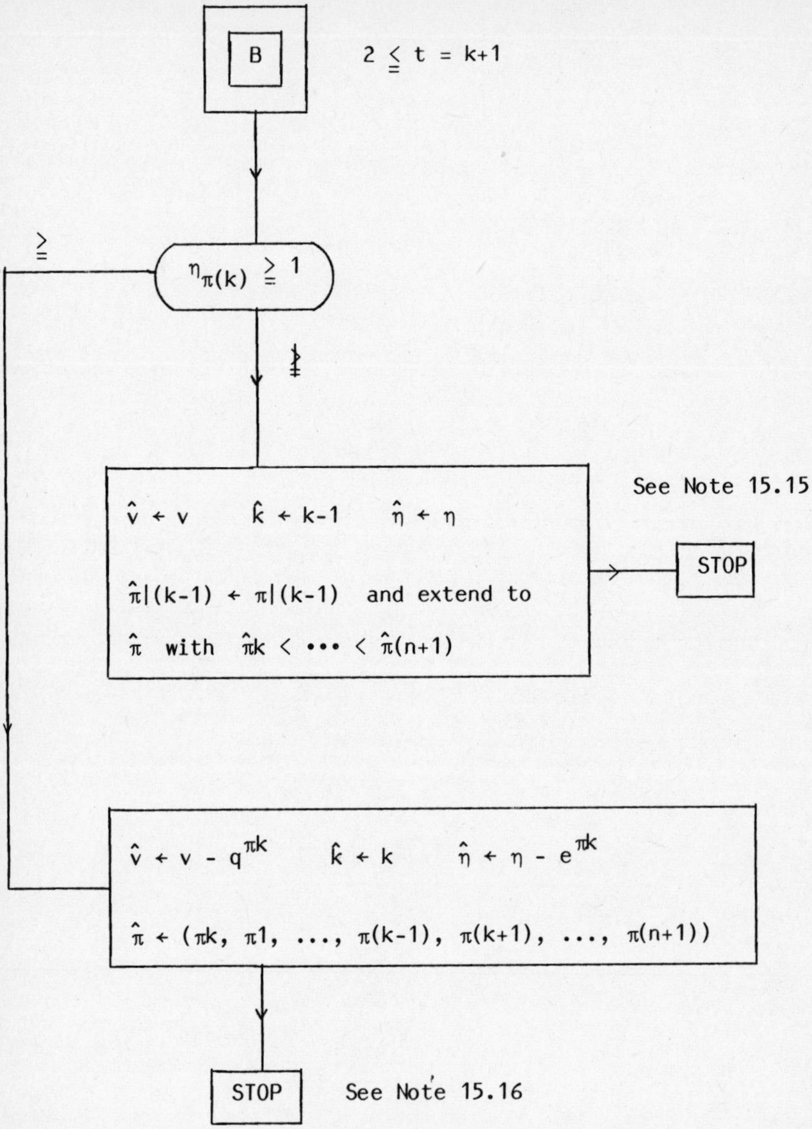
B
2 ≦ t = k+1
η_π(k) ≧ 1
≧
≱
v̂ ← v   k̂ ← k-1   η̂ ← η
π̂|(k-1) ← π|(k-1) and extend to
π̂ with π̂k < ••• < π̂(n+1)
See Note 15.15
STOP
v̂ ← v - q^πk   k̂ ← k   η̂ ← η - e^πk
π̂ ← (πk, π1, ..., π(k-1), π(k+1), ..., π(n+1))
STOP
See Note 15.16

Figure 16.5

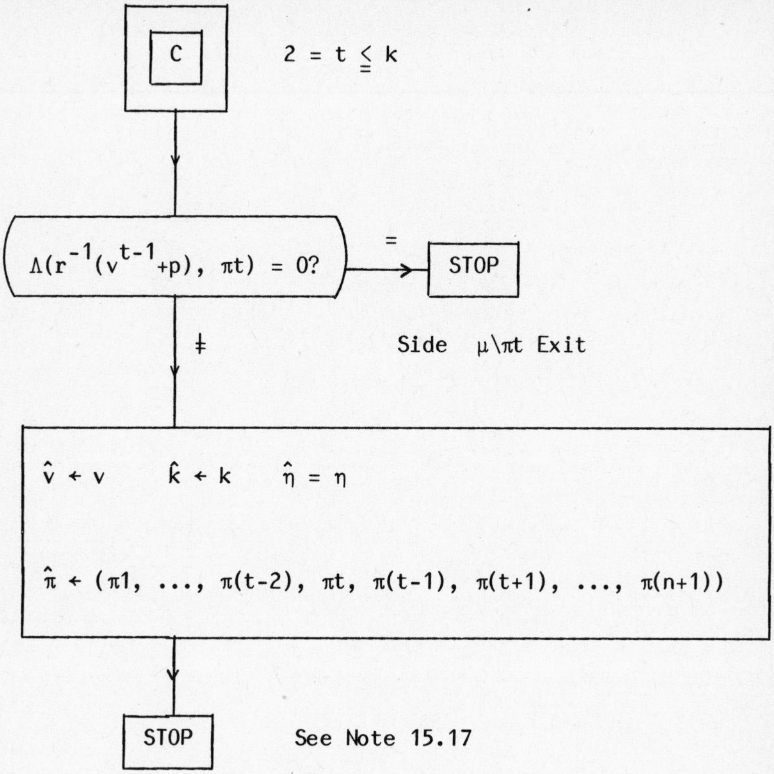
C
2 = t ≦ k
Λ(r⁻¹(vᵗ⁻¹+p), πt) = 0?
=
STOP
≠
Side μ\πt Exit
v̂ ← v   k̂ ← k   η̂ = η
π̂ ← (π1, ..., π(t-2), πt, π(t-1), π(t+1), ..., π(n+1))
STOP
See Note 15.17

Figure 16.6

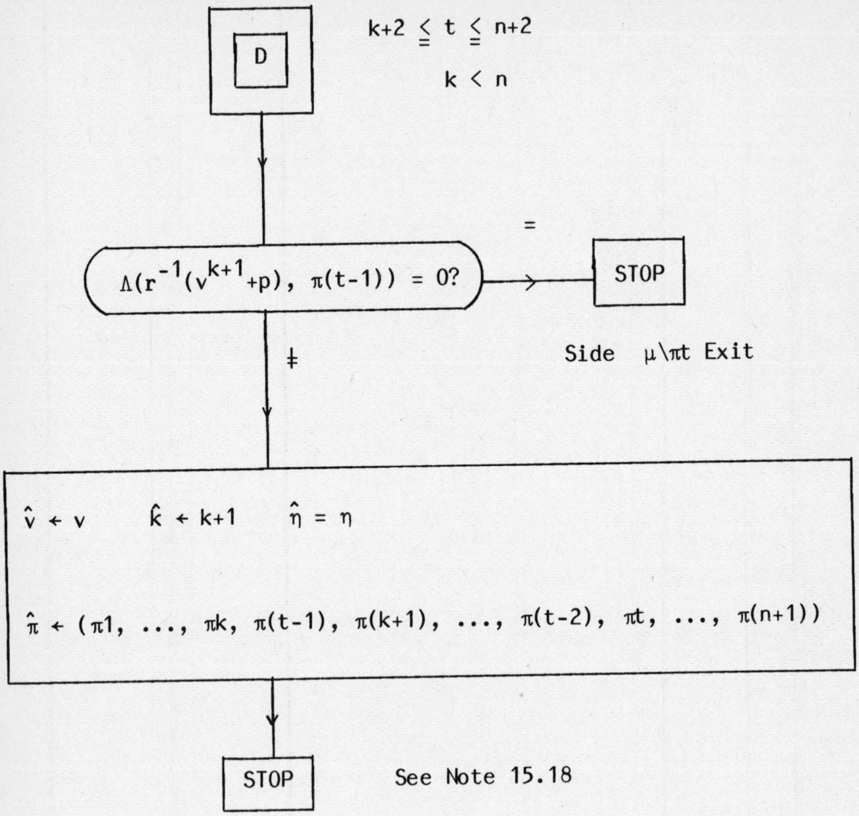
D
k+2 ≦ t ≦ n+2
k < n
Λ(r⁻¹(v^{k+1}+p), π(t-1)) = 0?
=
STOP
Side μ\πt Exit
≠
v̂ ← v k̂ ← k+1 η̂ = η
π̂ ← (π1, ..., πk, π(t-1), π(k+1), ..., π(t-2), πt, ..., π(n+1))
STOP
See Note 15.18

Figure 16.7

16.5 Lemma: Given $(v;\pi;k;\eta)$ in $\mathbf{V}(r,p)$ with $k \geq 1$ suppose $\Lambda(r^{-1}(v^{k+1}+p), \pi 1) = 0$, then all vertices of $(v;\pi;k;\eta)$ except $(v^1,1)$ lie in the facet $\mathrm{cvx}((rS_{\mu\setminus\pi 1}-p) \times 1) \cup (S_{\mu\setminus\pi 1} \times 0))$ of $V(r,p)$.

Proof: For $r^{-1}(v+p)$ in S, $\Lambda(r^{-1}(v+p) + \theta q^i, \pi 1)$ decreases in θ, if and only if $i = \pi(1)$, see Lemma 8.1. Thus $\Lambda(r^{-1}(v^i+p), \pi(1) = 0$ for $i = 2, \ldots, k+1$ and $v^i \in rS_{\mu\setminus\pi 1}-p$ for $i = 2, \ldots, k+1$. Since $\pi 1 \in (\pi|k)\mu$ $(s^{\pi 1},0)$ is not a vertex of $(v;\pi;k;\eta)$. □

Case B $(2 \leqq t = k+1)$: Let $(\hat{v},\hat{\pi},\hat{k},\hat{\eta})$ be the replacement in $\mathbf{iV}$ for (v,π,k,η), see Figures 15.9 through 15.11. If $\eta_{\pi k} < 1$, that is, $\eta_{\pi k} = 0$ then $(\hat{v};\hat{\pi};\hat{k};\hat{\eta})$ clearly lies in $V(r,p)$ as the new vertex is of form $(s^i,0)$. Thus $(\hat{v};\hat{\pi};\hat{k};\hat{\eta})$ is the replacement in $\mathbf{iV}(r,p)$. For $\eta_{\pi k} \geqq 1$, $(\hat{v},\hat{\pi},\hat{k},\hat{\eta})$ is the replacement in $\mathbf{iV}(r,p)$, but we must confirm that $(\hat{v};\hat{\pi};\hat{k};\hat{\eta})$ lies in $V(r,p)$. We address this point. As the adjoined vertex is $(v^1-q^{\pi k}, 1)$ the question becomes, is $\Lambda(r^{-1}(v^1+p), \pi k+1)$ positive? Note $\pi k+1 = (\pi k)+1$. See Lemma 4.1. If so, the new vertex $v^1-q^{\pi k}$ is in $rS-p$, $(\hat{v};\hat{\pi};\hat{k};\hat{\eta})$ is in $V(r,p)$ and $(\hat{v},\hat{\pi},\hat{k},\hat{\eta})$ is the replacement for (v,π,k,η) in $\mathbf{iV}(r,p)$. The following lemma establishes that $\Lambda(r^{-1}(v^1+p), \pi k+1) > 0$ and we see that Case B of $\mathbf{V}(r,p)$ and Case B of $\mathbf{V}$ are identical. □

16.6 Lemma: If (v,π,k,η) is in $\mathbf{iV}(r,p)^{n+1}$ and $\eta_{\pi k} \geqq 1$, then $\Lambda(r^{-1}(v+p), \pi k+1) > 0$.

Proof: Suppose $\Lambda(r^{-1}(v+p), \pi k+1) = 0$. We claim $\pi k+1$ is not in $(\pi|k)\mu$. To see this observe that $\pi(k)$ is the last element of $\pi|k$, and hence, if $\pi k+1$ occurred in $(\pi|k)\mu$ it would have to be before πk in π. But $\Lambda(r^{-}(v^{1}+p) + r^{-1}q^{j}, \pi k+1)$ is zero, positive, and negative for $j \in \mu\backslash\{\pi k, \pi k+1\}$, $j = \pi k$, and $j = \pi k+1$, respectively. This same vertex v^{i} for $i = 2, \ldots, k$ is not in $rS-p$. Thus, $\pi k+1$ is not in $(\pi|k)\mu$. Therefore $\eta_{\pi+1} = 0$ since, $v \in C_{(\pi|k)\mu}$. Setting $j = \pi k+1$ and using $v = q\eta$ we have $\Lambda(r^{-1}(v+p), j) = \Lambda(r^{-1}(q\eta+p), j)$ equal to

$$\begin{aligned}
&1 - r^{-1}p_1 + r^{-1}(\eta_{n+1} - \eta_1) , && j = 1 \\
&r^{-1}(p_{j-} - p_j) + r^{-1}(\eta_{j-1} - \eta_j) , && 2 \leq j \leqq n \\
&r^{-1}p_n = r^{-1}(\eta_n - \eta_{n+1}) , && j = n+1 .
\end{aligned}$$

As $r^{-1}p$ is in S and $\eta_j = \eta_{\pi k+1} = 0$ we see that $\Lambda(r^{-1}(v+p),j)$ exceeds

$$\begin{aligned}
&r^{-1}\eta_{n+1} , && \text{for } j = 1 \\
&r^{-1}\eta_{j-1} , && \text{for } 2 \leqq j \leqq n \\
&r^{-1}\eta_n , && \text{for } j = n+1
\end{aligned}$$

or that $\Lambda(r^{-1}(v+p), \pi k+1)$ exceeds $r\eta_{\pi k}$ since $\pi k = n+1$, $j-1$, and n for $j = 1$, $2 \leq j \leqq n$, and $j = n+1$, respectively. As $r^{-1}\eta_{\pi k}$ is positive so is $\Lambda(r^{-1}(v+p), \pi k+1)$ which contradicts the supposition. □

Case C $(2 \leqq t \leqq k)$: $w^t = (v^t,1)$ is dropped. First suppose $\Lambda(r^{-1}(v^{t-1}+p), \pi t)$ is positive. In this case $v^{t-1} + q^{\pi t}$ is in $rS-p$, and thus $(\hat{v};\hat{\pi};\hat{k})$ is in $rS-p$ where $\hat{v} = v$, $\hat{\pi} = (\pi 1, \ldots, \pi(t-2), \pi t, \pi(t-1), \pi t, \pi(t-1), \pi(t+1), \ldots, \pi k, \pi(k+1), \ldots, \pi(n+1))$ and $\hat{k} = k$. Thus $(\hat{v};\hat{\pi};\hat{k};\hat{\eta})$ where $\hat{\eta} = \eta$ is the replacement in $\mathbf{V}(r,p)$ as well as $\mathbf{V}$.

Second, suppose $\Lambda(r^{-1}(v^{t-1}+p), \pi t)$ is zero. The following lemma states that a side exit from $V(r,p)$ through facet $\mathrm{cvx}(((rS_{\mu\setminus\pi t} - p) \times 1) \cup (S_{\mu\setminus\pi t} \times 0))$ has occurred. Under the assumptions of Case C we have exhibited the replacement or proved none exists, and our analysis of Case C is complete. □

16.7 Lemma: Given $(v;\pi;k;\eta)$ in $\mathbf{V}(r,p)$ with $2 \leqq t \leqq k$ and $\Lambda(r^{-1}(v^{t-1}+p), \pi t) = 0$, then all vertices of $(v;\pi;k;\eta)$ except $(v^t,1)$ lie in the facet $\mathrm{cvx}(((rS_\beta-p) \times 1) \cup (S_\beta \times 0))$ of $V(r,p)$ with $\beta = \mu\setminus\pi t$.

Proof: We have $\Lambda(r^{-1}(v^{t-1}+p), \pi t) = 0$ and $\Lambda(r^{-1}(v^t+p), \pi t) > 0$. Thus $\Lambda(r^{-1}(v^{t+1}+p), \pi t) = 0$, since $v^{t+1} = v^t+q^{\pi t}$, see Lemma 8.1. Therefore $\Lambda(r^{-1}(v^i+p), \pi t) = 0$ for i in $\{1, \ldots, k+1\}\setminus t$; so all vertices of $(v;\pi;k)$ lie in $rS_{\mu\setminus\pi t}-p$ except v^t. Since $\pi t \in (\pi|k)\mu$; $(s^{\pi t},0)$ is not a vertex of $(v;\pi;k;\eta)$, and therefore, every vertex of $(v;\pi;k;\eta)$ except $(v^t,1)$ lies in the vertical facet $\mathrm{cvx}(((rS_{\mu\setminus\pi t}-p) \times 1) \cup (S_{\mu\setminus\pi t} \times 0).)$ □

Case D $(k+2 \leqq t \leqq n+2,\ k < n)$: Vertex $w^t = (s^{\pi(t-1)}, 0)$ is being dropped. First, let us suppose that $\Lambda(r^{-1}(v^{k+1}+p), \pi(t-1))$ is positive, then $v^{k+1} + q^{\pi(t-1)}$ is in $rS-p$ and the $(k+1)$-simplex

$(\hat{v};\hat{\pi};\hat{k})$ is in $rS-p$ where $\hat{v} = v$, $\pi = (\pi 1, \ldots, \pi k, \pi(t-1), \pi(k+1), \ldots, \pi(t-2), \pi t, \ldots, \pi(n+1))$ and $\hat{k} = k+1$. From the analysis for **V** we see that $(\hat{v},\hat{\pi},\hat{k},\hat{\eta})$ where $\hat{\eta} = \eta$ is the replacement in **V**(r,p). Second, let us suppose $\Lambda(r^{-1}(v^{k+1}+p), \pi(t-1))$ is zero. The following lemma states that all vertices of $(v;\pi;k;\eta)$ except w^t lie in the facet $\mathrm{cvx}(((rS_\beta-p) \times 1) \cup (S_\beta \times 0))$ of $V(r,p)$ with $\beta = \mu\backslash\pi(t-1)$.

Thus, under the assumptions of Case D we have produced the replacement or shown none exists and our analysis of Case D is complete. □

16.8 Lemma: Given $(v;\pi;k;\eta)$ in **V** with $k+2 \leqq t \leqq n+r$, $k < n$, and $\Lambda(r^{-1}(v^{k+1}+p), \pi(t-1)) = 0$, then all vertices of $(v;\pi;k;\eta)$ except $(s^{\pi(t-1)}, 0)$ lie in the facet $\mathrm{cvx}(((rS_\beta-p) \times 1) \cup (S_\beta \times 0))$ of $V(r,p)$ with $\beta = \mu\backslash\pi(t-1)$.

Proof: For v in $rS-p$, $\Lambda(r^{-1}(v+p) + \theta q^i, \pi(t-1)$ decreases in θ only for $i = \pi(t-1)$, see Lemma 4.1. Thus $\Lambda(r^{-1}(v^i+p), \pi(t-1)) = 0$ for $i = 1, \ldots, k+1$ since $\pi t \notin (\pi|k)\mu$. □

In the following charts the replacement rules for **V** are fully specified.

We now refine and shear **V**(r,p) to obtain a triangulation **V**$[r,p]$ of $S \times [0,1]$ with the property that **V**$[r,p]$ naturally restricted to $\mathbb{C}^n \times 1$ is $(r^{-1}F|S) \times 1$. It is **V**$[r,p]$ that we repeatedly juxtapose and stack to construct the homotopy triangulation **S**.

Define the $\mathbf{V}(r,p)$-PL map $g : V(r,p) \to \mathbb{C}^n \times [0,1]$ by setting

$$g(v,1) = (r^{-1}v + r^{-1}p,\ 1)$$

$$g(v,0) = (v,0)$$

for vertices (v,s) of $\mathbf{V}(r,p)$. We define $\mathbf{V}[r,p]$ to be the triangulation $g(\mathbf{V}(r,p))$ which is obtained by squeezing (r^{-1}) and shearing $(+r^{-1}p)$ the triangulation $\mathbf{V}(r,p)$. As we see in the next lemma $\mathbf{V}[r,p]$ triangulates $S \times [0,1]$, see Figures 16.8 and 16.9.

16.9 Lemma: $g(V(r,p)) = S \times [0,1]$.

Proof: g is the identity on $S \times 0$. Since $g(rx-p,1) = (r^{-1}(rx-p) + r^{-1}p,\ 1) = (x,1)$ we see that $g(rS-p,\ 1) = S \times 1$. As $\mathbf{V}(r,p)$ triangulates the convex set $V(r,p)$, we see that $g(V(r,p)) = S \times [0,1]$ by Lemma 9.3. □

As $r^{-1}\mathbf{F} + r^{-1}p = r^{-1}\mathbf{F}$ from Lemma 6.20, it is clear that $\mathbf{V}[r,p]|(S \times 1) = (r^{-1}\mathbf{F}|S) \times 1$.

The representation set and replacement rules of $\mathbf{V}(r,p)$ and $\mathbf{V}[r,p]$ are identical; define $i\mathbf{V}[r,p]^{n+1} = i\mathbf{V}(r,p)^{n+1}$, see Figure 16.10. If (v,π,k,η) is in $i\mathbf{V}[r,p]^{n+1}$ we define $[v;\pi;k;\eta]$ to be the $(n+1)$-simplex in $\mathbf{V}[r,p]^{n+1}$ with vertices

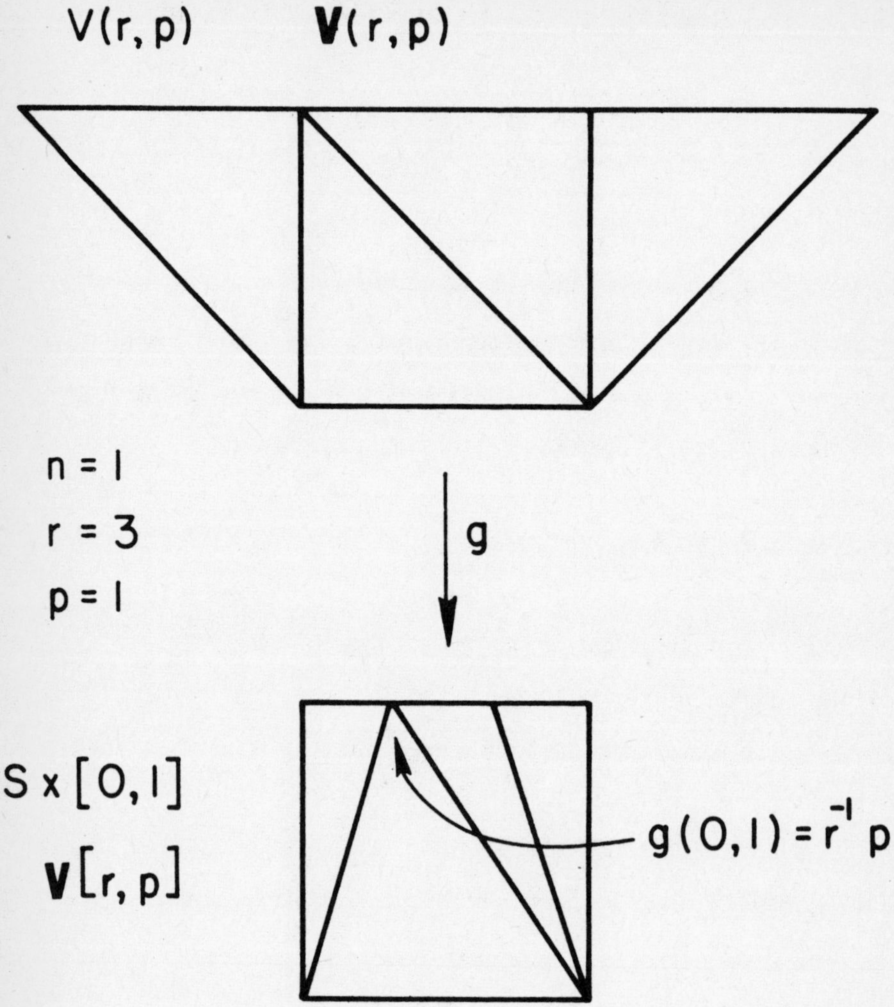

Figure 16.8

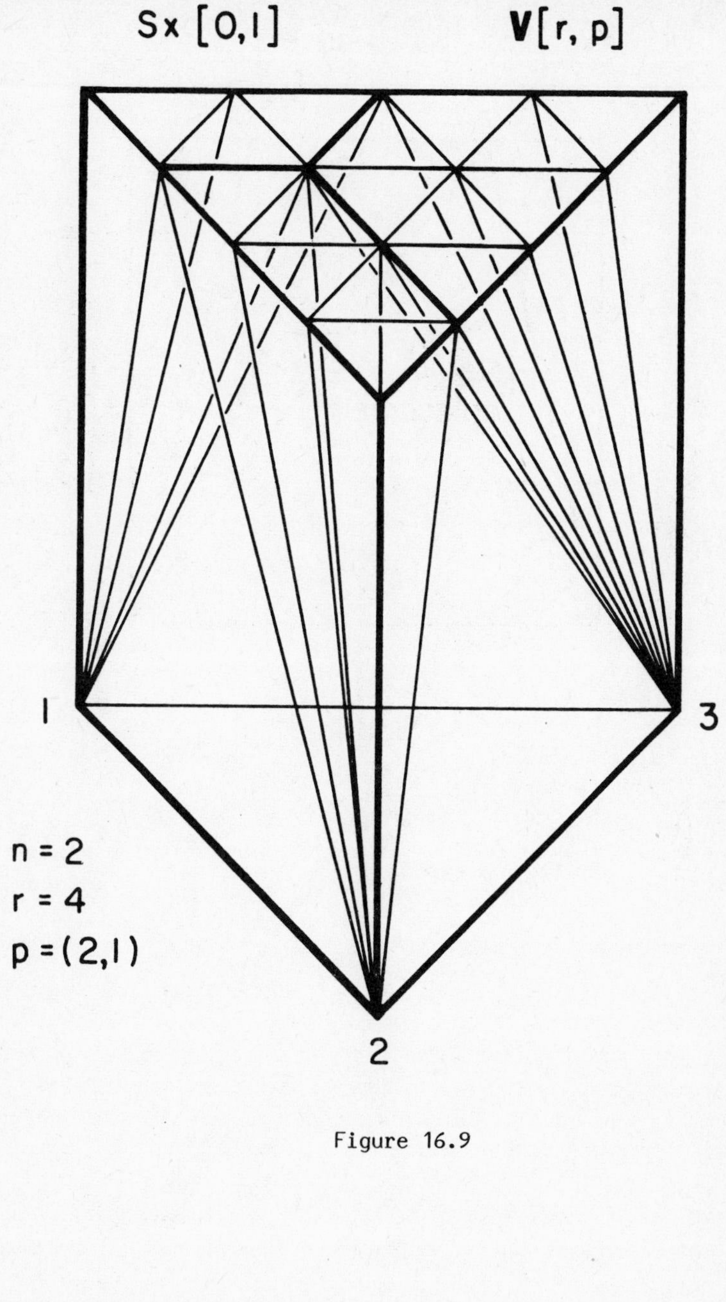

Figure 16.9

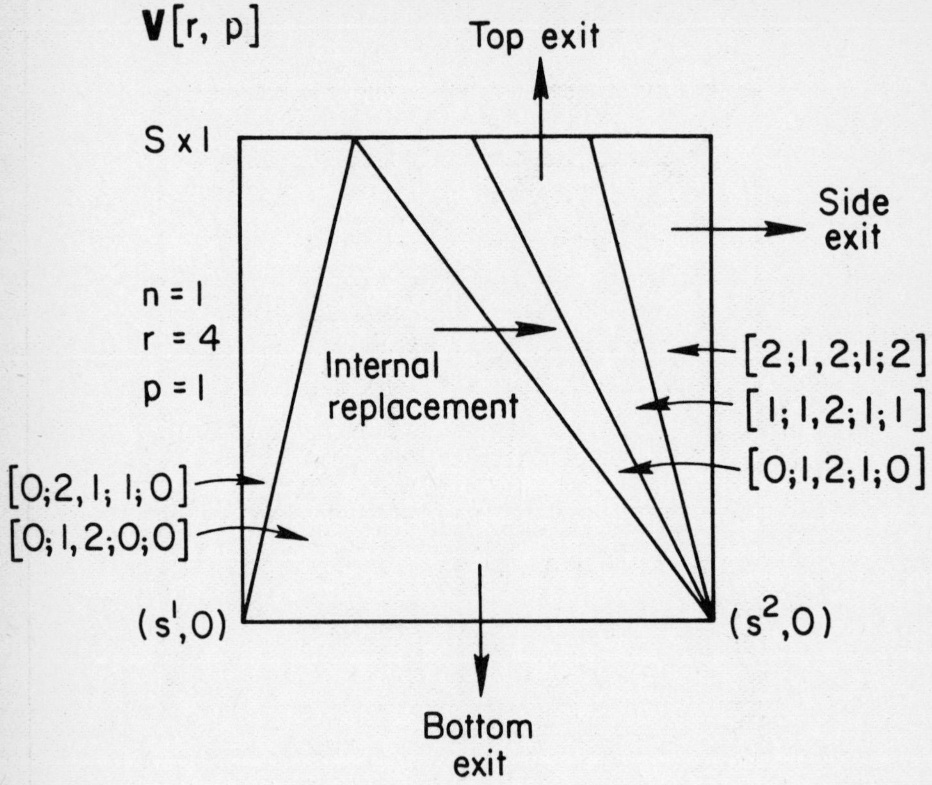
V[r, p]
Top exit
S x I
Side exit
n = 1
r = 4
p = 1
Internal replacement
[2;1,2;1;2]
[1;1,2;1;1]
[0;1,2;1;0]
[0;2,1;1;0]
[0;1,2;0;0]
(s¹,0)
(s²,0)
Bottom exit

Figure 16.10

$$(r^{-1}v^1 + r^{-1}p,\ 1),\ \ldots,\ (r^{-1}v^{k+1} + r^{-1}p,\ 1)$$

$$(s^{\pi(k+1)},\ 0),\ \ldots,\ (s^{\pi(n+1)},\ 0)$$

where

$$(v^1,\ 1),\ \ldots,\ (v^{k+1},\ 1)$$

$$(s^{\pi(k+1)},\ 0)\ ,\ \ldots,\ (s^{\pi(n+1)},\ 0)$$

are the vertices of $(v;\pi;k;\eta)$ in $\mathbf{V}(r,p)$. The replacement charts for $\mathbf{V}(r,p)$ and $\mathbf{V}[r,p]$ are identical, but, of course, the representation rules, and hence, notes, vary. The representation rules differ in an obvious way, namely

$\underline{\mathbf{V}(r,p)}$		$\underline{\mathbf{V}[r,p]}$
$(v^i,1)$	$\rightarrow$	$(r^{-1}v^i + r^{-1}p,\ 1)$
$(s^i,0)$	$\rightarrow$	$(s^i,0)$.

A side exit through facet $\text{cvx}((rS_\beta - p) \times 1) \cup (S_\beta \times 0)$ in $\mathbf{V}(r,p)$ corresponds to a side exit through $S_\beta \times [0,1]$ in $\mathbf{V}[r,p]$, and a top exit through the simplex $(v;\pi;k)$ in $\mathbf{V}(r,p)$ corresponds to a top exit through the simplex $r^{-1}((v;\pi;k) + r^{-1}p$ in $\mathbf{V}[r,p]$.

16.10 Bibliographical Remarks:

The triangulations **V**[r,p] were used by both Shamir [1979] and van der Laan and Talman [1980b] in the construction of the variable rate refining triangulation **S**. □

17. VARIABLE RATE REFINING TRIANGULATION **S** OF $\mathbb{C}^n \times [0,+\infty)$ BY JUXTAPOSITIONING **V**[r,p]'s

Using Freudenthal's triangulation **F** of $\mathbb{C}^n$, the triangulation **V**[r,p] of $S \times [0,1]$, we construct the variable rate refining homotopy triangulation **S** of $\mathbb{C}^n \times [0, +\infty)$. As a preview of such a triangulation see Figure 17.1.

The triangulation **S** depends upon two infinite sequences, the integral refinement factors $r_1, r_2, \ldots,$ and the integral translators $p^1, p^2, \ldots$. The r_i must be positive integers, the p_i integral vectors in $\mathbb{C}^n$, and the focal point $r_i^{-1} p_i$ must lie in S. As we shall see, **S** naturally restricted to $\mathbb{C}^n \times h$ is $((r_1, \ldots, r_h)^{-1}$ **F**$) \times h$ for $h = 0, 1, 2, \ldots$ where **F**, as always, is the Freudenthal triangulation of $\mathbb{C}^n$.

In Figure 17.1 we have

$$r_1 = 3, \; r_2 = 3, \; r_3 = 2, \; \ldots \; .$$

The role of the translators p^i is related to the structure of **S** in the regions $\mathbb{C}^n \times (h-1,h)$ just as the translator is in **V**[r,p]. In Figure 17.1 we have $p^1 = 1, \; p^2 = 2, \; p^3 = 2, \; \ldots$.

Throughout our discussion in this section we assume that the sequence of scalars r_i and translators p_i is known. However, it is important to note as we progress that **S** in $\mathbb{C}^n \times [0,h]$ depends only upon $r_1, \ldots,$ r_h, and $p^1, \ldots, p^h$. In actual computations the sequences r_i and p^i are generated dynamically as the equation solving algorithm proceeds. From this view we shall describe the triangulation **S** first restricted to $\mathbb{C}^n \times [0,1]$, then to $\mathbb{C}^n \times [0,2]$, then to $\mathbb{C}^n \times [0,3]$, etc.

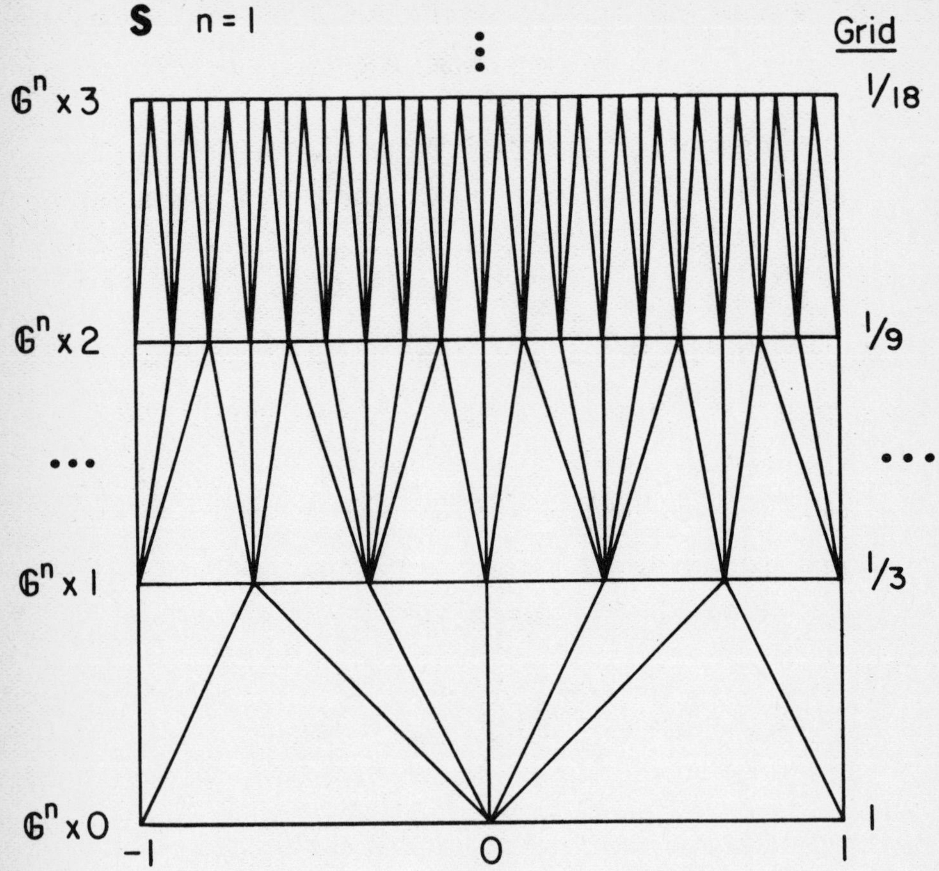
S n = 1
Grid
$\mathbb{G}^n \times 3$
$\mathbb{G}^n \times 2$
$\mathbb{G}^n \times 1$
$\mathbb{G}^n \times 0$
1/18
1/9
1/3
1
-1
0
1

Figure 17.1

As in juxtapositioning for any n-simplex σ in $\mathbf{F}$ we define the linear map $T_\sigma : \mathbb{C}^n \to \mathbb{C}^n$ by $T(s^i) = u^i$ where u^i is the vertex of σ with $\underline{\ell}(u^i) = \underline{\ell}(s^i)$ and s^i is the ith vertex of S for i in μ. Recall that each n-simplex σ in $\mathbf{F}$ can be represented uniquely by $(u;\gamma;n)$ with $\underline{\ell}(u^i) = \underline{\ell}(s^i)$ for i in μ, and we have $T_\sigma(x) = q^{\gamma|n}x + u$, see Lemma 12.1.

Let $I_h : \mathbb{C} \to \mathbb{C}$ be the map defined by $I_h(t) = t+h$ and we shall employ the notation $(T_\sigma, I_h)(x,t) = (T_\sigma(x), t+h)$.

We describe $\mathbf{S}_1$ which is a triangulation of $\mathbb{C} \times [0,1]$; $\mathbf{S}_1$ is the natural restriction of $\mathbf{S}$ to $\mathbb{C} \times [0,1]$. Let $\mathbf{V}[r_1,p^1]$ be the van der Laan-Talman triangulation of $S \times [0,1]$ with integral factor r_1, integral translator p^1, and $r_1^{-1}p_1 \in S$. Define $\mathbf{S}_1$ to be the collection of all simplexes ρ of form

$$(T_\sigma, I_0)(\tau),$$

for σ in $\mathbf{F}$ and τ in $\mathbf{V}[r_1,p^1]$; that the collection $\mathbf{S}_1$ triangulates $\mathbb{C}^n \times [0,1]$ is proved in Section 12. Furthermore, it is proved there that the restriction of $\mathbf{S}_1$ to $\mathbb{C}^n \times 1$ is $r_1^{-1}\mathbf{F} \times 1$, see Figure 17.2.

Now suppose that we have described triangulation $\mathbf{S}_i$ of $\mathbb{C}^n \times [i-1,i]$ for $i = 1, \ldots, h$, that their union $\cup_{i=1}^h \mathbf{S}_i$ triangulates $\mathbb{C}^n \times [0,h]$, and that $\mathbf{S}_h$ restricted to $\mathbb{C}^n \times h$ is $(r_1, \ldots, r_h)^{-1}\mathbf{F} \times h$. We proceed to define $\mathbf{S}_{h+1}$ a triangulation of $\mathbb{C}^n \times [h, h+1]$.

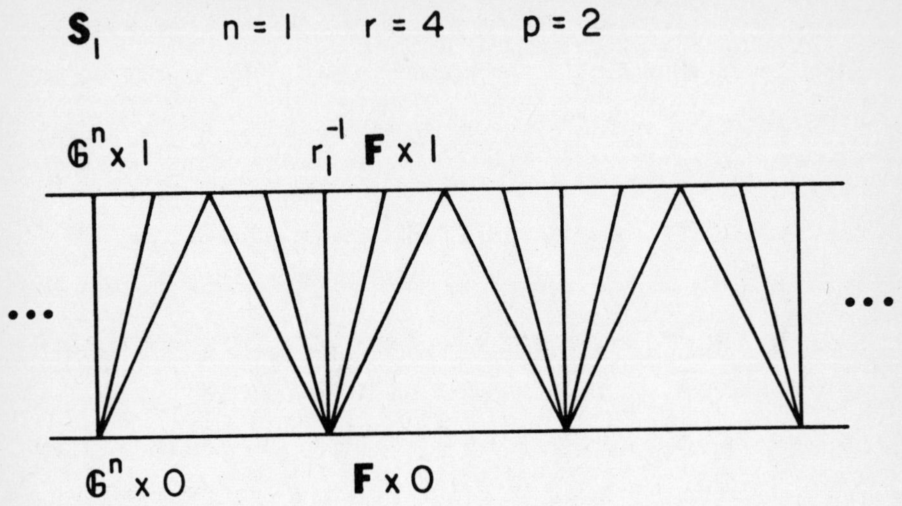
$\mathbf{S}_1$ n = 1 r = 4 p = 2
$\mathbb{G}^n \times 1$
$r_1^{-1}\,\mathbf{F} \times 1$
$\mathbb{G}^n \times 0$
$\mathbf{F} \times 0$

Figure 17.2

Let $\mathbf{V}[r_{h+1}, p^{h+1}]$ be the van der Laan-Talman triangulation of $S \times [0,1]$ with factor r_{h+1} and translator p^{h+1}. Define $\mathbf{S}_{h+1}$ to be the collection of all simplexes ρ of form

$$((r_1, \ldots, r_h)^{-1} T_\sigma, I_h)(\tau)$$

for σ in $\mathbf{F}$ and τ in $\mathbf{V}[r_{h+1}, p^{h+1}]$, see Figure 17.3. It follows immediately from Section 12 that $\mathbf{S}_{h+1}$ is a triangulation of $\mathbb{C}^n \times [h,h+1]$, and that $\mathbf{S}_{h+1}$ restricted to $\mathbb{C}^n \times h$ and $\mathbb{C}^n \times (h+1)$ is, respectively, $(r_1, \ldots, r_h)^{-1} \mathbf{F} \times h$ and $(r_1, \ldots, r_{h+1})^{-1} \mathbf{F} \times (h+1)$. Thus $\cup_{i=1}^{h+1} \mathbf{S}_i$ is a triangulation of $\mathbb{C}^n \times [0, h+1]$. The triangulation $\mathbf{S}$ is defined to be the union $\cup_{i=1}^{\infty} \mathbf{S}_i$; we refer to $\mathbf{S}$ as the variable rate refining triangulation.

17.1 Theorem: $\mathbf{S}$ is a locally finite triangulation of $\mathbb{C}^n \times [0,+\infty)$.

Proof: Since any element and any two simplexes lie in some $\mathbb{C}^n \times [0,h]$ the results follow from the same conclusion of $\cup_{i=1}^{h} \mathbf{S}_i$. □

Now that the variable rate refining triangulation $\mathbf{S}$ of $\mathbb{C}^n \times [0,+\infty)$ has been defined and proved, our next task is to develop a representation for simplexes in $\mathbf{S}$ and to describe the replacement rules.

Define $i\mathbf{S}^{n+1}$ to be the collection of 7-tuples of form $(h,u,\gamma,v,\pi,k,\eta)$ where

(a) $h = 0, 1, 2, \ldots$

(b) $(u,\gamma,n) \in i_*\mathbf{F}$ and $\underline{\ell}(u^i) = \underline{\ell}(s^i)$ for $i \in \mu$

(c) $(v,\pi,k,\eta) \in i\mathbf{V}[r_{h+1}, p^{h+1}]^{n+1}$.

For a 7-tuple $(h,u,\gamma,v,\pi,k,\eta)$ in $\mathbf{iS}^{n+1}$ we define $(h;u;\gamma;v;\pi;k;\eta)$ to be the (n+1)-simplex

$$\rho = \left((r_1, \ldots, r_h)^{-1} T_{(u;\gamma;n)}, I_h\right) [v;\pi;k;\eta] .$$

It is evident that these (n+1)-simplex, are all of and only those of $\mathbf{S}^{n+1}$. The next lemma states that each (n+1)-simplex of $\mathbf{S}$ can be referenced by only one element in $\mathbf{iS}^{n+1}$.

17.2 Lemma: The representation $(h;u;\gamma;v;\pi;k;\eta)$ is unique.

Proof: If $(h;u;\gamma;v;\pi;k;\eta) = (\bar{h};\bar{u};\bar{\gamma};\bar{v};\bar{\pi};\bar{k};\bar{\eta})$ then $h = \bar{h}$, then $(u,v) = (\bar{u},\bar{v})$ from uniqueness in $\mathbf{F}$, and then $(u,\pi,k,\eta) = (\bar{u},\bar{\pi},\bar{k},\bar{\eta})$ from uniqueness in $\mathbf{V}[r_{h+1}, p^{h+1}]$. □

As before the representation $(h;u;\gamma;v;\pi;k;\eta)$ is integral which permits the replacement operation to be made with no numerical error.

The vertices of $\rho = (h;u;\gamma;v;\pi;k;\eta)$ are defined to be in order

$$W^i = \left((r_1 \cdots r_h)^{-1} T_{(u;\gamma;n)}, I_h\right)(w^i) ,$$

for $i = 1, \ldots, n+2$ where $w^1, \ldots, w^{n+2}$ are the vertices of $[v;\pi;k;\eta]$ in $\mathbf{V}[r_{h+1}, p^{h+1}]$; namely,

$$w^1 = (r^{-1}_{h+1}v^1 + r^{-1}_{h+1}p^{h+1}, 1), \ldots, w^{k+1} = (r^{-1}_{h+1}v^{k+1} + r^{-1}_{h+1}p^{h+1}, 1)$$

$$w^{k+2} = (s^{\pi(k+1)}, 0), \ldots, w^{n+2} = (s^{\pi(n+1)}, 0)$$

where $v^1 = v$, $v^{i+1} = v^i + q^{\pi i}$ for $i = 1, \ldots, k$.

Thus, for example, upon reference to vertex t of $\rho = (h;u;\gamma;v;\pi;k;\eta)$ we mean

$$w^t = \left((r_1, \ldots, r_h)^{-1} T_{(u;\gamma;n)}, I_h\right)(w^t)$$

where w^t is vertex t of $[v;\pi;k;\eta]$ in $\mathbf{V}[r_{h+1}, p^{h+1}]$.

In moving to an n-simplex $\hat{\rho} = (\hat{h};\hat{u};\hat{\gamma};\hat{v};\hat{\pi};\hat{k};\hat{\eta})$ adjacent to the n-simplex $\rho = (h;u;\gamma;v;\pi;k;\eta)$ there are five cases, namely, internal, traverse, ascent, descent replacement, and bottom exit corresponding to where $\hat{\rho}$ is in relation to ρ, if $\hat{\rho}$ exists, see Figure 17.4. The bottom exit can only occur in $\mathbb{C}^n \times [0,1]$.

If $(h,u,v) = (\hat{h},\hat{u},\hat{v})$ the replacement is termed "internal" for both simplexes ρ and $\hat{\rho}$ lie in the cylinder $(r_1 \cdots r_h)^{-1} (u;\gamma;n) \times [h,h+1]$. If $h = \hat{h}$ but $(u,v) \neq (\hat{u},\hat{v})$ the replacement is termed "traverse" for "sideways" movement. If $\hat{h} = h+1$ or $h-1$ the replacement is called "ascent" or descent", respectively.

A significant portion of the replacement rules for **S** is contained verbatim in the replacement rules $\mathbf{V}(r,p)$. Indeed the tests for which of the four cases has been encountered and the rules for internal replacement are contained in the chart for $\mathbf{V}(r,p)$

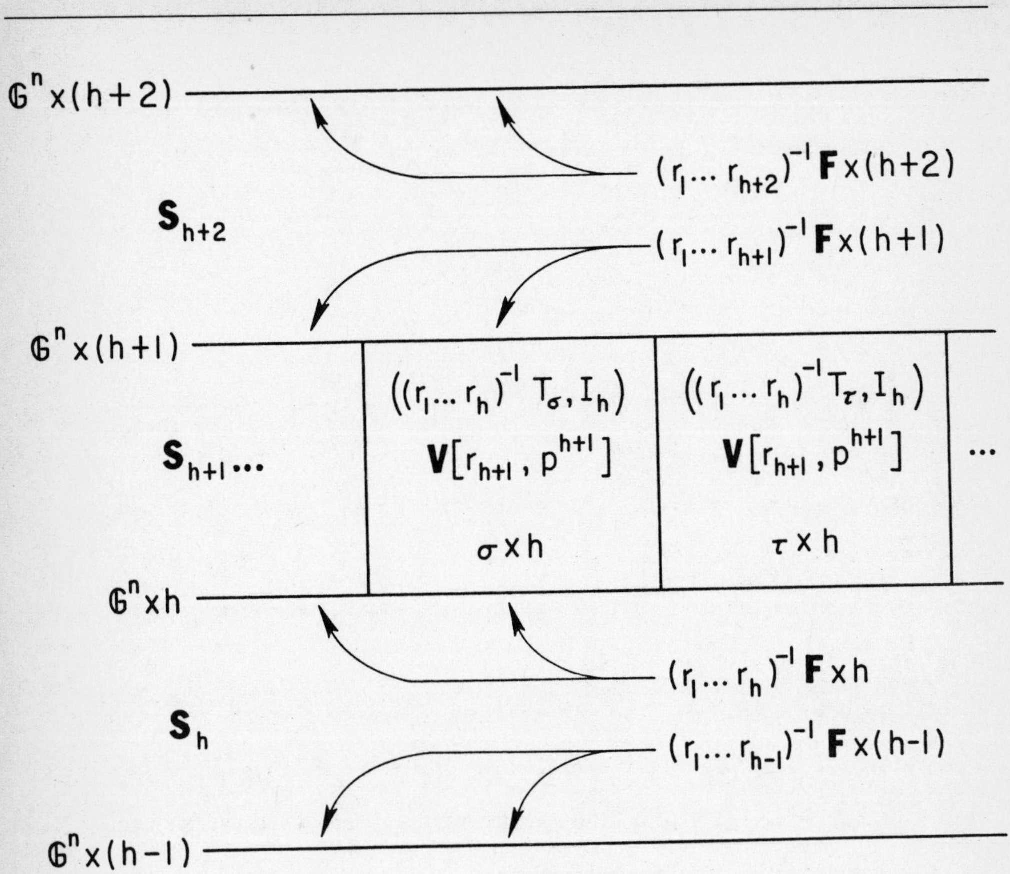

$\mathbb{G}^n \times (h+2)$
$(r_1 \dots r_{h+2})^{-1} \mathbf{F} \times (h+2)$
$\mathbf{S}_{h+2}$
$(r_1 \dots r_{h+1})^{-1} \mathbf{F} \times (h+1)$
$\mathbb{G}^n \times (h+1)$
$((r_1 \dots r_h)^{-1} T_\sigma, I_h)$
$\mathbf{V}[r_{h+1}, p^{h+1}]$
$\sigma \times h$
$((r_1 \dots r_h)^{-1} T_\tau, I_h)$
$\mathbf{V}[r_{h+1}, p^{h+1}]$
$\tau \times h$
$\mathbf{S}_{h+1} \dots$
...
$\mathbb{G}^n \times h$
$(r_1 \dots r_h)^{-1} \mathbf{F} \times h$
$\mathbf{S}_h$
$(r_1 \dots r_{h-1})^{-1} \mathbf{F} \times (h-1)$
$\mathbb{G}^n \times (h-1)$

Figure 17.3

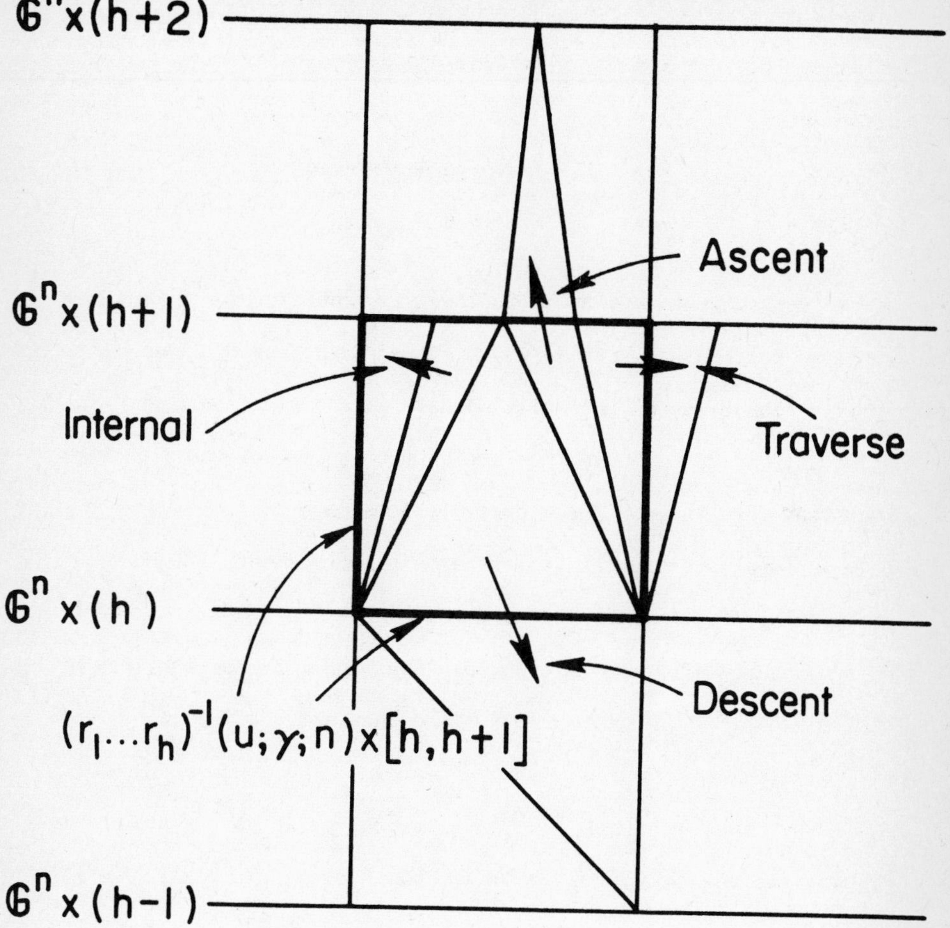

$\mathbb{G}^n \times (h+2)$
Ascent
$\mathbb{G}^n \times (h+1)$
Internal
Traverse
$\mathbb{G}^n \times (h)$
Descent
$(r_1 \ldots r_h)^{-1}(u;\gamma;n) \times [h,h+1]$
$\mathbb{G}^n \times (h-1)$

Figure 17.4

$\mathbf{V}(r,p)$		$\mathbf{S}$
Replacement	• • •	Internal Replacement
Bottom Exit	• • •	Descent Replacement (and Bottom Exit)
Side Exit	• • •	Traverse Replacement
Top Exit	• • •	Ascent Replacement

The traverse replacements are as in Section 12. Therefore, the only new ideas required for replacements in $\mathbf{S}$ are the ascent and descent cases which we will soon describe. Finally we develop a grand flow chart which encompasses all cases, but note that a significant part of the flow chart duplicates that for $\mathbf{V}(r,p)$, see Figures 16.3 through 16.7.

First we turn our attention to the ascent replacement. Given $\rho = (h;u;\gamma;v;\pi;k;\eta)$ assume $k = n$ and that vertex $t = n+2$ is to be dropped, see Figures 17.5 and note 15.13. Thus $n+1$ of the vertices of ρ are in $\left((r_1 \cdots r_h)^{-1}(u;\gamma;h)\right) \times (h+1)$ and one vertex of σ is in $\left((r_1 \cdots r_h)^{-1}(u;\gamma;h)\right) \times h$.

We seek an $(n+1)$-simplex $\hat{\rho} = (\hat{h};\hat{u};\hat{\gamma};\hat{v};\hat{\pi};\hat{k};\hat{\eta})$ in $\mathbf{S}^{n+1}$ which shows exactly the top $n+1$ vertices with ρ. First it is clear that $\hat{h} = h+1$ which is to say that $\hat{\rho}$ lies in $G^n \times [h+1, h+2]$. The new $[\hat{v};\hat{\pi};\hat{k};\hat{\eta}]$ in $\mathbf{V}[r_{h+2}, p^{h+2}]$ must have $n+1$ vertices in $S \times 0$ for $\hat{\rho}$ to have $n+1$ vertices in $\mathbb{C}^n \times (h+1)$. There is exactly one such simplex $[\hat{v};\hat{\pi};\hat{k};\hat{\eta}]$, namely, that given by $\hat{v} = 0$, $\hat{\pi} = (1, \ldots, n+1)$, $\hat{k} = 0$, and $\hat{\eta} = 0$, see Figure 17.6. With $(\hat{h},\hat{v},\hat{\pi},\hat{k},\hat{\eta})$ determined we turn our attention to the generation of $(\hat{u},\hat{\gamma})$.

A $(\hat{u},\hat{\gamma})$ is sought such that

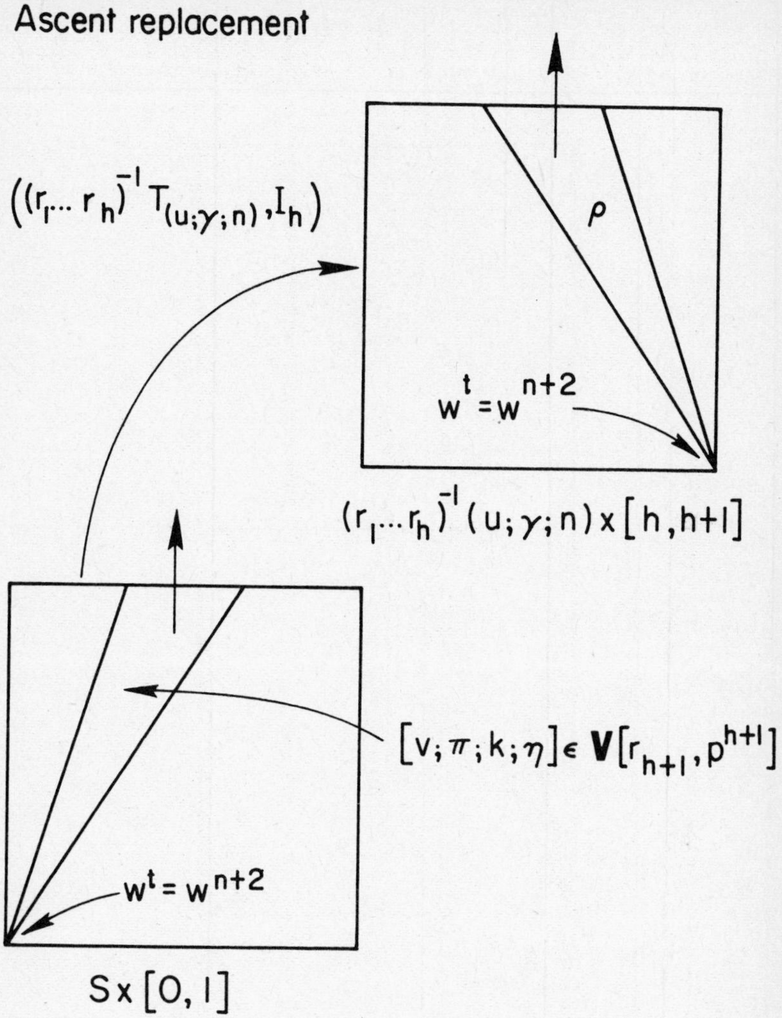
Ascent replacement
((r1... rh)-1 T(u;γ;n), Ih)
ρ
wt = wn+2
(r1...rh)-1 (u;γ;n) x [h, h+1]
[v;π;k;η] ∈ V[rh+1, ph+1]
wt = wn+2
S x [0, 1]

Figure 17.5

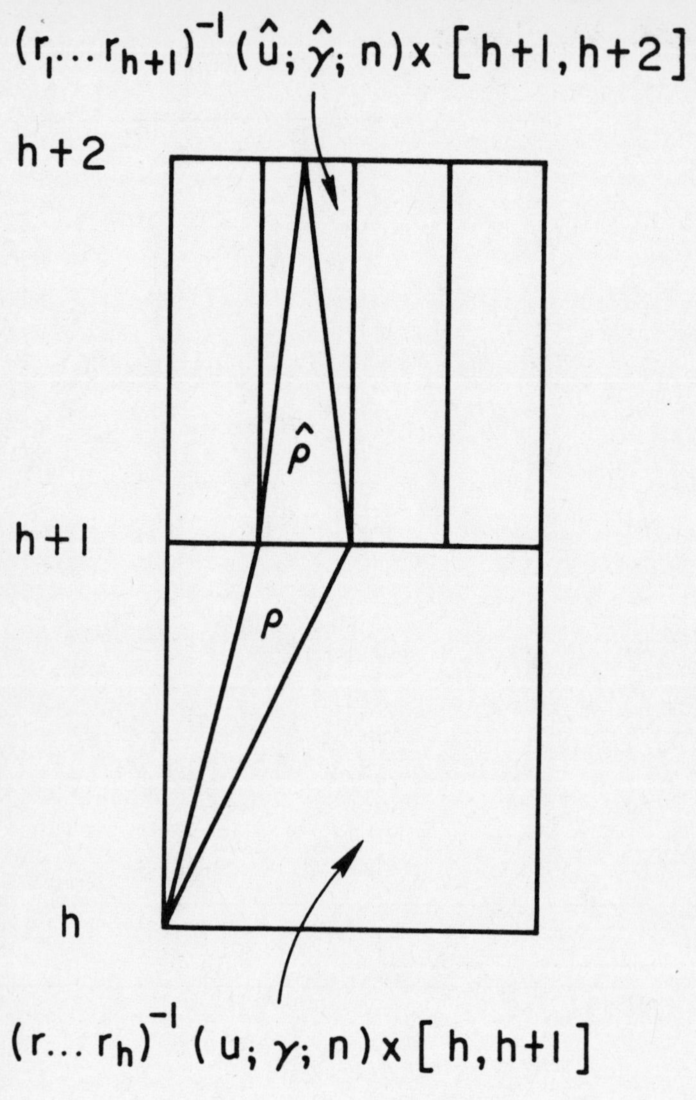
$(r_1 \ldots r_{h+1})^{-1}(\hat{u}; \hat{\gamma}; n) \times [h+1, h+2]$
h+2
$\hat{\rho}$
h+1
ρ
h
$(r \ldots r_h)^{-1}(u; \gamma; n) \times [h, h+1]$

Figure 17.6

$$(r_1 \cdots r_{h+1})^{-1}\ (\hat{u};\hat{\gamma};n) = (r_1 \cdots r_h)^{-1}\ T_{(u;\gamma;n)}(r_{h+1}^{-1}(v;\pi;n) + r_{h+1}^{-1}p^{h+1}),$$

see Figure 17.7. Using Lemma 12.1, it is sufficient to solve for $(\hat{u},\hat{\gamma})$ in

$$(1) \qquad (\hat{u};\hat{\gamma};n) = q^{\gamma|n}(v;\pi;n) + r_{h+1}u + q^{\gamma|n}\ p^{h+1}\ .$$

As the ordering of the vertices in $(v;\pi;n)$ and $(\hat{u};\hat{\gamma};n)$ is irrelevant in equation (1), using 6.2 it is sufficient to solve

$$\hat{u} = q^{\gamma|n}(v + p^{h+1}) + r_{h+1}u$$

$$q^{\hat{\gamma}|n} = q^{\gamma|n}\ q^{\pi|n}$$

for $(\hat{u},\hat{\gamma})$. The efficient manner to obtain $(\hat{u},\hat{\gamma})$ is according to Lemmas 4.2 and 4.5, to wit:

$$\hat{u}_i = r_{h+1}u_i + (v + p^{h+1})_{\xi i} - (v + p^{h+1})_{\xi(n+1)}$$

where $\xi = \gamma^{-1}$ and $(\cdot)_{n+1} \triangleq 0$, and $\hat{\gamma} = \gamma\pi$. Thus $(\hat{u},\hat{\gamma})$ solves (1) but we are not quite done for it is required that $(\hat{u};\hat{\gamma};n)$ be coordinated with $\underline{\ell}$, that is, $\underline{\ell}(u^i) = \underline{\ell}(s^i)$ for i in μ.

However, using Figure 10.4 it is a simple matter to rerepresent $(\hat{u};\hat{\gamma};n)$ to obtain $\underline{\ell}(\hat{u}^i) = \underline{\ell}(s^i)$ for i in μ. Once this is done we have the simplex $\hat{\rho} = (\hat{h};\hat{u};\hat{\gamma};\hat{v};\hat{\pi};\hat{k};\hat{\eta})$ in $\mathbb{C}^n \times [h+1, h+2]$ that shares the top $n+1$ vertices with $\rho = (h;u;\gamma;v;\pi;k;\eta)$ in $\mathbb{C}^n \times [h,h+1]$; and our analysis of the ascent replacement is complete.

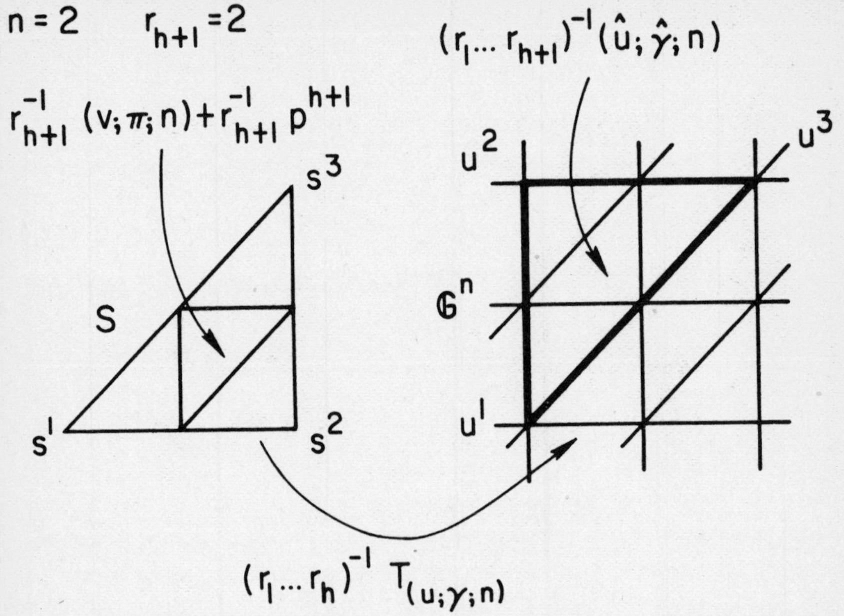
n = 2
$r_{h+1} = 2$
$r_{h+1}^{-1}(v;\pi;n) + r_{h+1}^{-1}p^{h+1}$
s^3
S
s^1
s^2
$(r_1 \dots r_{h+1})^{-1}(\hat{u};\hat{\gamma};n)$
u^2
u^3
$\mathbb{G}^n$
u^1
$(r_1 \dots r_h)^{-1} T_{(u;\gamma;n)}$

Figure 17.7

We turn our attention to the descent replacement. Given $\rho = (h;u;\gamma;v;\pi;k;\eta)$ assume $k = 0$ and that vertex 1 is to be dropped, see Figures 17.8 and Note 15.2.

We seek an $(n+1)$-simplex $\hat{\rho} = (\hat{h};\hat{u};\hat{\gamma};\hat{v};\hat{\pi};\hat{k};\hat{\eta})$ in **S** which shares, exactly, the bottom $(n+1)$- vertices with ρ. It is clear that $\hat{h} = h-1$ and $\hat{k} = n$ since $\hat{\rho}$ will lie in $\mathbb{C}^n \times [h-1,h]$ and have a facet in $\mathbb{C}^n \times h$. Next we must solve the following equation for $(\hat{u},\hat{\gamma},\hat{v},\hat{\pi})$,

$$(r_1 \cdots r_{h-1})^{-1}\, T_{(\hat{u};\hat{\gamma};n)}(r_h^{-1}((\hat{v};\hat{\pi},n) + p^h))$$

$$= (r_1 \cdots r_h)^{-1}\,(u;\gamma;n) \tag{2}$$

$$r_h^{-1}(\hat{v};\hat{\pi};n) + r_h^{-1}p \subseteq S \quad .$$

Our first step is to find $(\hat{u};\hat{\gamma};n)$ which contains $r_h^{-1}(u;\gamma;n)$ which is done as in Figure 6.7.

Rerepresent $(u;\gamma;n)$ as $(\bar{u};\bar{\gamma};n)$ with $\bar{\gamma}(n+1) = n+1$, see Figure 10.2. Let

$$\hat{u} = \lfloor r_h^{-1}\bar{u} \rfloor$$

and $\hat{\gamma}$ order $(\bar{u}_i - r_h\hat{u}_i, -\bar{\gamma}^{-1}(i))$ for i in ν lexico decreasing and set $\hat{\gamma}(n+1) = n+1$. Now $(\hat{u};\hat{\gamma};n)$ contains $r_h^{-1}(u;\gamma;n)$.

Rerepresent $(\hat{u};\hat{\gamma};n)$ according to Figure 10.4 so that $\underline{\ell}(\hat{u}^i) = \underline{\ell}(s^i)$ for i in ν. Returning to equation (2) we now have the correct $(\hat{u};\hat{\gamma};n)$. Since $T^{-1}_{(\hat{u};\hat{\gamma};n)}(r_h^{-1}(u;\gamma;n))$ lies in S, system (2) can be solved by solving (3)

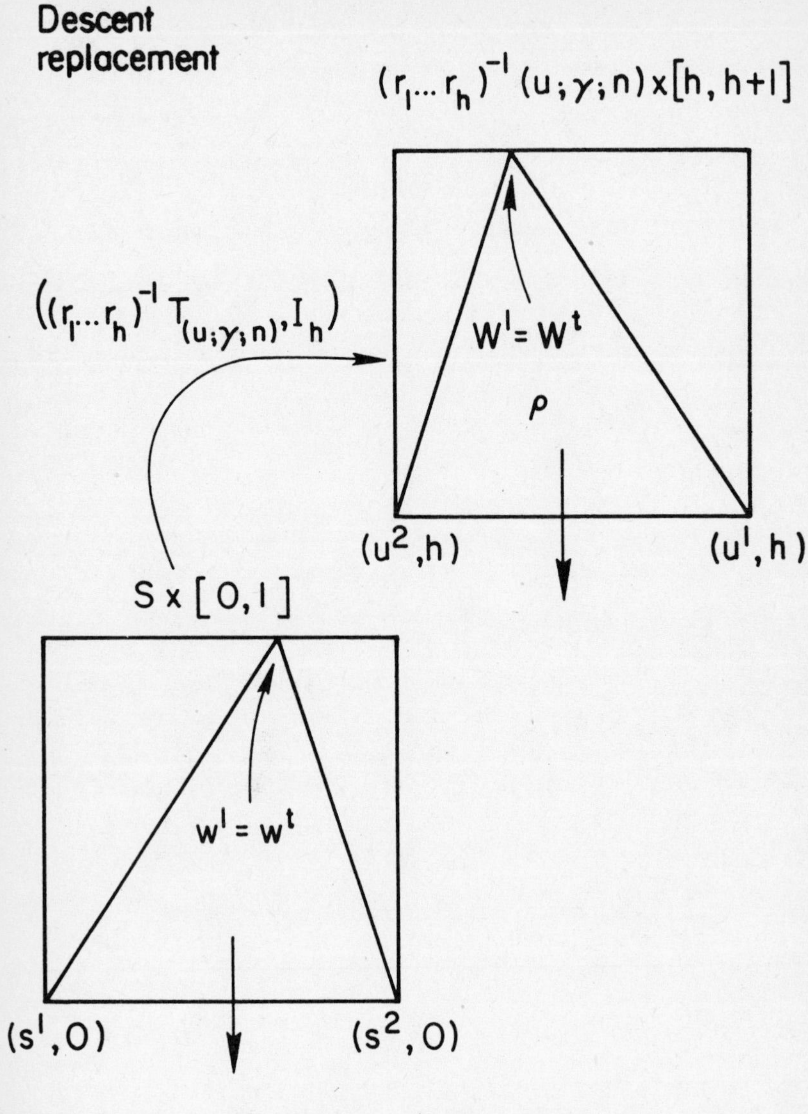
Descent
replacement
(r1... rh)^-1 (u; γ; n) x [h, h+1]
((r1... rh)^-1 T(u;γ;n), Ih)
W1 = Wt
ρ
(u2, h)
(u1, h)
S x [0, 1]
w1 = wt
(s1, 0)
(s2, 0)

Figure 17.8

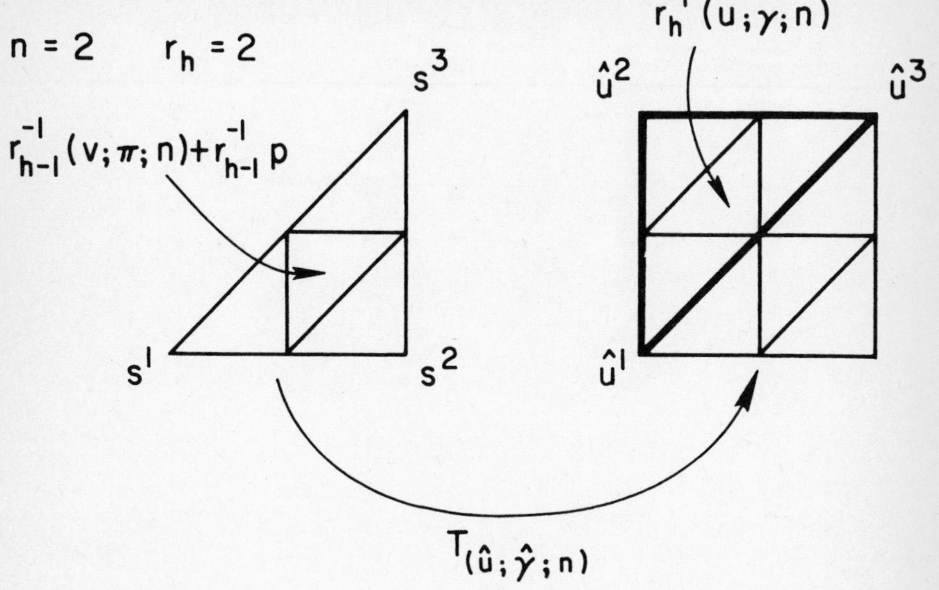

Figure 17.9

$$q^{\hat{\gamma}|n}(\hat{v} + p^h) + r_h\hat{u} = u$$

(3)

$$q^{\hat{\gamma}|n}\, q^{\hat{\pi}|n} = q^{\gamma|n} .$$

By Lemmas 4.1 and 4.5 if $\hat{v}$ is defined by

$$\hat{v}_i = (u-r_h\hat{u})_{\hat{\gamma}i} - (u-r_h\hat{u})_{\hat{\gamma}(n+1)} - p_i^h$$

for i in ν and $(\cdot)_{n+1} \triangleq 0$, and if $\hat{\pi} = \hat{\gamma}^{-1}\gamma$ then $(\hat{v},\hat{\pi})$ solves (3). Consequently $(\hat{u};\hat{\gamma};\hat{v};\hat{\pi})$ solves (2). There is one small difficulty remaining which is related to $\hat{\eta}$.

Such $(\hat{v};\hat{\pi};n)$ may not lie in $Q_{(\hat{\pi}|n)\mu}$ as is required of a simplex $(\hat{v};\hat{\pi};\hat{k};\hat{\eta})$ in $\mathbf{V}(r_h,p^h)$. To attend to this matter we proceed as in Figure 6.8. First rerepresent $(\hat{v};\hat{\pi};n)$ to obtain $\hat{\pi}(n+1) = n+1$. Let j index the lexico smallest element of $(\hat{v}_i,-\hat{\pi}^{-1}(i))$ for i in ν. Let $\ell = n+1$ or $\ell = j$ if $\hat{v}_j \geqq 0$ or $\hat{v}_j < 0$, respectively, and $(\hat{v};\hat{\pi};n)$ is contained in $Q_{\mu\backslash\ell}$. Finally rerepresent $(\hat{v};\hat{\pi};n)$ so that $\hat{\pi}(n+1) = \ell$. To obtain $\hat{\eta}$ solve $q\hat{\eta} = \hat{v}$ for $0 \nleq \hat{\eta} \geqq 0$ by proceeding as in Figure 5.2. That is, for $\ell = n+1$ let $(\hat{\eta}_\nu, \eta_{n+1}) = (\hat{v},0)$ and for $\ell < n+1$ let $\hat{\eta} = (\hat{v},0) - v_\ell e$ where $e = (1, \ldots, 1) \in \mathbb{C}^{n+1}$.

We now have an $(n+1)$-simplex $\hat{\rho} = (\hat{h};\hat{u};\hat{\gamma};\hat{v};\hat{\pi};\hat{k};\hat{\eta})$ of $\mathbf{S}$ which shares the bottom $n+1$ vertices with $\rho = (h;u;\gamma;v;\pi;k;\eta)$.

In the following charts the replacement rules for $\mathbf{S}$ are summarized. For $(h;u;\gamma;v;\pi;k;\eta)$ the vertices of $(v;\pi;k)$ are represented by v^1, $\ldots$, v^{k+1}.

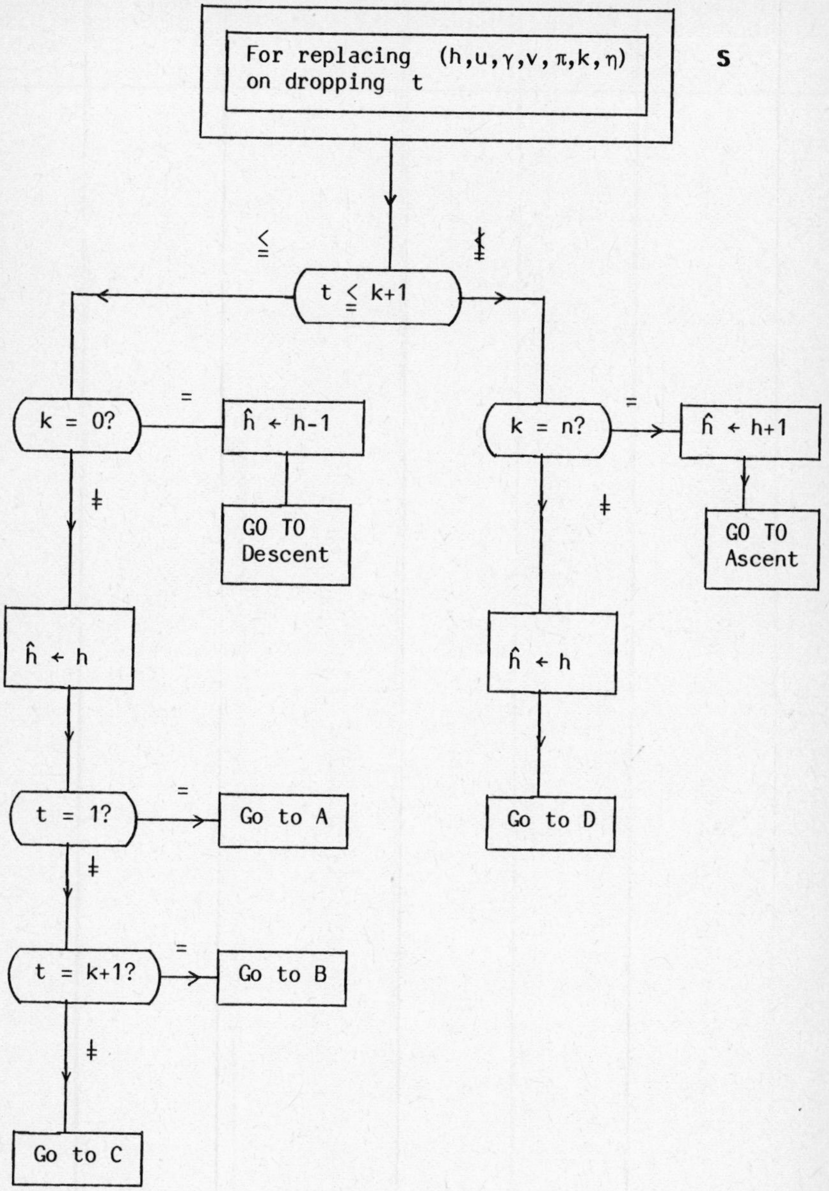
For replacing (ħ,u,γ,v,π,k,η)
on dropping t
S
≦
≰
t ≦ k+1
k = 0?
=
ĥ ← h-1
GO TO
Descent
≠
ĥ ← h
t = 1?
=
Go to A
≠
t = k+1?
=
Go to B
≠
Go to C
k = n?
=
ĥ ← h+1
GO TO
Ascent
≠
ĥ ← h
Go to D

Figure 17.10

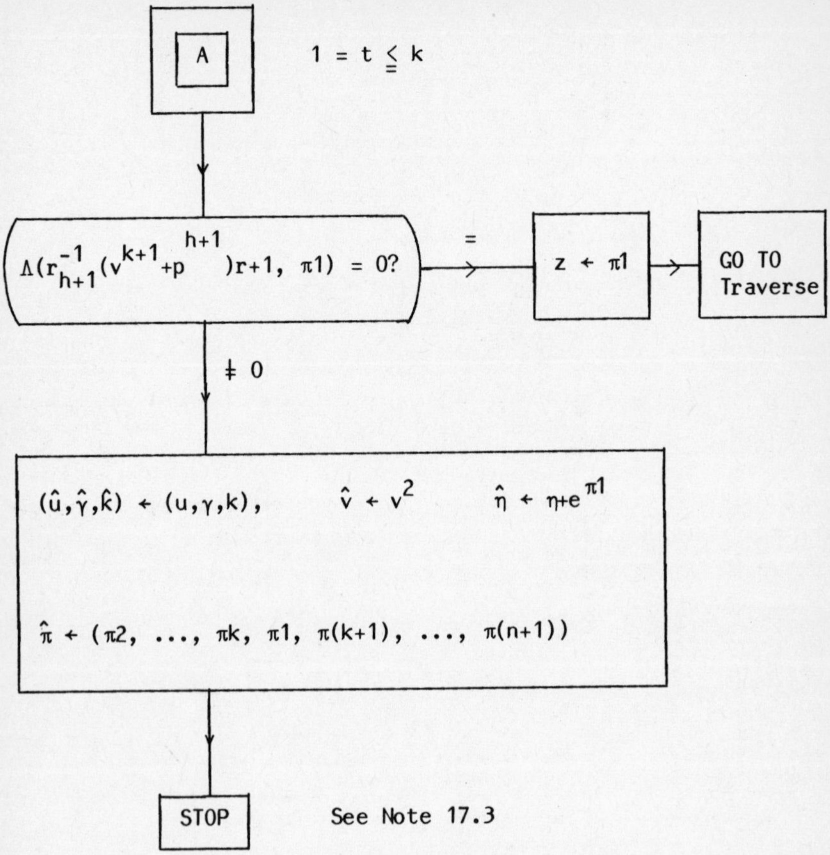
A
1 = t ≦ k
Λ(r⁻¹ₕ₊₁(v^{k+1}+p^{h+1})r+1, π1) = 0?
=
z ← π1
GO TO Traverse
≠ 0
(û,γ̂,k̂) ← (u,γ,k), v̂ ← v² η̂ ← η+e^{π1}
π̂ ← (π2, ..., πk, π1, π(k+1), ..., π(n+1))
STOP
See Note 17.3

Figure 17.11

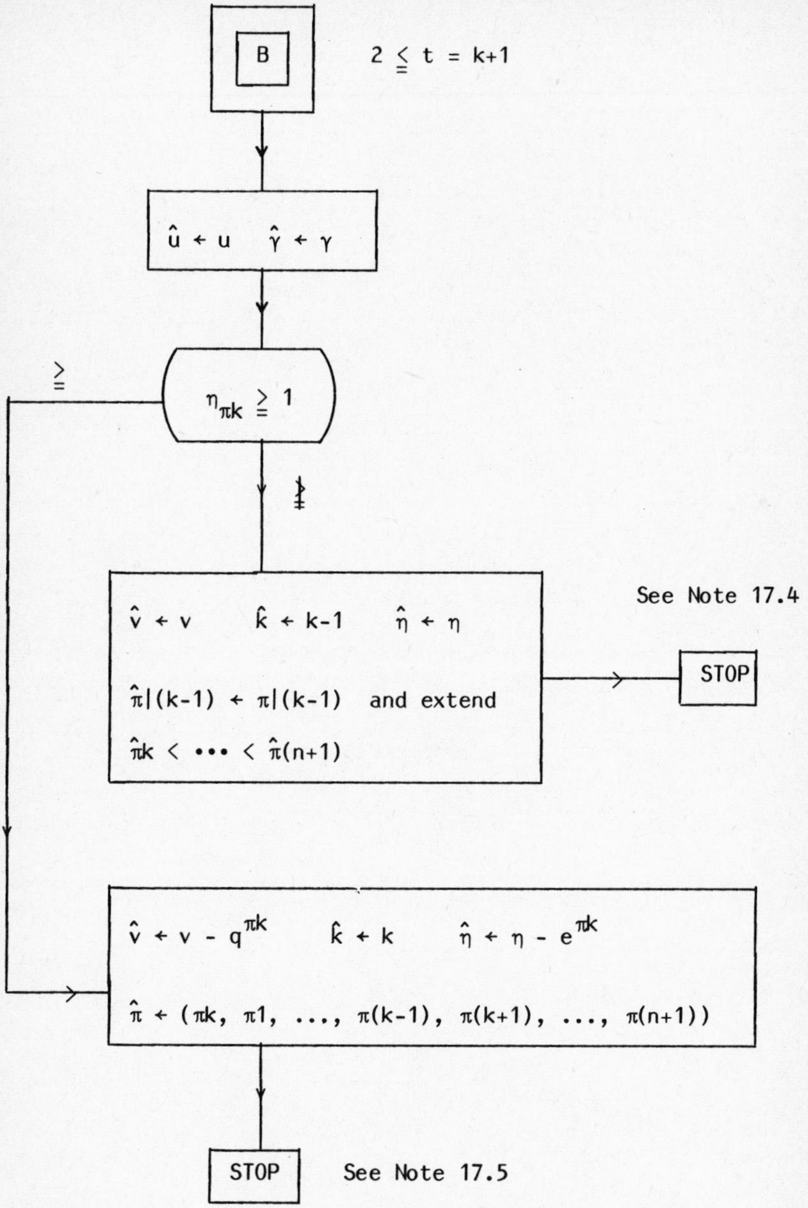
B
2 ≦ t = k+1
û ← u   γ̂ ← γ
η_πk ≧ 1
≧
≱
v̂ ← v   k̂ ← k-1   η̂ ← η
π̂|(k-1) ← π|(k-1) and extend
π̂k < ••• < π̂(n+1)
See Note 17.4
STOP
v̂ ← v - q^πk   k̂ ← k   η̂ ← η - e^πk
π̂ ← (πk, π1, ..., π(k-1), π(k+1), ..., π(n+1))
STOP
See Note 17.5

Figure 17.12

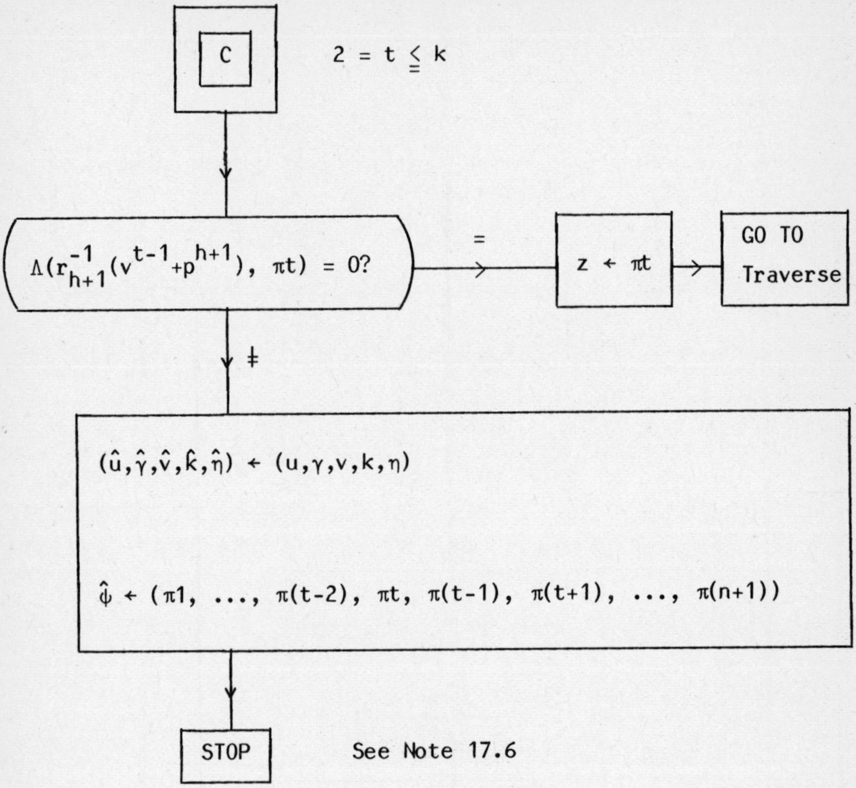
C
2 = t ≦ k
Λ(r⁻¹ₕ₊₁(vᵗ⁻¹+pʰ⁺¹), πt) = 0?
=
z ← πt
GO TO Traverse
≠
(û,γ̂,v̂,k̂,η̂) ← (u,γ,v,k,η)
φ̂ ← (π1, ..., π(t-2), πt, π(t-1), π(t+1), ..., π(n+1))
STOP
See Note 17.6

Figure 17.13

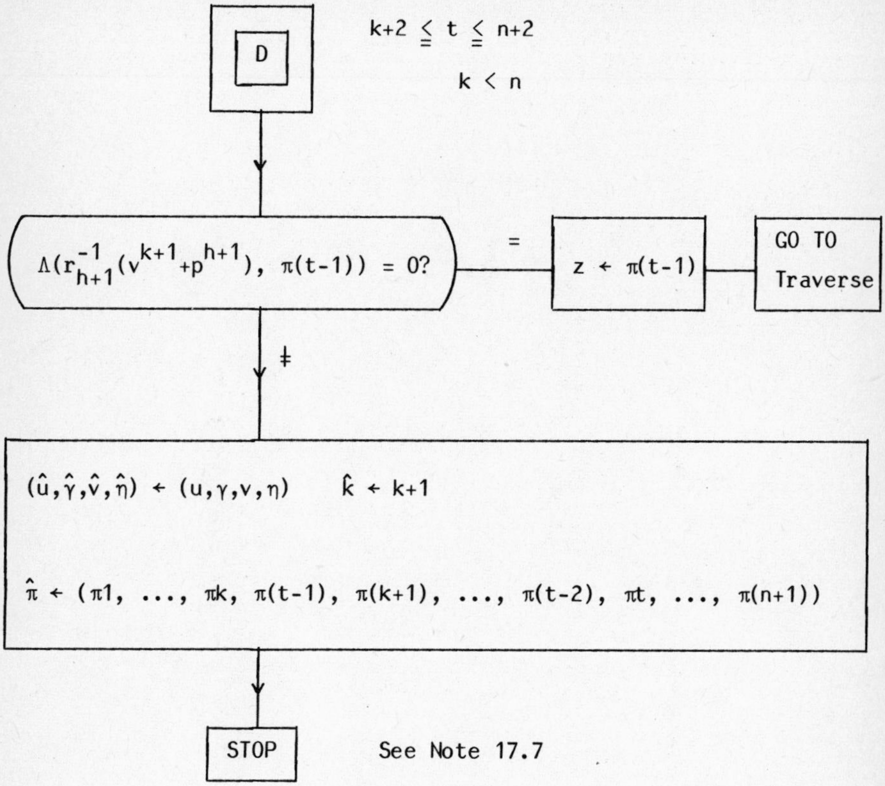
D
$k+2 \leq t \leq n+2$
$k < n$
$\Lambda(r_{h+1}^{-1}(v^{k+1}+p^{h+1}), \pi(t-1)) = 0?$
=
$z \leftarrow \pi(t-1)$
GO TO Traverse
≠
$(\hat{u},\hat{\gamma},\hat{v},\hat{\eta}) \leftarrow (u,\gamma,v,\eta)$
$\hat{k} \leftarrow k+1$
$\hat{\pi} \leftarrow (\pi 1, \ldots, \pi k, \pi(t-1), \pi(k+1), \ldots, \pi(t-2), \pi t, \ldots, \pi(n+1))$
STOP
See Note 17.7

Figure 17.14

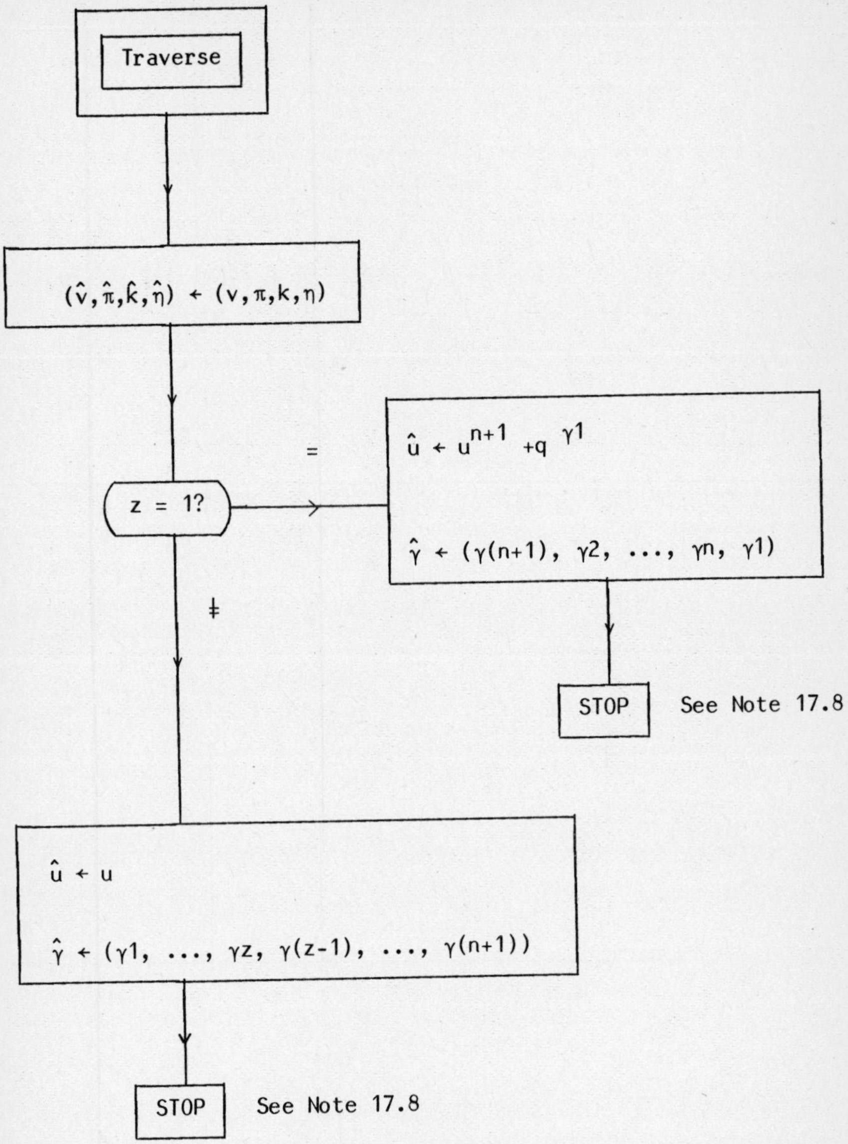
Traverse
(v̂, π̂, k̂, η̂) ← (v, π, k, η)
z = 1?
=
û ← u^{n+1} +q γ1
γ̂ ← (γ(n+1), γ2, ..., γn, γ1)
STOP
See Note 17.8
≠
û ← u
γ̂ ← (γ1, ..., γz, γ(z-1), ..., γ(n+1))
STOP
See Note 17.8

Figure 17.15

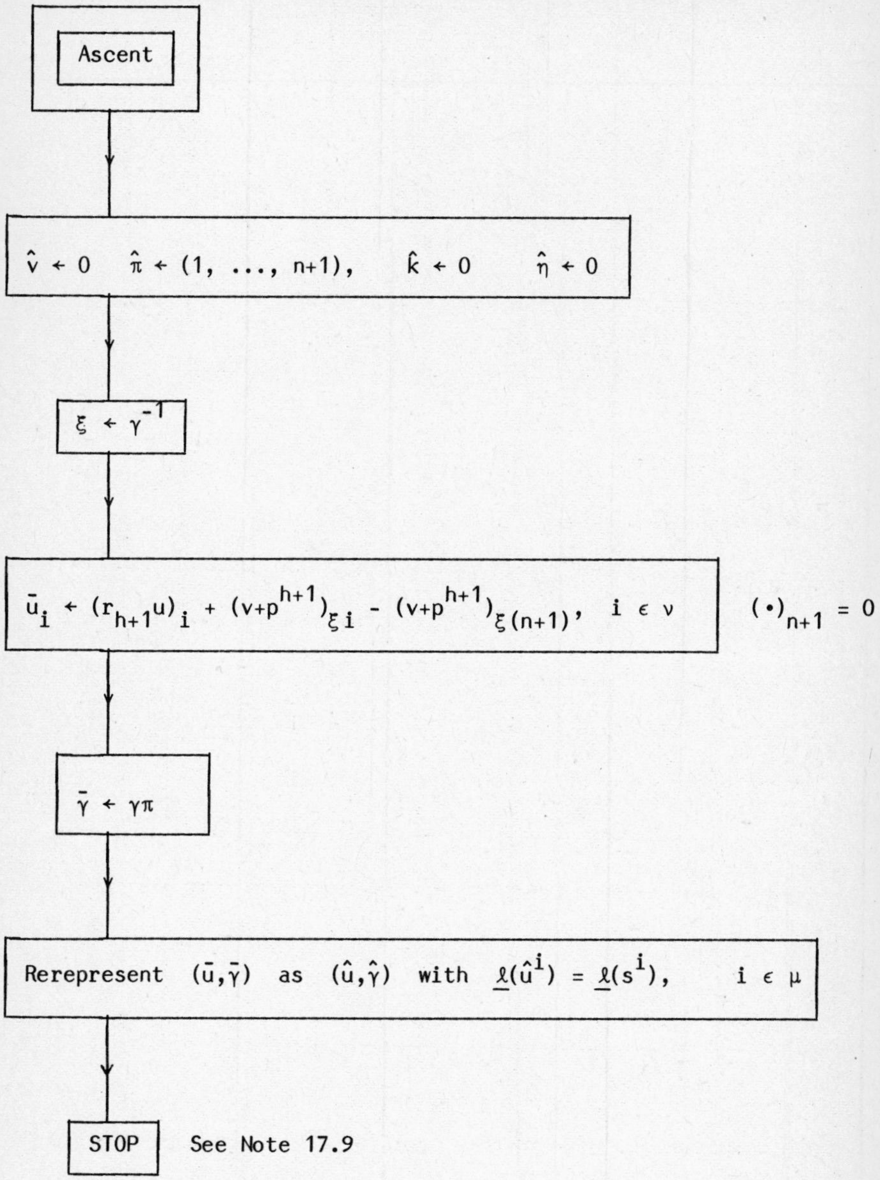
Ascent
$\hat{v} \leftarrow 0 \quad \hat{\pi} \leftarrow (1, \ldots, n+1), \quad \hat{k} \leftarrow 0 \quad \hat{\eta} \leftarrow 0$
$\xi \leftarrow \gamma^{-1}$
$\bar{u}_i \leftarrow (r_{h+1}u)_i + (v+p^{h+1})_{\xi i} - (v+p^{h+1})_{\xi(n+1)}, \quad i \in \nu$
$(\cdot)_{n+1} = 0$
$\bar{\gamma} \leftarrow \gamma\pi$
Rerepresent $(\bar{u},\bar{\gamma})$ as $(\hat{u},\hat{\gamma})$ with $\underline{\ell}(\hat{u}^i) = \underline{\ell}(s^i), \quad i \in \mu$
STOP
See Note 17.9

Figure 17.16

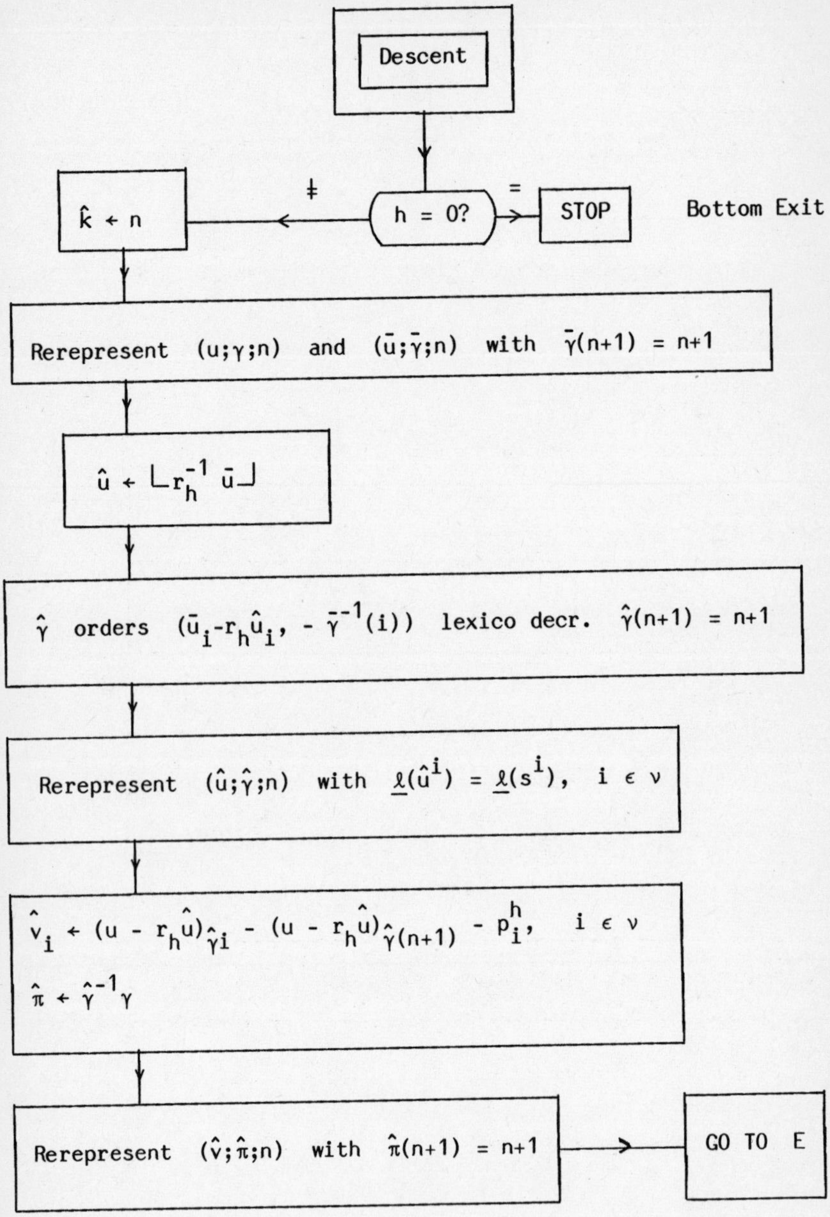
Descent
≠
h = 0?
=
STOP
Bottom Exit
$\hat{k} \leftarrow n$
Rerepresent $(u;\gamma;n)$ and $(\bar{u};\bar{\gamma};n)$ with $\bar{\gamma}(n+1) = n+1$
$\hat{u} \leftarrow \lfloor r_h^{-1} \bar{u} \rfloor$
$\hat{\gamma}$ orders $(\bar{u}_i - r_h \hat{u}_i, -\bar{\gamma}^{-1}(i))$ lexico decr. $\hat{\gamma}(n+1) = n+1$
Rerepresent $(\hat{u};\hat{\gamma};n)$ with $\underline{\ell}(\hat{u}^i) = \underline{\ell}(s^i)$, $i \in \nu$
$\hat{v}_i \leftarrow (u - r_h\hat{u})_{\hat{\gamma}i} - (u - r_h\hat{u})_{\hat{\gamma}(n+1)} - p_i^h$, $i \in \nu$
$\hat{\pi} \leftarrow \hat{\gamma}^{-1}\gamma$
Rerepresent $(\hat{v};\hat{\pi};n)$ with $\hat{\pi}(n+1) = n+1$
GO TO E

Figure 17.17

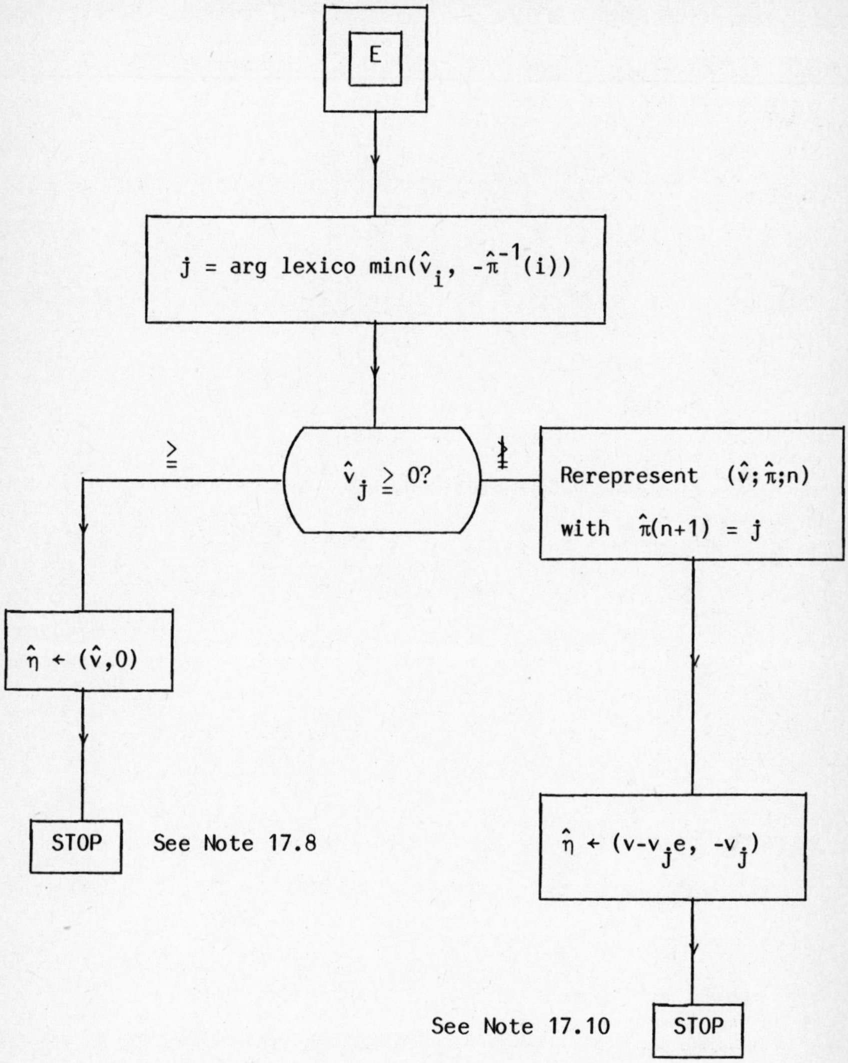
E
ĵ = arg lexico min(v̂ᵢ, -π̂⁻¹(i))
≧
v̂ⱼ ≧ 0?
≱
Rerepresent (v̂;π̂;n)
with π̂(n+1) = ĵ
η̂ ← (v̂,0)
STOP
See Note 17.8
η̂ ← (v-vⱼe, -vⱼ)
See Note 17.10
STOP

Figure 17.18

17.3 Note (for Figure 17.11): Internal replacement. The correspondence between vertices of $\rho = (h;u;\gamma;v;\pi;k;\eta)$ and $\hat{\rho} = (\hat{h};\hat{u};\hat{\gamma};\hat{v};\hat{\pi};\hat{k};\hat{\eta})$ is

$$\begin{array}{ccccccccccccc} \boxed{1} & & 2 & & 3 & & 4 & & & & k+1 & k+2 & n+2 \\ & / & & / & & / & & \cdots & & / & & | & | \\ 1 & & 2 & & 3 & & & & k & & \boxed{k+1} & k+2 & n+2 \end{array}$$

where the squared numbers indicate the dropped vertex and the new vertex. Thus the first vertex was dropped from ρ and the adjoined vertex is vertex $k+1$ of $\hat{\rho}$. The second vertex W^2 of ρ is the first vertex $\hat{W}^{-1}$ of $\hat{\rho}$, that is, $W^2 = \hat{W}^1$, etc. The adjoined vertex is

$$\hat{W}^{k+1} = \left((r_1 \cdots r_h)^{-1} T_{(u;\gamma;n)}(r_{h+1}^{-1}(\hat{v}^{k+1} + p^{h+1})), h+1\right) .$$

We have and shall let $W^1, \ldots, W^{n+2}$ indicate the vertices of ρ and $\hat{W}^1, \ldots, \hat{W}^{n+2}$ those for $\hat{\rho}$. □

17.4 Note (for Figure 17.12): Internal Replacement. Let πk be the ith smallest element in $\{\pi k, \ldots, \pi(n+1)\}$ where $i = 1, \ldots, n-k+2$. Then $\hat{\pi} = (\pi 1, \ldots, \pi(k-1), \pi(k+1), \ldots, \pi(k+i-2), \pi k, \pi(k+i-1), \ldots \pi(n+1))$. The correspondence between vertices of ρ and $\hat{\rho}$ is

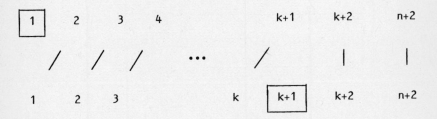

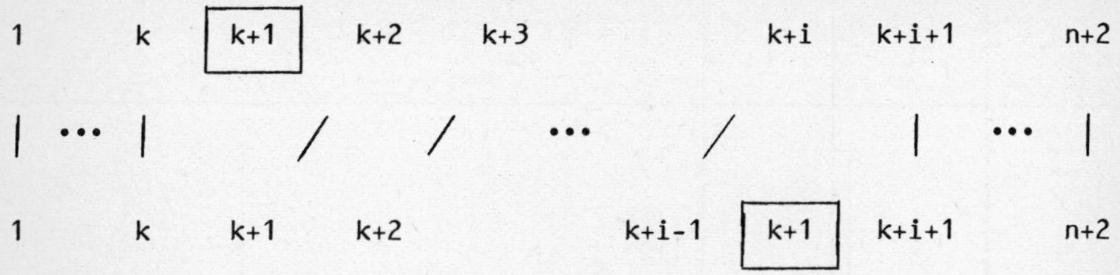

1		k	$\boxed{k+1}$		k+2		k+3				k+i	k+i+1		n+2
|	⋯	|		/		/		⋯		/		|	⋯	|
1		k	k+1		k+2				k+i-1		$\boxed{k+1}$	k+i+1		n+2

For more explicit treatment see Note 15.15. The adjoined vertex is

$$\hat{W}^{k+i} = \big((r_1 \cdots r_h^{-1})\ T_{(u;\gamma;n)}(s^{\pi k}),\ h\big)\ . \qquad \square$$

17.5 Note (for Figure 17.12): Internal Replacement. The correspondence between vertices of ρ and $\hat{\rho}$ is

1		2		3				k		$\boxed{k+1}$	k+2		n+2
	\		\		\		⋯		\		|	⋯	|
$\boxed{1}$		2		3		k				k+1	k+2		n+2

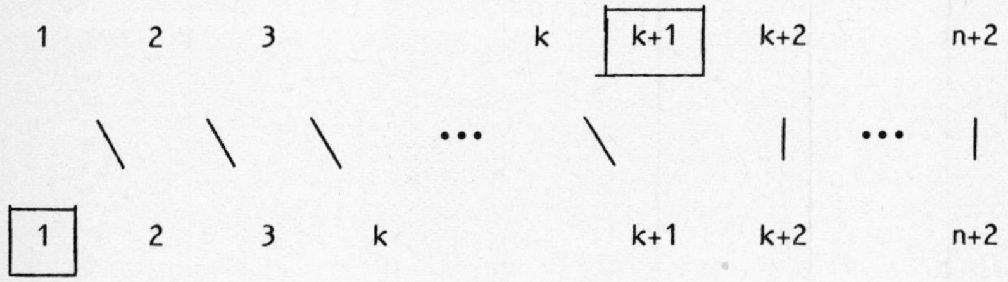

The adjoined vertex is

$$\hat{W}^{1} = \big((r_1 \cdots r_h)^{-1}\ T_{(u;\gamma;n)}(r_{h+1}^{-1}(\hat{v} + p^{h+1})),\ h+1\big)\ . \qquad \square$$

17.6 Note (for Figure 17.13): Internal replacement. The correspondence between vertices of ρ and $\hat{\rho}$ is

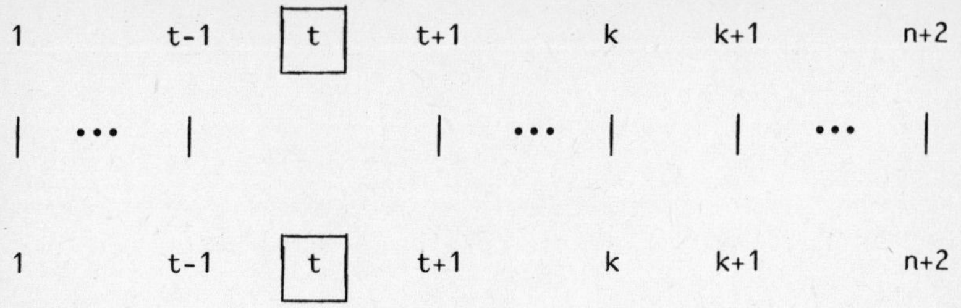
1
t-1
t
t+1
k
k+1
n+2
1
t-1
t
t+1
k
k+1
n+2

The new vertex is

$$\hat{W}^t = \left((r_1 \cdots r_h)^{-1} T_{(u;\gamma;n)}(r_{h+1}^{-1}(\hat{v}^t + p^{h+1})), h+1\right) . \quad \square$$

17.7 Note (for Figure 17.14): Internal Replacement. The correspondence between vertices of ρ and $\hat{\rho}$ is

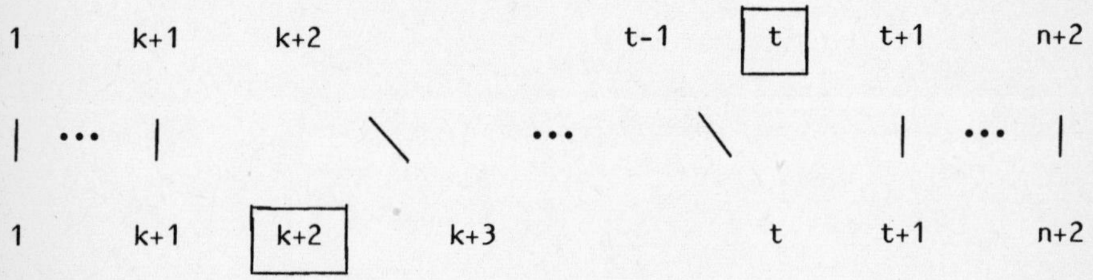
1
k+1
k+2
t-1
t
t+1
n+2
1
k+1
k+2
k+3
t
t+1
n+2

The adjoined vertex is

$$\hat{W}^{k+2} = \left((r_1 \cdots r_h)^{-1} T_{(u;\gamma;n)}(r_{h+1}^{-1}(\hat{v}^{k+2} + p^{h+1})), h+1\right) .$$

Note that $\hat{\pi} = \left(\pi(t-1)), \pi 1, \ldots, \pi(t-2), \pi t, \ldots, \pi(n+1)\right)$ for $k = 0$. $\square$

17.8 Note (for Figure 17.15): Traverse Replacement. The correspondence between vertices of ρ and $\hat{\rho}$ is

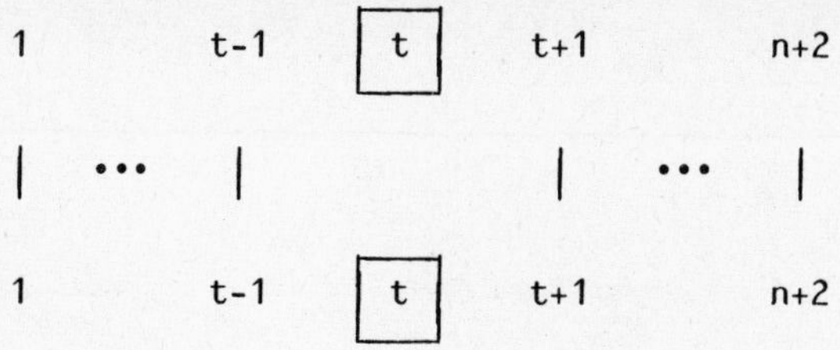

The adjoined vertex is

$$\hat{W}^t = \left((r_1 \cdots r_h)^{-1}\, T_{(\hat{u};\hat{\pi};n)}(r_{h+1}^{-1}(v^t + p^{h+1})),\ h+1\right) \quad ,$$

if $1 \leqq t \leqq k+1$ and

$$\hat{W}^t = \left((r_1 \cdots r_h)^{-1}\, T_{(\hat{u};\hat{\pi};n)}(s^{\pi(t-1)}),\ h\right) ,$$

if $k+2 \leqq t \leqq n+2$. □

17.9 Note (for Figure 17.16): Ascent Replacement. The top vertices $W^{-1}, \ldots, W^{n+1}$ of ρ are, in order,

$$\left((r_1 \cdots r_h)^{-1}\, T_{(u;\gamma;n)}(r_{h+1}^{-1}(v^i + p^{h+1})),\ h+1\right)$$

for $i = 1, \ldots, n+1$, and the bottom vertices $\hat{W}^2, \ldots, \hat{W}^{n+2}$ of $\hat{\rho}$ are, in order,

$$\left((r_1 \cdots r_{h+1})^{-1}\, T_{(\hat{u};\hat{\gamma};n)}(s^i),\ h+1\right)$$

for $i = 1, \ldots, n+1$. Therefore

$$q^{\gamma|n}\big((v;\pi;n) + p^{h+1}\big) + r_{h+1}u = (\hat{u};\hat{\gamma};n) \ .$$

The left hand side is $(\bar{u};\bar{\gamma};n)$, see Figure 17.4, where the order of vertices is preserved. Thus we have $(\bar{u};\bar{\gamma};n) = (\hat{u};\hat{\gamma};n)$ but where the ordering of the vertices may differ. Compute $\underline{\ell}(\bar{u}) - \underline{\ell}(\hat{u}) = \underline{\ell}(\bar{u}) = z$ modulo $n+1$ where z is chosen in $\{0, \ldots, n\}$. Then $\bar{u}^i = \hat{u}^{i+z}$ for i in u where the vertices are regarded modulo $n+1$. Thus the correspondence between vertices of ρ and $\hat{\rho}$ is

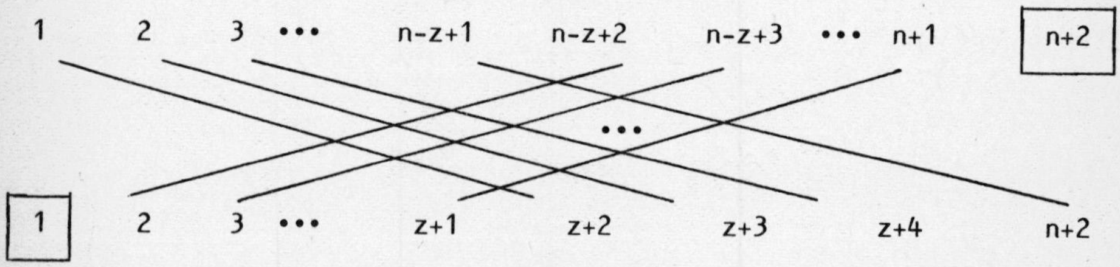

That is,

$$W^i = \begin{cases} \hat{W}^{i+z+1}, & 1 \leqq i \leqq n-z+1 \\ \hat{W}^{i-n+z} & n-z+2 \leqq i \leqq n+1 \end{cases}$$

The adjoined vertex is

$$\hat{W}^1 = \big((r_1 \cdots r_{h+1})^{-1}\, T_{(\hat{u};\hat{\gamma};n)}(r_{h+2}^{-1}(\hat{v} + p^{h+2})),\ h+2\big) \ . \qquad \square$$

17.10 Note (for Figure 17.18): Descent Replacement. The bottom vertices $W^2, \ldots, W^{n+2}$ of ρ are, in order,

$$\left((r_1 \cdots r_h)^{-1} T_{(u;\gamma;n)}(s^i), h\right) = (r_1, \ldots, r_h)^{-1} (u;\gamma;n)$$

for $i = 1, \ldots, n+1$, and the top vertices $\hat{W}^1, \ldots, \hat{W}^{n+1}$ of $\hat{\rho}$ are, in order,

$$\left((r_1 \cdots r_{h-1})^{-1} T_{(\hat{u};\hat{\gamma};n)}(r_h^{-1}(\hat{v}^i + p^h)), h\right)$$

for $i = 1, \ldots, n+1$. We thus have

$$(u;\gamma;n) = q^{\hat{\gamma}|n}((\hat{v};\hat{\pi};n) + p^h) + r_h\hat{u}$$

or $(u;\gamma;n) = (\bar{u};\bar{\gamma};n)$ where

$$\bar{u}_i = r_h\hat{u}_i + (\hat{v} + p^h)_{\xi i} - (\hat{v} + p^h)_{\xi(n+1)}$$

$$\xi = \hat{\gamma}^{-1} \qquad \text{and} \qquad \bar{\gamma} = \hat{\gamma}\,\hat{\pi} \quad .$$

Letting $z = \underline{\ell}(\bar{u}) - \underline{\ell}(u)$ we have

$$z = \underline{\ell}(\bar{u}) = \ell(\hat{v} + p^h) - (n+1)\,(\hat{v}_j + p_j)$$

where $\hat{\gamma}j = n+1$ and $(\bullet)_{n+1} \triangleq 0$, and we get $(\bar{u}^i) = u^{i+z}$ for i in μ where indices are regarded modulo $n+1$. Thus the correspondence between vertices of ρ and $\hat{\rho}$ is:

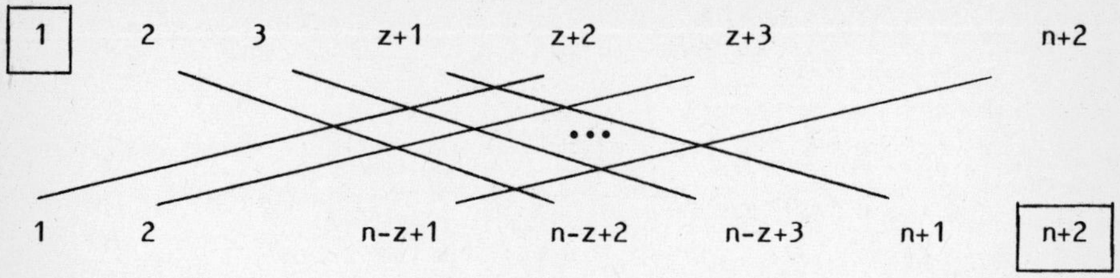

That is,

$$w^i = \begin{cases} \hat{w}^{i+n-z}, & 2 \leqq i \leqq z+1 \\ \hat{w}^{i-z-1} & z+2 \leqq i \leqq n+2 \,. \end{cases}$$

The adjoined vertex is

$$\hat{w}^{n+2} = \left((r_1, \ldots, r_{h-1})^{-1}\, T_{(\hat{u};\hat{\gamma};n)}(s^{\pi(n+1)}),\ h-1\right)$$

$$= \left((r_1, \ldots, r_{h-1})^{-1}\, \hat{u}^{\pi(n+1)},\ h-1\right) \qquad \square$$

17.11 Example: For $r_i = 2, 3, 2$ and $p^i = 1,2,0$ for $i = 1,2,3$ we show a portion of **S** in Figure 17.19 with a path. Beginning with simplex $i = 1$ the path is followed using the charts of Figures 17.10 to 17.18. The sequence of $(h,u,\gamma,v,\psi,k,\eta)$ generated is shown in Figure 17.20. The $\mathbf{V}[r_i,p_i]$ and $\mathbf{V}(r_i,p_i)$ are shown in Figures 17.21 and 17.22. The t to be dropped in order to follow the path must be determined by inspection at each replacement. $\square$

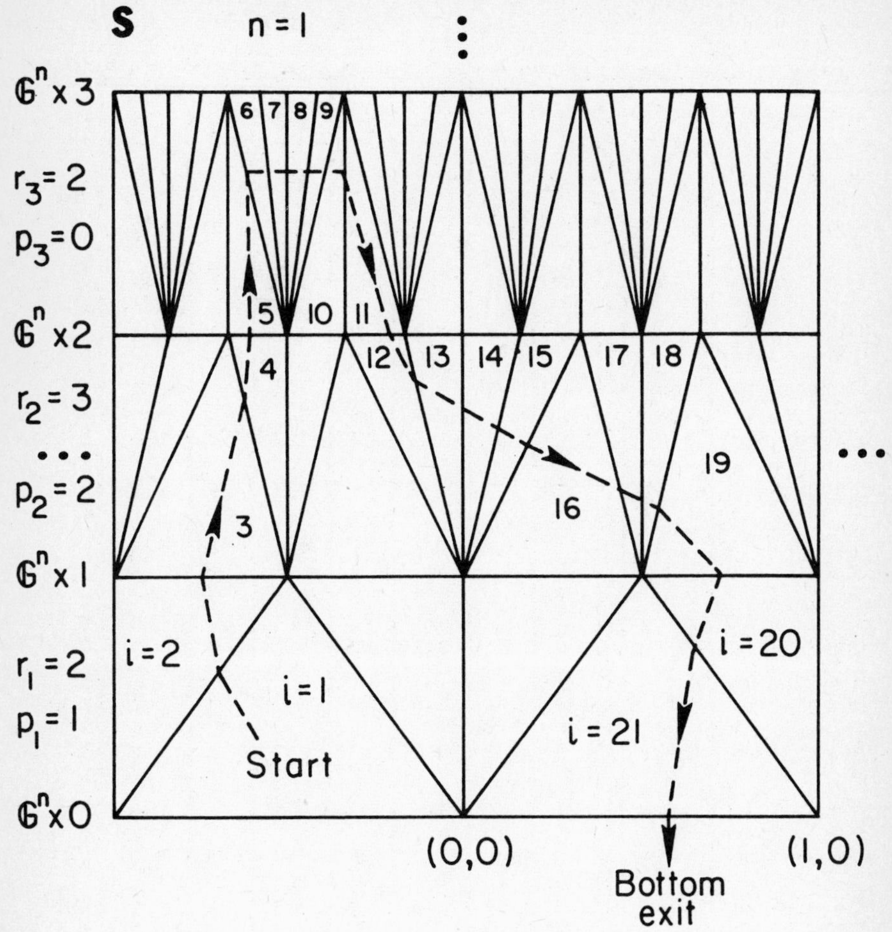
S
n = 1
$\mathbb{G}^n \times 3$
$r_3 = 2$
$p_3 = 0$
$\mathbb{G}^n \times 2$
$r_2 = 3$
$p_2 = 2$
$\mathbb{G}^n \times 1$
$r_1 = 2$
$p_1 = 1$
$\mathbb{G}^n \times 0$
6 7 8 9
5 10 11
4
12 13 14 15 17 18
3
16
19
i = 2
i = 1
i = 20
i = 21
Start
(0,0)
(1,0)
Bottom exit

Figure 17.19

	i	h	u	γ	ν	π	k	η	t
START	1	0	0	2 1	0	1 2	0	0 0	2
Internal to	2	0	0	2 1	0	1 2	1	0 0	3
Ascend to	3	1	-2	1 2	0	1 2	0	0 0	2
Internal to	4	1	-2	1 2	0	1 2	1	0 0	3
Ascend to	5	2	-4	1 2	0	1 2	0	0 0	2
Internal to	6	2	-4	1 2	0	1 2	1	0 0	1
Internal to	7	2	-4	1 2	1	1 2	1	1 0	1
Traverse to	8	2	-2	2 1	1	1 2	1	1 0	2
Internal to	9	2	-2	2 1	0	1 2	1	0 0	2
Internal to	10	2	-2	2 1	0	1 2	0	0 0	3
Traverse to	11	2	-2	1 2	0	1 2	0	0 0	1
Descend to	12	1	0	2 1	0	2 1	1	0 0	1
Internal to	13	1	0	2 1	-1	2 1	1	0 1	1
Traverse to	14	1	0	1 2	-1	2 1	1	0 1	2
Internal to	15	1	0	1 2	0	2 1	1	0 0	2
Internal to	16	1	0	1 2	0	1 2	0	0 0	2
Internal to	17	1	0	1 2	0	1 2	1	0 0	1
Traverse to	18	1	2	2 1	0	1 2	1	0 0	2
Internal to	19	1	2	2 1	0	1 2	0	0 0	1
Descend to	20	0	0	1 2	0	1 2	1	0 0	2
Internal to	21	0	0	1 2	0	1 2	0	0 0	1
Bottom Exit	22	0							

Figure 17.20

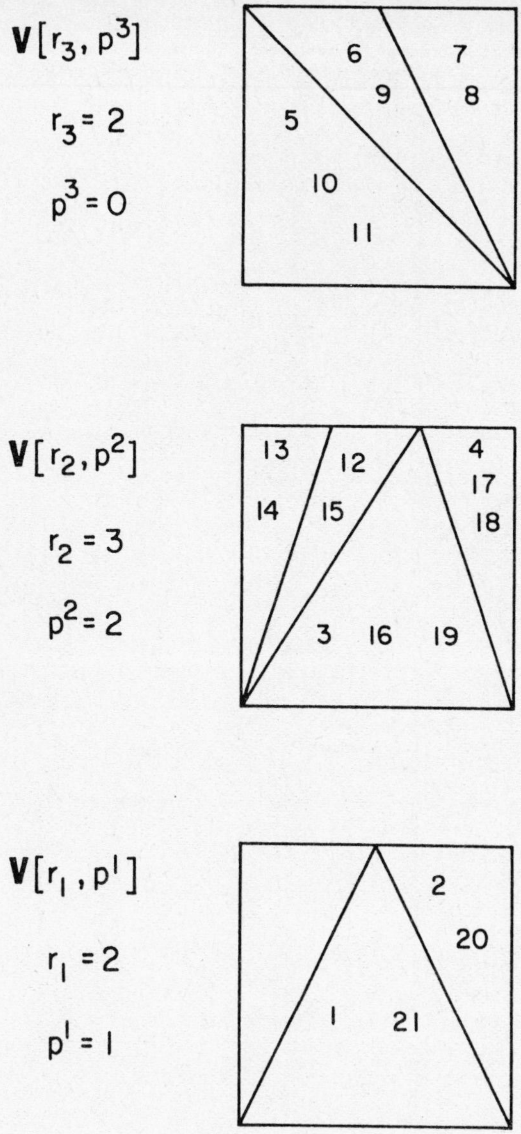
V[r3, p3]
r3 = 2
p3 = 0
6
7
9
8
5
10
11
V[r2, p2]
r2 = 3
p2 = 2
13
12
4
17
14
15
18
3
16
19
V[r1, p1]
r1 = 2
p1 = 1
2
20
1
21

Figure 17.21

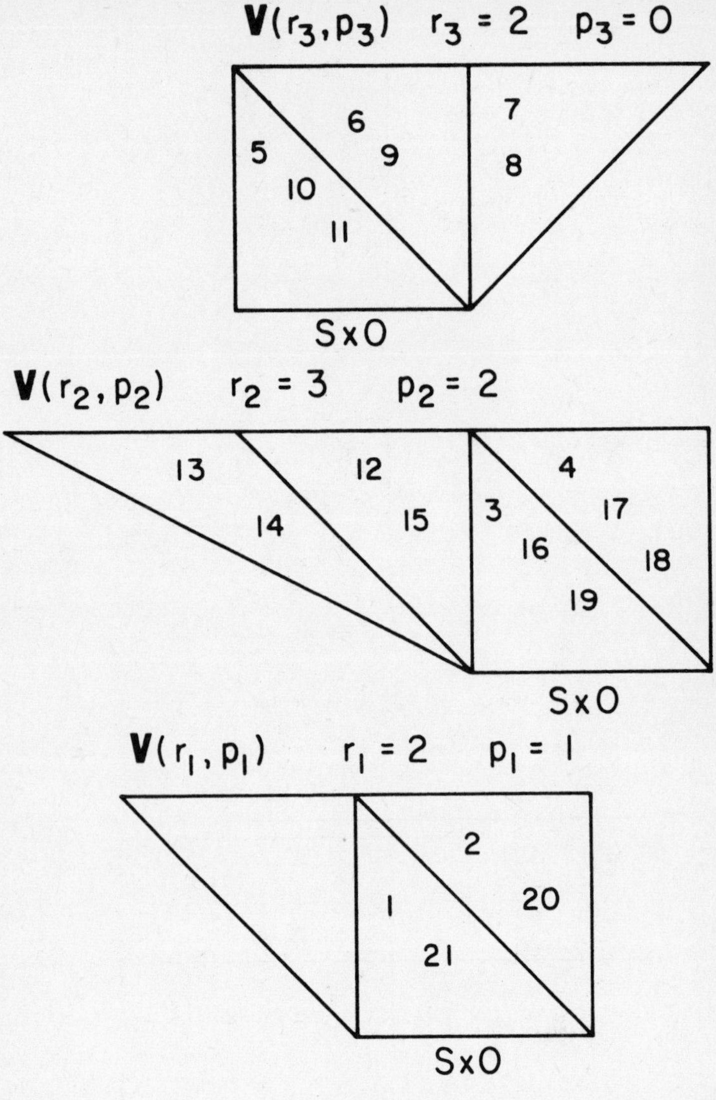
V(r3,p3) r3 = 2 p3 = 0
5
6
9
10
11
7
8
SxO
V(r2,p2) r2 = 3 p2 = 2
13
14
12
15
3
16
4
17
18
19
SxO
V(r1,p1) r1 = 2 p1 = 1
2
1
20
21
SxO

Figure 17.22

S depends upon the infinite sequence $(r_i,p^i) \in Z \times Z^n$ with $i = 1,2,\ldots$ where r_i is positive and $r_i^{-1}p^i$ is in S. Now let $(\bar{r}_i,\bar{p}_i) \in \mathbb{G} \times \mathbb{G}^n$ with $i = 0, 1, 2, \ldots$ be another infinite sequence with $\bar{r}_i$ positive. Using the second sequence we can squeeze and shear **S**, see Section 9. Of course, the representation set and replacement rules for the new subdivision of $\mathbb{G}^n \times [0,+\infty)$ are those of **S**. Clearly, in the selection of the parameters (r_i,p^i) and $(\bar{r}_i,\bar{p}^i)$ there is much latitude to accomplishing good things and bad.

17.12 Bibliographical Notes

Isomorphisms of **S** were constructed independently by Shamir [1979] and van der Laan and Talman [1980]. Neither supplied a complete proof that **S** was a triangulation, however, van der Laan and Talman did allude to the necessary line of argument. Shamir stated a full set of representative and replacement rules for **S** and conducted computational experiments with **S**. The presentations of **S** by van der Laan and Talman and Shamir appear quite different from that given here, in part, because we chose the standard n-simplex S in $\mathbb{G}^n$ rather than in $\mathbb{G}^{n+1}$. □

18. $\mathbf{S}_+$ AN AUGMENTATION OF $\mathbf{S}$

As our last major topic in the construction of triangulations we consider the triangulation $\mathbf{S}_+$ which is obtained by modifying $\mathbf{S}$ in the region $\mathbb{C}^n \times [0,1]$. $\mathbf{S}_+$ and $\mathbf{S}$ restricted to $\mathbb{C}^n \times [1,+\infty)$ agree. However, $\mathbf{S}_+$ naturally restricted to $\mathbb{C}^n \times [0,1]$ is the triangulation $\mathbf{V}$ which has been squeezed and sheared on level $\mathbb{C}^n \times 1$ to match $\mathbf{S}$ on $\mathbb{C}^n \times 1$, see Figure 18.1.

$\mathbf{S}_+$ will also be parametrized by an infinite sequence (r_i, p^i) in $\mathbb{Z} \times \mathbb{Z}^n$ with $i = 1, 2, \ldots$. The parameters r_1, $(r_2, p^2), \ldots, (r_3, p^3), \ldots$ play exactly the same role in $\mathbf{S}_+$ as they did in $\mathbf{S}$, and in particular, it is required that $r_i^{-1} p^i \in \mathbf{S}$ for $i = 2, 3, \ldots$. The integral parameter p_1 is related to the structure of $\mathbf{S}_+$ in the region $\mathbb{C}^n \times [0,1]$ or more specifically, in the region $V = \mathrm{cvx}(\mathbf{S} \times 0) \cup (\mathbb{C}^n \times 1)$. Although, it is, most likely, always best to select $r_1^{-1} p^1$ in S, it is not required.

Now let us define $\mathbf{S}_+$ formally. The natural restriction of $\mathbf{S}_+$ to $\mathbb{C}^n \times [1,+\infty)$ is defined to be $\mathbf{S}|(\mathbb{C}^n \times [1,+\infty))$, that is, the natural restriction of $\mathbf{S}$ to $\mathbb{C}^n \times [1,+\infty)$. The natural restriction of $\mathbf{S}_+$ to $\mathbb{C}^n \times [0,1]$ is defined to be $g(\mathbf{V})$ where $g : V \to V$ is $\mathbf{V}$-PL and

$$g(x,1) = \left(r_1^{-1}(x + p_1),\ 1\right) ,$$

$$g(x,0) = (x,\ 0) .$$

As shown in Section 15, $g(\mathbf{V})$ is a triangulation of V and the restriction of $g(\mathbf{V})$ to $\mathbb{C}^n \times 1$ is $r_1^{-1}\mathbf{F} \times 1$. Therefore, $g(\mathbf{V}|(\mathbb{C}^n \times 1) = \mathbf{S}|(\mathbb{C}^n \times 1)$ and $\mathbf{S}_+$ is a triangulation of $V \cup \mathbb{C}^n \times [1,+\infty)$.

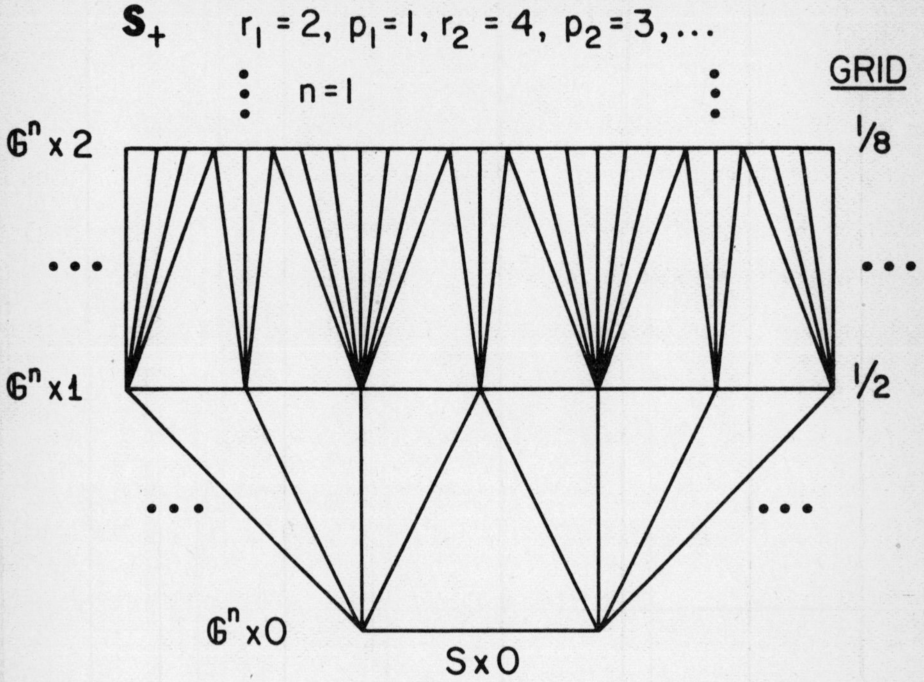

Figure 18.1

As the reader might anticipate our next step is to describe the representation rules of $\mathbf{S}_+$. We set $i\mathbf{S}_+^{n+1}$ to be the set of all $(h,u,\gamma,v,\pi,k,\eta)$ such that either a) or b)

a) $(h,u,\gamma,v,\pi,k,\eta) \in i\mathbf{S}^{n+1}$ and $h > 0$.

b) $h = 0$, $u = 0$, $\gamma = (1, \ldots, n+1)$ and $(v,\pi,k,\eta) \in i\mathbf{V}^{n+1}$.

For a) $(h;u;\gamma;v;\pi;k;\eta)$ represents that $(n+1)$-simplex in $\mathbf{S}$. For b) $(h;u;\gamma;v;\pi;k;\eta)$ represents that $(n+1)$-simplex in $g(\mathbf{V})$ with vertices $(r_1^{-1}(v^1 + p^1), 1), \ldots, (r_1^{-1}(v^{k+1} + p^1), 1), (s^{\pi(k+1)}, 0), \ldots, (s^{\pi(n+1)}, 0)$ where $(v^1,1), \ldots, (v^{k+1},1)$ and $(s^{\pi(k+1)}, 0), \ldots, (s^{\pi(n+1)}, 0)$ are the vertices of $(v;\pi;k;\eta)$ in $\mathbf{V}$. Equivalently, $(0;u;\gamma;v\pi;k;\eta)$ represents the simplex $(v;\pi;k;\eta)$ in $\mathbf{V}$ after it has been squeezed (r_1^{-1}) and sheared $(r_1^{-1}\, p_1)$ on top. The terms (u,γ) could be regarded as defunct for $h = 0$, however, to be consistent with $\mathbf{S}$ we set $u = 0$ and $\gamma = (1, \ldots, n+1)$ and we see that the vertices of $(0;u;\gamma;v;\pi;k;\eta)$ are

$$(T_{(u;\gamma;n)}(r_1^{-1}(v^i + p^1), 1) = (r_1^{-1}v^i + p^1, 1)$$

for $i = 1, \ldots, k+1$ and

$$(T_{(u;\gamma;n)}(s^{\pi(i-1)}), 0) = (u^{\pi(i-1)}, 0) = (s^{\pi(i-1)}, 0)$$

for $i = k+2, \ldots, n+2$.

As for the replacement rules, they are as $\mathbf{S}$ for $h = 1, 2, 3, \ldots$ and as $\mathbf{V}$ for $h = 0$. The only remaining consideration is to patch them together at $h = 1$. Thus there are two cases which we must consider, namely, a top exit from $g(\mathbf{V})$ into $\mathbf{S}|(\mathbb{C}^n \times [1,+\infty))$ which we call

"Ascent$_+$" and a bottom exit from $\mathbf{S}|\mathbb{C}^n \times [1,+\infty)$ into $g(\mathbf{V})$ which we call "Descent$_+$".

Ascent$_+$: Given $\rho = (0;u;\gamma;v;\pi;k;\eta)$ in $g(\mathbf{V})$ suppose we are dropping $t = n+2$ and $k = n$. These are the conditions which yield a top exit from $g(\mathbf{V})$ and we seek a simplex in $\mathbf{S}|(\mathbb{C}^n \times [1,+\infty))$ which shares $n+1$ vertices with $r_1^{-1}(v+p^1; \pi; n)$. Let $\hat{h} = 1$, $\hat{u} = v+p^1$, $\hat{\gamma} = \pi$, $\hat{v} = 0$, $\hat{\pi} = (1, 2, \ldots, n+1)$, $\hat{k} = 0$, and $\hat{\eta} = 0$. Finally, we must rerepresent $(\hat{u};\hat{\gamma};n)$ to coordinate it by $\underline{\ell}$. The $\hat{\rho} = (\hat{h};\hat{u};\hat{\gamma};\hat{v};\hat{\pi};\hat{k};\hat{\eta})$ is an $(n+1)$-simplex in $\mathbf{S}|\mathbb{C}^n \times [1,2])$ that has as a facet $r_1^{-1}(v+p^1; \pi; n) \times 1$ and the replacement is found. □

Descent$_+$: Next given $\rho = (1;u;\gamma;v;\pi;k;\eta)$ with $(1,u,\gamma,v,\pi,k,\eta)$ in $i\mathbf{S}^{n+1}$ suppose $k = 0$ and we are dropping $t = 1$. These are the conditions which yield a descent in $\mathbf{S}$ which crosses $\mathbb{C}^n \times 1$. We seek a simplex in $g(\mathbf{V})$ which shares $n+1$ vertices with ρ. Let $\hat{u} = 0$, $\hat{\gamma} = (1, \ldots, n+1)$, and $\hat{k} = n$. Next we solve $r_1^{-1}(\hat{v}+p^1; \hat{\pi}; n) = r_1^{-1}(u;\gamma;n)$ for $(\hat{v},\hat{\pi})$. Thus let $\hat{v} = u-p^1$ and $\hat{\pi} = \gamma$. Next we discern which cone Q_α contains $(\hat{v};\hat{\pi};n)$, see Figure 6.8. First rerepresent $(\hat{v};\hat{\pi};n)$ so that $\hat{\pi}(n+1) = n+1$. Let j index the lexico minimum of $(\hat{v}_i, -\hat{\pi}_i^{-1})$ for i in ν. If $v_j \geqq 0$ set $\eta = (\nu,0)$. If $v_j < 0$ rerepresent $(\hat{v};\hat{\pi};n)$ with $\hat{\pi}(n+1) = j$ and set $\eta = (\nu-e\nu_j, -\nu_j)$. The $(n+1)$-simplex $(\hat{v};\hat{\pi};\hat{k};\hat{\eta})$ is in $g(\mathbf{V})$ it shares the face $r_1^{-1}((\hat{v}_j;\hat{\pi};n) + p_1)$ with ρ. Thus the desired representation in $i\mathbf{S}_+^{n+1}$ is $(\hat{h};\hat{u};\hat{\gamma};\hat{v};\hat{\pi};\hat{k};\hat{\eta})$ and the replacement is complete. □

A temporary chart for $\mathbf{S}_+$ is given in the following figures. Observe that Ascent$_+$ is executed properly by Ascent and that Descent$_+$ differs only a little from Descent.

18.1 Note (for Figure 18.3): Ascent replacement crossing $\mathbb{C}^n \times 1$. The bottom vertices of $\hat{\rho}$ are those of $r_1^{-1}(\hat{u};\hat{\gamma};n)$ in order. The top vertices of ρ are those of $r_1^{-1}(r+p^1;\ \pi;\ n)$ in order. So $(\hat{u};\hat{\gamma};n) = (r+p^1;\pi;n)$. Let $z = \underline{\ell}(\hat{r}+\rho^1) - \underline{\ell}(\hat{u}) = \underline{\ell}(\hat{v}+p^1)$ and we have $\hat{u}^{i+z} = v^i+p^1$ for $i \in u$. Thus the correspondence between vertices of ρ and $\hat{\rho}$ is as given in Note 17.9 for $\mathbf{S}$. The adjoined vertex is

$$\hat{W}^1 = \left(r_1^{-1}\ T_{(\hat{u};\hat{\gamma};n)}(r_2^{-1}(\hat{v}+p^2)),\ 2\right)$$

which is that of Note 17.9 for $\mathbf{S}$ with $h = 0$. □

18.2 Note (for Figure 18.4): Descent replacement crossing

$\mathbb{C}^n \times 1$. The bottom vertices of ρ are those of $r_1^{-1}(u;\gamma;n)$ in order. The top vertices of $\hat{\rho}$ are those of $r_1^{-1}(\hat{v}+p^1;\hat{\pi};n)$. Thus $(u;\gamma;n) = (\hat{v}+p^1;\hat{\pi};n)$. Let $z = \underline{\ell}(\hat{v}+p^1) - \underline{\ell}(u) = \underline{\ell}(\bar{v}+p^1)$ and we have $u^{i+z} = \hat{v}^i+p^i$ for $i \in \mu$ and the correspondence between vertices of ρ and $\hat{\rho}$ is as given in Note 17.10 for $\mathbf{S}$. The adjoined vertex is $\hat{W}^{n+2} = (s^{\pi(n+1)},\ 0)$. □

Using the charts for $\mathbf{S}_+$ and $\mathbf{S}$ in Figures 18.2 to 18.4 and 17.10 to 17.18 we construct a more compact version for $\mathbf{S}_+$ in Figures 18.5 to 18.13.

In Figures 18.14 to 18.16 a version of $\mathbf{S}_+$ is exhibited and a path is followed.

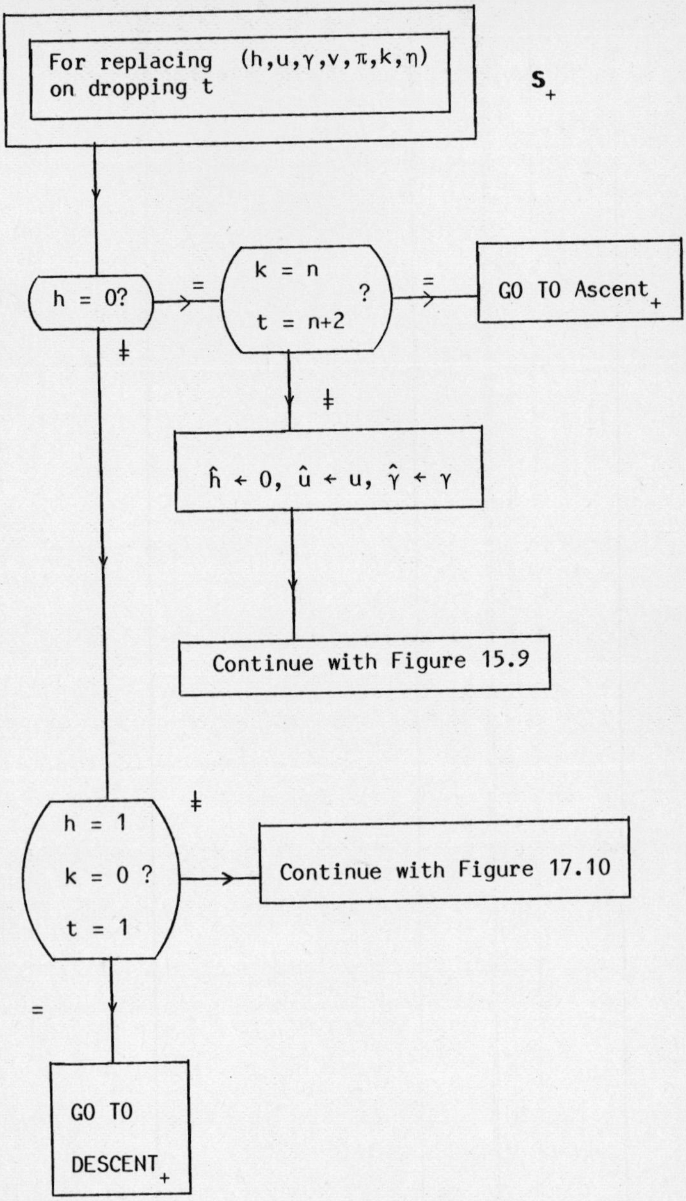
For replacing (h,u,γ,v,π,k,η) on dropping t
S+
h = 0?
=
k = n
?
t = n+2
=
GO TO Ascent+
≠
≠
ĥ ← 0, û ← u, γ̂ ← γ
Continue with Figure 15.9
≠
h = 1
k = 0 ?
t = 1
Continue with Figure 17.10
=
GO TO
DESCENT+

Figure 18.2

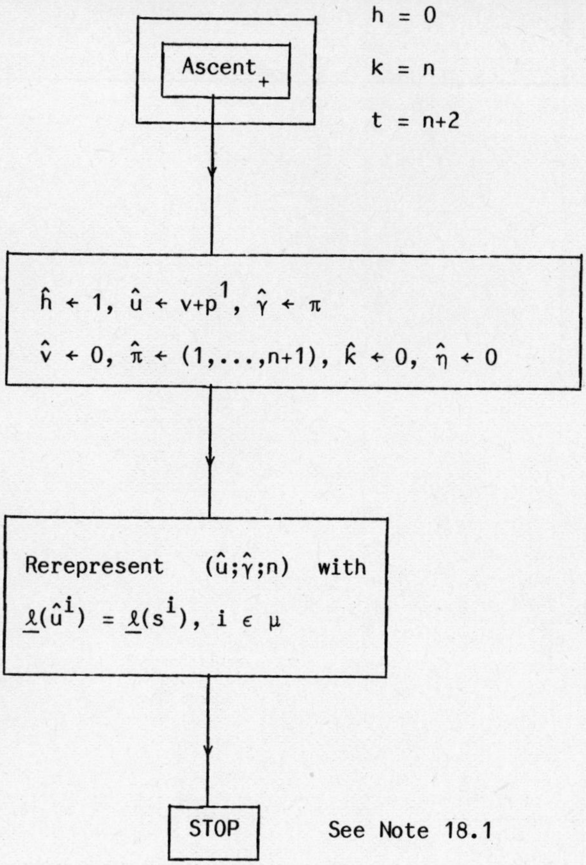
Ascent$_+$
h = 0
k = n
t = n+2
$\hat{h} \leftarrow 1,\ \hat{u} \leftarrow v+p^1,\ \hat{\gamma} \leftarrow \pi$
$\hat{v} \leftarrow 0,\ \hat{\pi} \leftarrow (1,\ldots,n+1),\ \hat{k} \leftarrow 0,\ \hat{\eta} \leftarrow 0$
Rerepresent $(\hat{u};\hat{\gamma};n)$ with
$\underline{\ell}(\hat{u}^i) = \underline{\ell}(s^i),\ i \in \mu$
STOP
See Note 18.1

Figure 18.3

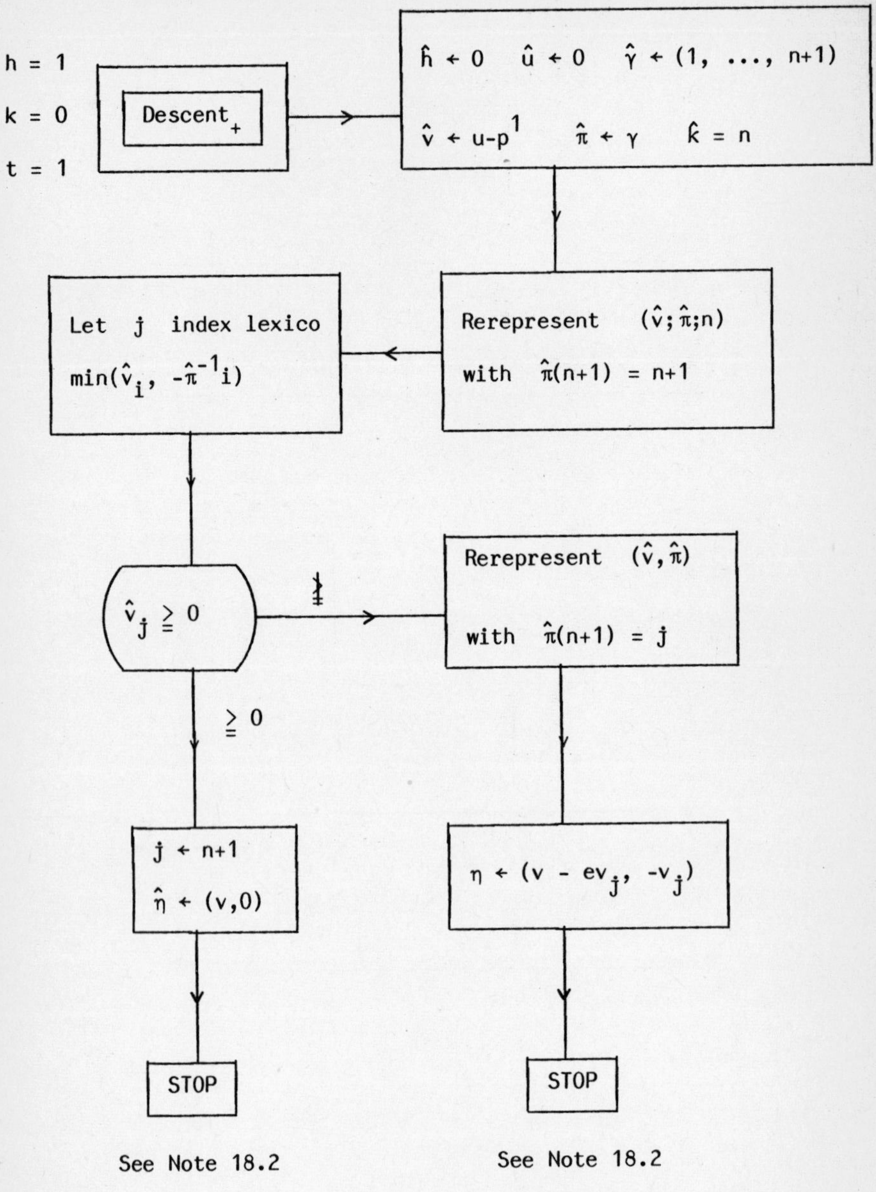
h = 1
k = 0
t = 1
Descent+
ĥ ← 0 û ← 0 γ̂ ← (1, ..., n+1)
v̂ ← u-p1 π̂ ← γ k̂ = n
Rerepresent (v̂; π̂;n)
with π̂(n+1) = n+1
Let j index lexico
min(v̂i, -π̂-1i)
v̂j ≧ 0
≱
Rerepresent (v̂, π̂)
with π̂(n+1) = j
≧ 0
j ← n+1
η̂ ← (v,0)
η ← (v - evj, -vj)
STOP
STOP
See Note 18.2
See Note 18.2

Figure 18.4

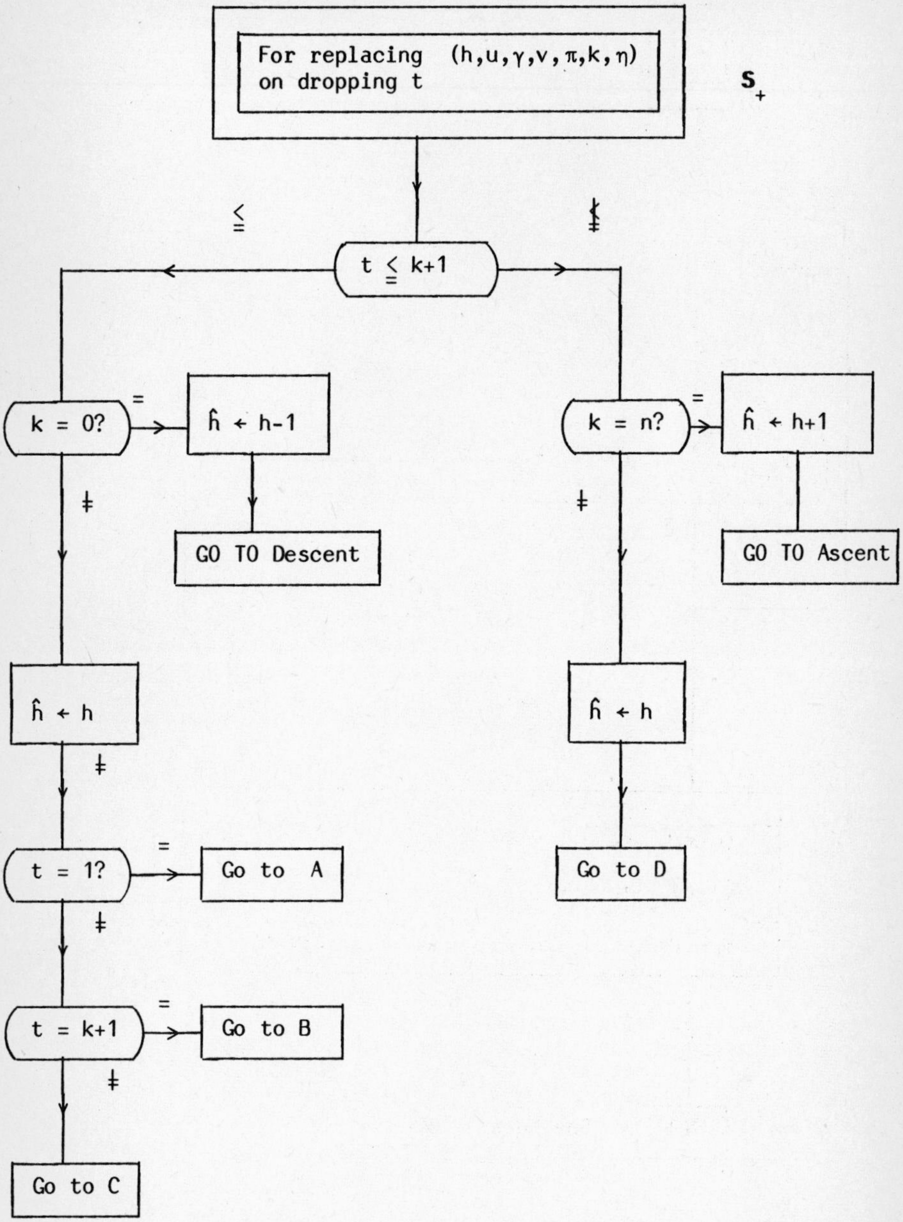
For replacing (h,u,γ,v,π,k,η) on dropping t
S+
≦
≰
t ≦ k+1
k = 0?
=
ĥ ← h-1
≠
GO TO Descent
k = n?
=
ĥ ← h+1
≠
GO TO Ascent
ĥ ← h
≠
t = 1?
=
Go to A
≠
t = k+1
=
Go to B
≠
Go to C
ĥ ← h
Go to D

Figure 18.5

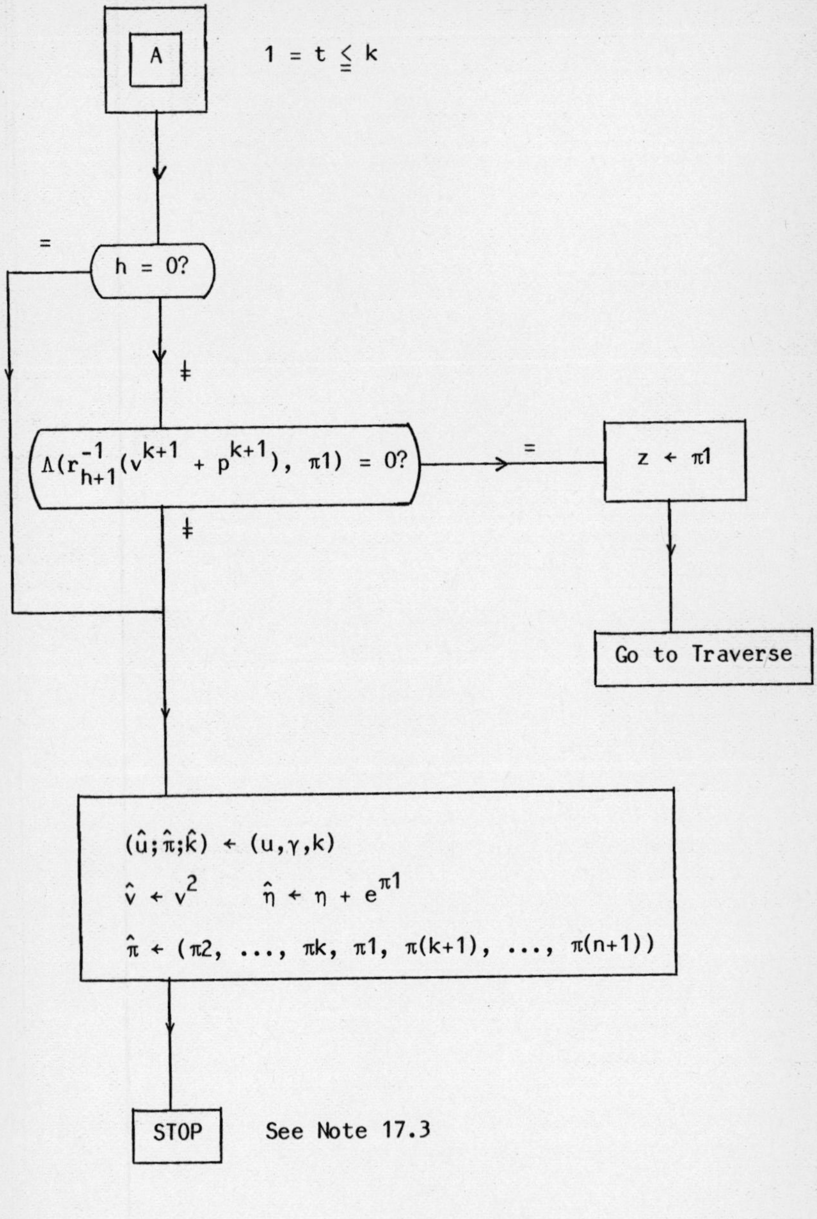
A
1 = t ≦ k
h = 0?
=
≠
Λ(r$^{-1}_{h+1}$(v^{k+1} + p^{k+1}), π1) = 0?
=
z ← π1
Go to Traverse
≠
(û;π̂;k̂) ← (u,γ,k)
v̂ ← v^{2}
η̂ ← η + e^{π1}
π̂ ← (π2, ..., πk, π1, π(k+1), ..., π(n+1))
STOP
See Note 17.3

Figure 18.6

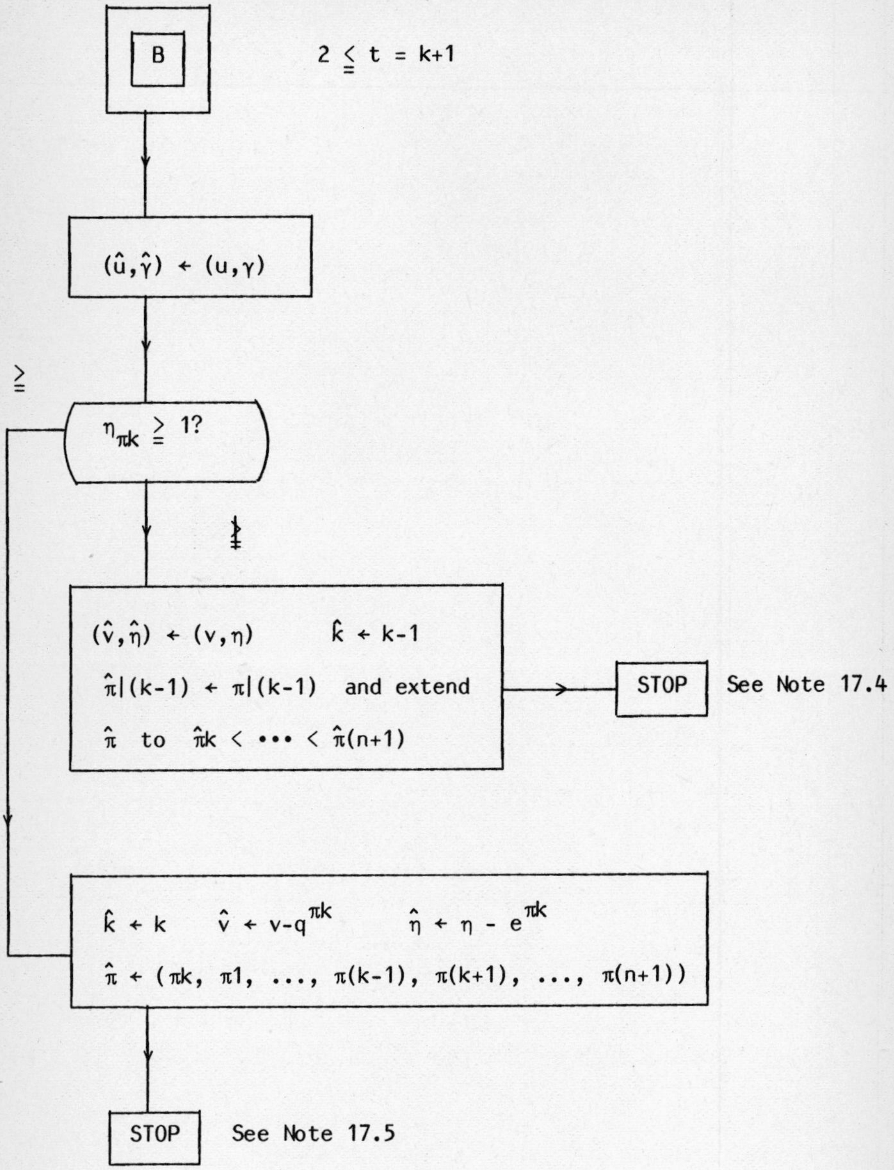
B
2 ≦ t = k+1
(û, γ̂) ← (u, γ)
≧
η_πk ≧ 1?
≱
(v̂, η̂) ← (v, η)   k̂ ← k-1
π̂|(k-1) ← π|(k-1) and extend
π̂ to π̂k < ••• < π̂(n+1)
STOP
See Note 17.4
k̂ ← k   v̂ ← v-q^πk   η̂ ← η - e^πk
π̂ ← (πk, π1, ..., π(k-1), π(k+1), ..., π(n+1))
STOP
See Note 17.5

Figure 18.7

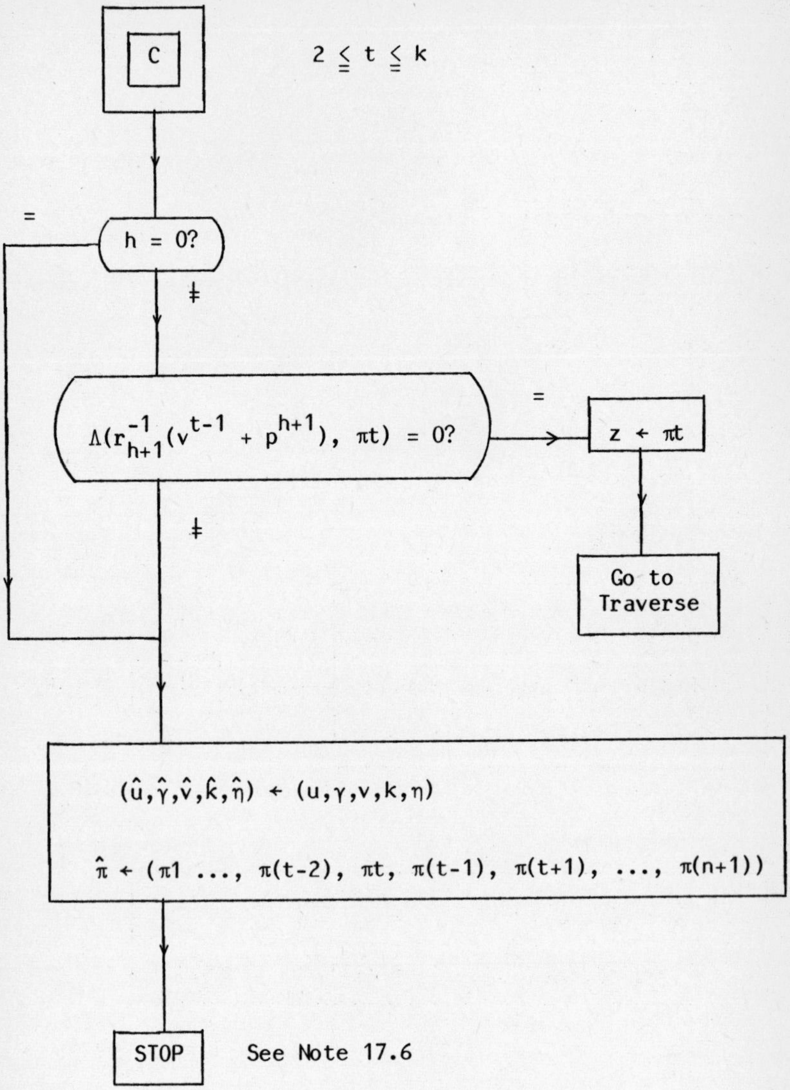
C
2 ≦ t ≦ k
=
h = 0?
≠
=
Λ(r^{-1}_{h+1}(v^{t-1} + p^{h+1}), πt) = 0?
z ← πt
≠
Go to Traverse
(û, γ̂, v̂, k̂, η̂) ← (u, γ, v, k, η)
π̂ ← (π1 ..., π(t-2), πt, π(t-1), π(t+1), ..., π(n+1))
STOP
See Note 17.6

Figure 18.8

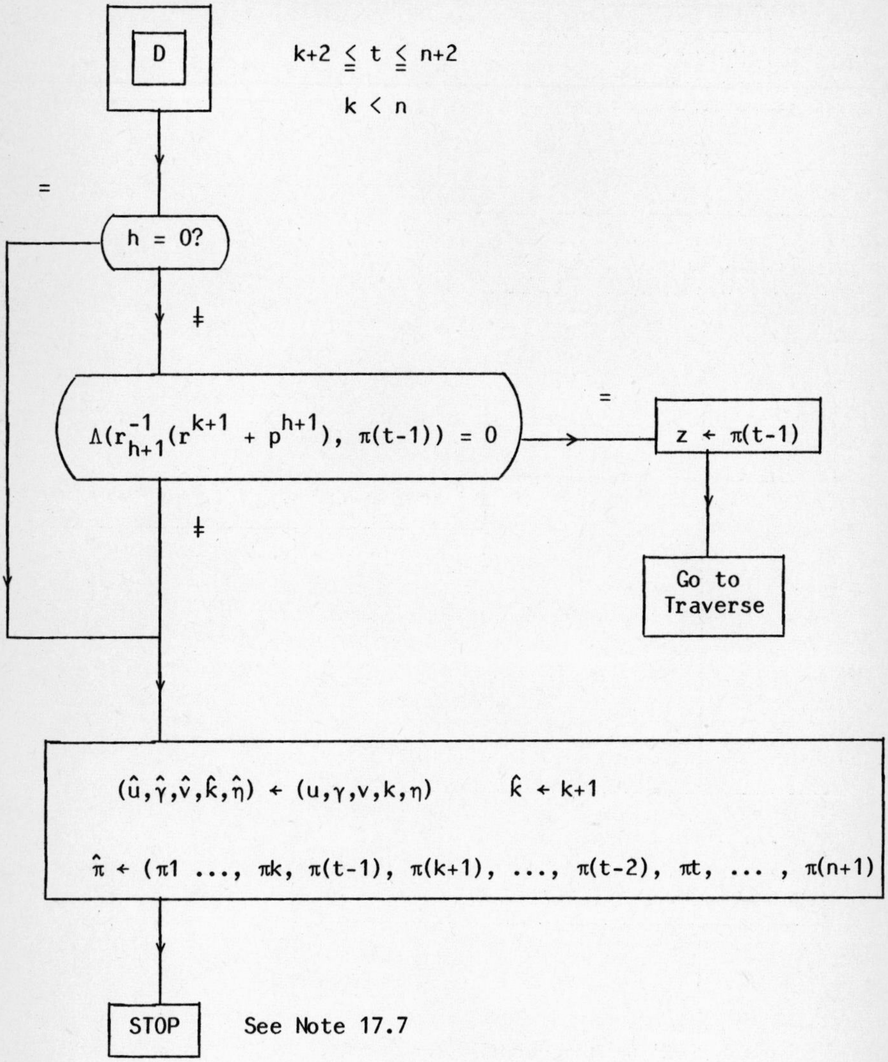
D
k+2 ≦ t ≦ n+2
k < n
=
h = 0?
≠
Λ(r⁻¹ₕ₊₁(rᵏ⁺¹ + pʰ⁺¹), π(t-1)) = 0
=
z ← π(t-1)
≠
Go to Traverse
(û,γ̂,v̂,k̂,η̂) ← (u,γ,v,k,η)   k̂ ← k+1
π̂ ← (π1 ..., πk, π(t-1), π(k+1), ..., π(t-2), πt, ... , π(n+1)
STOP
See Note 17.7

Figure 18.9

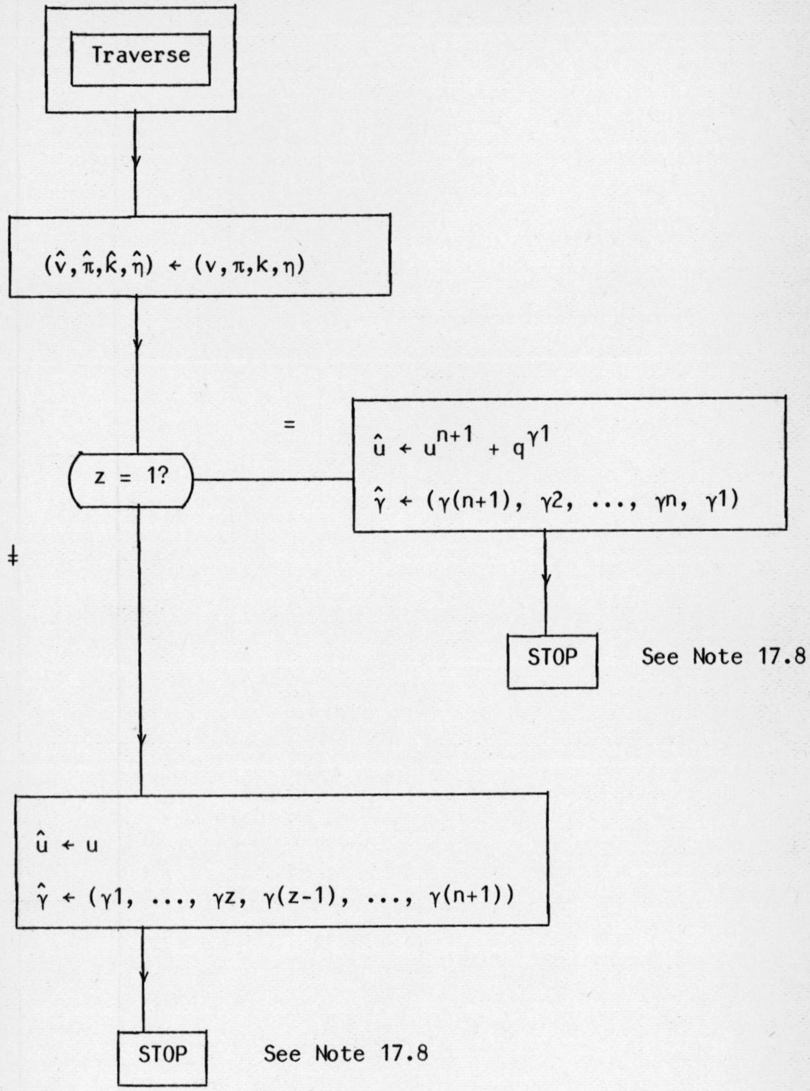
Traverse
(v̂, π̂, k̂, η̂) ← (v, π, k, η)
z = 1?
=
û ← u^{n+1} + q^{γ1}
γ̂ ← (γ(n+1), γ2, ..., γn, γ1)
STOP
See Note 17.8
≠
û ← u
γ̂ ← (γ1, ..., γz, γ(z-1), ..., γ(n+1))
STOP
See Note 17.8

Figure 18.10

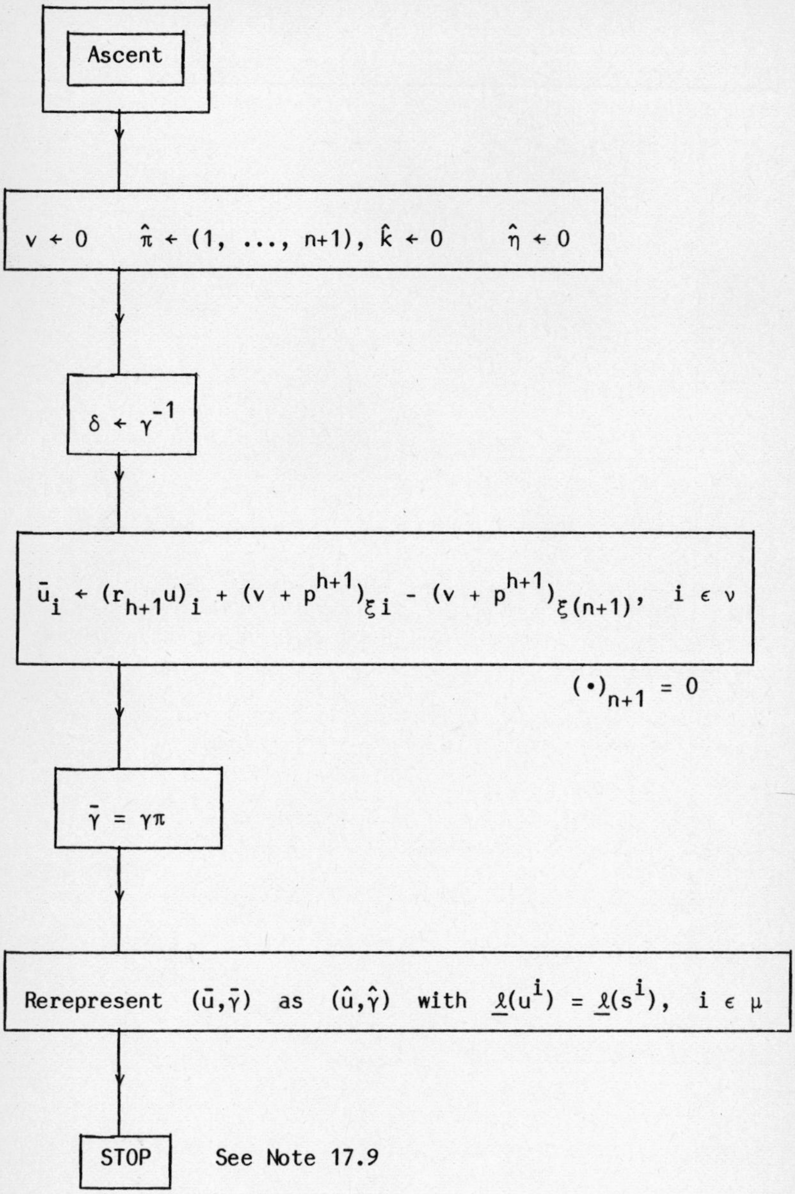
Ascent
$v \leftarrow 0 \quad \hat{\pi} \leftarrow (1, \ldots, n+1), \ \hat{k} \leftarrow 0 \quad \hat{\eta} \leftarrow 0$
$\delta \leftarrow \gamma^{-1}$
$\bar{u}_i \leftarrow (r_{h+1}u)_i + (v + p^{h+1})_{\xi i} - (v + p^{h+1})_{\xi(n+1)}, \quad i \in \nu$
$(\cdot)_{n+1} = 0$
$\bar{\gamma} = \gamma\pi$
Rerepresent $(\bar{u},\bar{\gamma})$ as $(\hat{u},\hat{\gamma})$ with $\underline{\ell}(u^i) = \underline{\ell}(s^i), \quad i \in \mu$
STOP
See Note 17.9

Figure 18.11

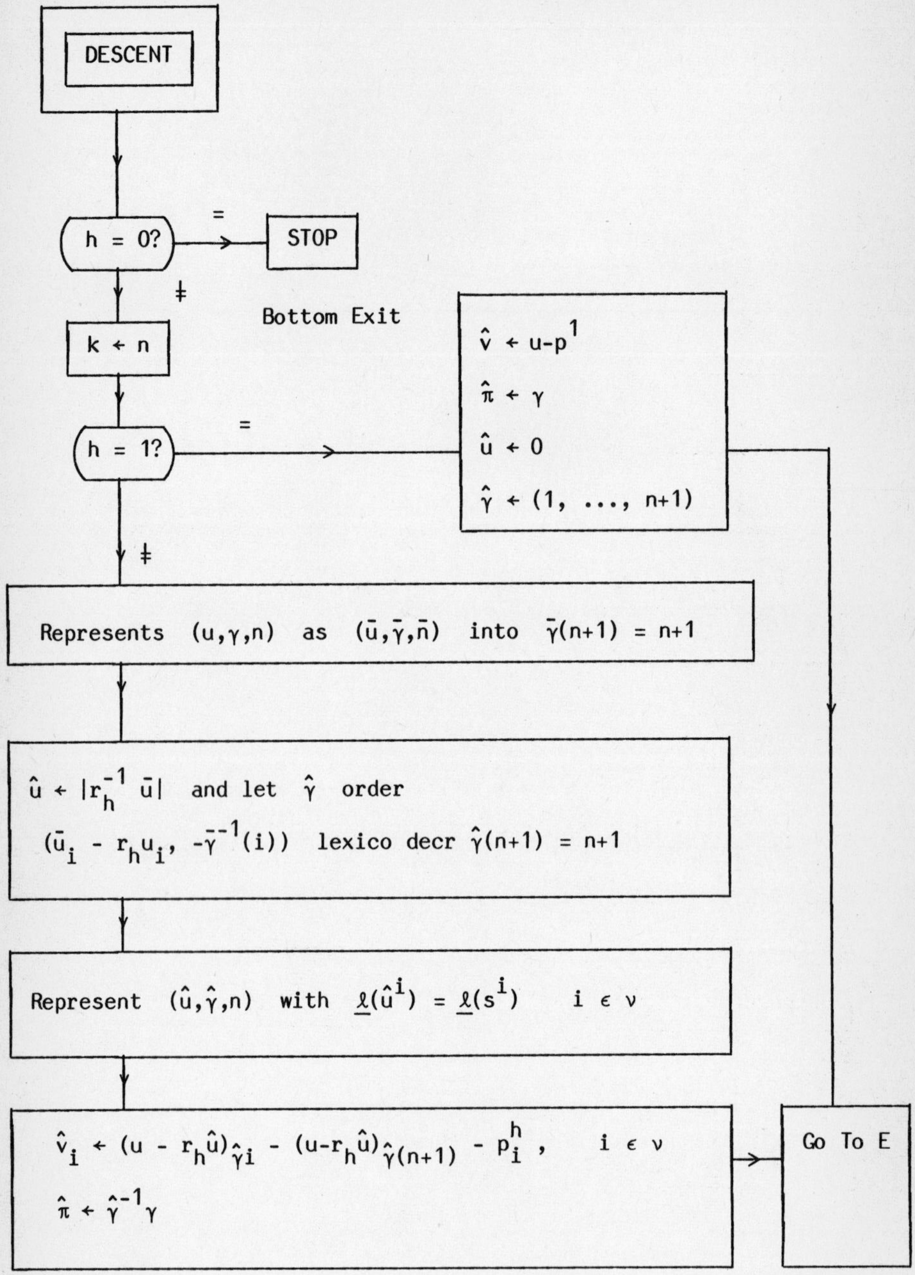
DESCENT
h = 0?
=
STOP
≠
Bottom Exit
k ← n
h = 1?
=
v̂ ← u-p^1
π̂ ← γ
û ← 0
γ̂ ← (1, ..., n+1)
≠
Represents (u,γ,n) as (ū,γ̄,n̄) into γ̄(n+1) = n+1
û ← |r_h^-1 ū| and let γ̂ order
(ū_i - r_h u_i, -γ̄^-1(i)) lexico decr γ̂(n+1) = n+1
Represent (û,γ̂,n) with ℓ(û^i) = ℓ(s^i) i ∈ ν
v̂_i ← (u - r_h û)_γ̂i - (u-r_h û)_γ̂(n+1) - p_i^h , i ∈ ν
π̂ ← γ̂^-1 γ
Go To E

Figure 18.12

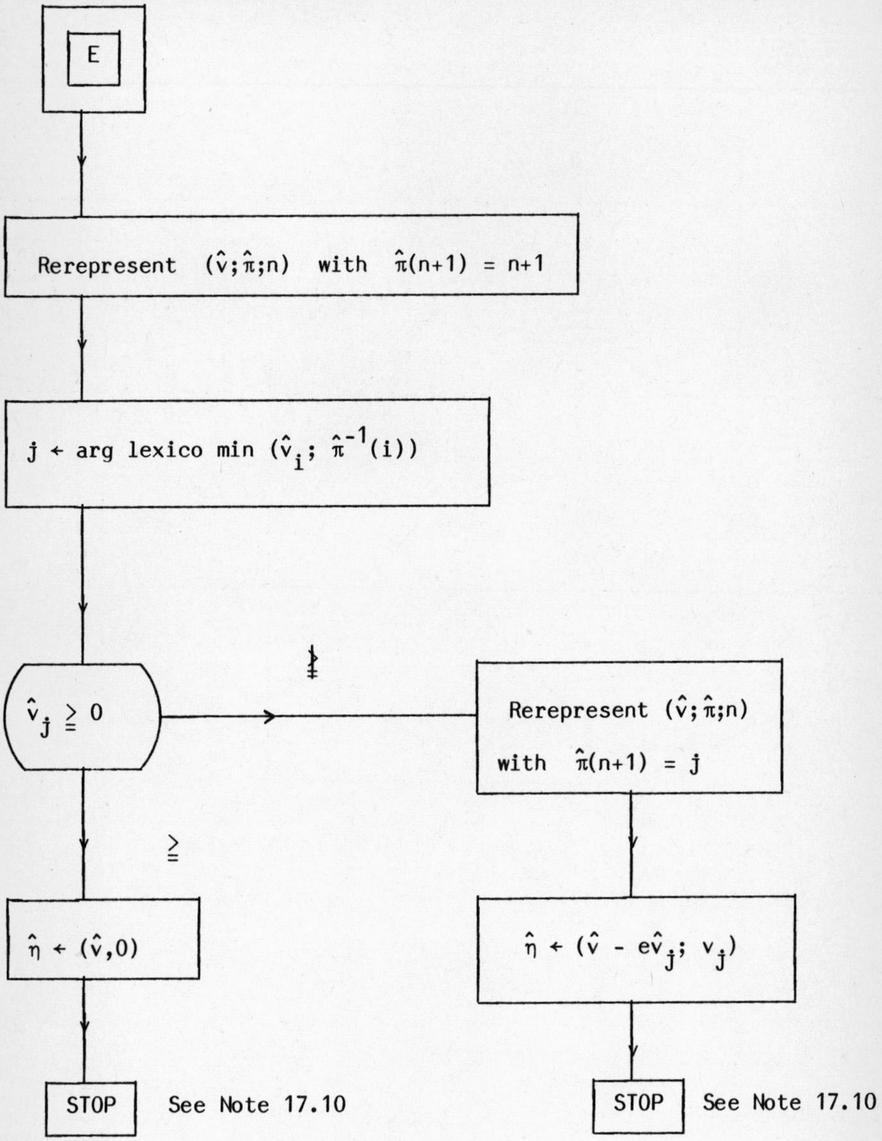
E
Rerepresent (v̂;π̂;n) with π̂(n+1) = n+1
j ← arg lexico min (v̂_i; π̂^-1(i))
v̂_j ≧ 0
≱
Rerepresent (v̂;π̂;n)
with π̂(n+1) = j
≧
η̂ ← (v̂,0)
η̂ ← (v̂ - ev̂_j; v_j)
STOP
See Note 17.10
STOP
See Note 17.10

Figure 18.13

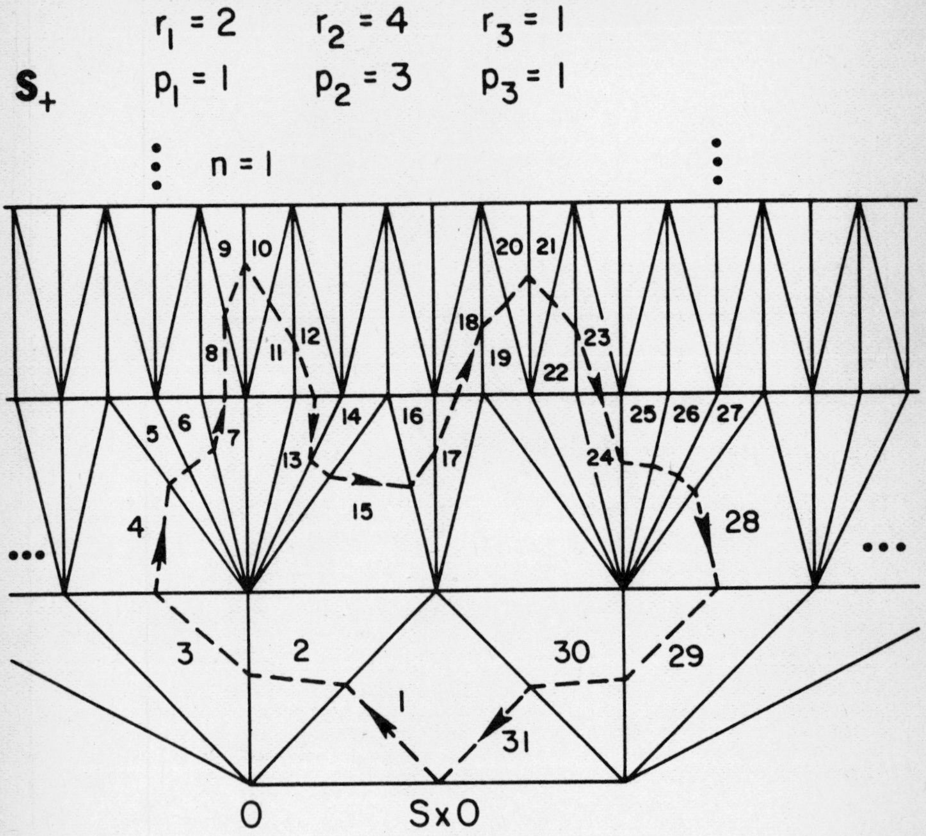

$r_1 = 2$
$r_2 = 4$
$r_3 = 1$
S_+
$p_1 = 1$
$p_2 = 3$
$p_3 = 1$
n = 1
1
2
3
4
5
6
7
8
9
10
11
12
13
14
15
16
17
18
19
20
21
22
23
24
25
26
27
28
29
30
31
0
SxO

Figure 18.14

	i	h	u	γ	ν	π	k	η	t
START	1	0	0	1 2	0	1 2	0	0 0	3
Internal to	2	0	0	1 2	0	2 1	1	0 0	1
Traverse to	3	0	0	1 2	-1	2 1	1	0 1	3
Ascend to	4	1	0	2 1	0	1 2	0	0 0	3
Internal to	5	1	0	2 1	0	2 1	1	0 0	1
Internal to	6	1	0	2 1	-1	2 1	1	0 1	1
Internal to	7	1	0	2 1	-2	2 1	1	0 2	3
Ascend to	8	2	0	2 1	0	1 2	0	0 0	3
Internal to	9	2	0	2 1	0	2 1	1	0 0	1
Traverse to	10	2	0	1 2	0	2 1	1	0 0	2
Internal to	11	2	0	1 2	0	1 2	0	0 0	2
Traverse to	12	2	2	2 1	0	1 2	0	0 0	1
Descend to	13	1	0	1 2	-1	2 2	2	0 1	2
Internal to	14	1	0	1 2	0	2 1	1	0 0	2
Internal to	15	1	0	1 2	0	1 2	0	0 0	2
Internal to	16	1	0	1 2	0	1 2	1	0 0	1
Traverse to	17	1	2	2 1	0	1 2	1	0 0	3
Ascend to	18	2	4	1 2	0	1 2	0	0 0	2
Traverse to	19	2	6	2 1	0	1 2	0	0 0	3
Internal to	20	2	6	2 1	0	2 1	1	0 0	1
Traverse to	21	2	6	1 2	0	2 1	1	0 0	2
Internal to	22	2	6	1 2	0	1 2	0	0 0	2
Traverse to	23	2	8	2 1	0	1 2	0	0 0	1
Descend to	24	1	2	2 1	-2	2 1	1	0 2	1
Traverse to	25	1	2	1 2	-2	2 1	1	0 2	2
Inernal to	26	1	2	1 2	-1	2 1	1	0 1	2
Internal to	27	1	2	1 2	0	2 1	1	0 0	2
Interval to	28	1	2	1 2	0	1 2	0	0 0	1
Descend to	29	0	0	1 2	1	1 2	1	1 0	2
Traverse to	30	0	0	1 2	0	1 2	1	0 0	2
Internal to	31	0	0	1 2	0	1 2	0	0 0	1

Figure 18.15

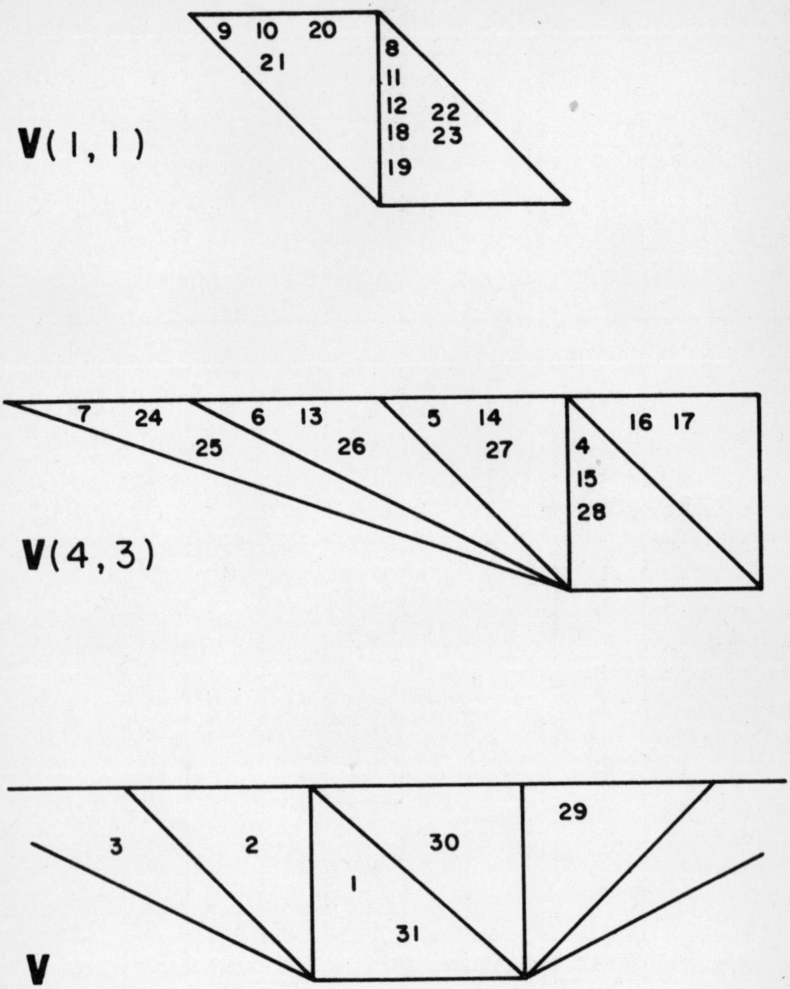
V(1,1)
9 10 20
21
8
11
12 22
18 23
19
V(4,3)
7 24
25
6 13
26
5 14
27
4
15
28
16 17
V
3
2
1
30
31
29

Figure 18.16

18.3 Exercise: Select (r_i, p^i) for $i = 1, 2, 3$ and draw a portion of $\mathbf{S}_+$. Draw a path in $\mathbf{S}_+$ and use the charts to generate the representatives along the path as in the previous figures. □

The proper selection of the parameters (r_i, p^i) is quite an interesting, involved, and unresolved one, and one which will not be discussed in this edition of this manuscript.

As a final remark, to take best advantage of $\mathbf{S}_+$ one must be willing to work with isomorphisms, for example, the squeeze and shear, of it.

The replacement rules for $\mathbf{S}_+$ have been programmed for a computer and their validity verified; such testing is conducted by repeatedly generating long sequences of random replacements, reversing the replacements, and checking that the starting and ending simplexes are, in fact, one and the same.

References

Allgower, E.L., and K. Georg, Simplicial and Continuation Methods for Approximating Fixed Points and Solutions to Systems of Equations, **SIAM Review 22** (1980) 28-85.

Bárány, I., Subdivisions and Triangulations in Fixed Point Algorithms, International Research Institute for Management Science, ul. Ryleeva, Moskou, U.S.S.R. (1979).

Broadie, M.N., and B.C. Eaves, A Variable Rate Refining Triangulations for PL Homotopies, Department of Operations Research, Stanford University, in progress.

Cohen, D.I.A., On the Sperner Lemma, **Journal of Combinatorial Theory 2** (1967) 585-587.

Cottle, R.W., Nonlinear Programs with Positively Bounded Jacobians, Ph.D. Dissertation, Department of Mathematics, University of California, Berkeley (1964).

Cottle, R.W., Minimal Triangulation of the 4-Cube, **Discrete Mathematics 40** (1982) 25-29.

Davidenko, D., On the Approximate Solution of a System of Nonlinear Equations, **Ukraine Mat Zurnal 5** (1953) 196-206.

Eaves, B.C., An Odd Theorem, **Proceedings of the American Mathematical Society 26** (1970) 509-513.

Eaves, B.C., On the Basic Theorem of Complementarity, **Mathematical Programming 1** (1971a) 68-75.

Eaves, B.C., Computing Kakutani Fixed Points, **SIAM Journal of Applied Mathematics 21** (1971b) 236-244.

Eaves, B.C., The Linear Complementarity Problem, **Management Science 17** 9 (1971c) 612-634.

Eaves, B.C., Homotopies for the Computation of Fixed Points, **Mathematical Programming 3** (1972) 1-22.

Eaves, B.C., A Short Course in Solving Equations with PL Homotopies, (eds. Cottle, R.W., and Lemke, C.E.) **Nonlinear Programming, SIAM-AMS Proceedings,** Volume 9, American Mathematical Society, Providence, Rhode Island (1976) 73-143.

Eaves, B.C., Permutation Congruent Transformations of the Freudenthal Triangulation with Minimal Surface Density, Department of Operations Research, Stanford University, (March 1982). **Mathematical Programming 29** (1984) 77-99.

Eaves, B.C., Subdivisions from Primal and Dual Cones and Polytopes, **Linear Algebra,** to appear.

Eaves, B.C., and M.N. Broadie, A Variable Rate Refining Triangulations for PL Homotopies, Department of Operations Research, Stanford University, in progress. (1983)

Eaves, B.C., and U. Rothblum, Homotopies for Solving Equations in Ordered Fields with Application to an Invariant Curve Theorem, in progress.

Eaves, B.C., and R. Saigal, Homotopies for Computation of Fixed Points on Unbounded Regions, **Mathematical Programming 3** (1972) 225-237.

Eaves, B.C., and H. Scarf, The Solution of Systems of Piecewise Linear Equations, **Mathematics of Operations Research 1** (1976) 1-27.

Eaves, B.C., and J.A. Yorke, Equivalence of Surface Density and Average Directional Density, Department of Operations Research, Stanford University, (January 1982). **Mathematics of Operations Research,** (to appear).

Engles, C., Economic Equilibrium Under Deformation of the Economy, Ph.D. Thesis, Department of Operations Research, Stanford University, (October 1978).

Freudenthal, H., Simplizialzerlegungen von Beschrankter Flachheit, **Annals of Mathematics 43** (1942) 580-582.

Freund, R.M., Variable-Dimension Complexes with Applications, Ph.D. Thesis, Department of Operations Research, Stanford University, (June 1980).

Garcia, C.B., Soime Classes of Matrices in Linear Complementarity Theory, **Mathematical Programming 5** 3 (1973) 299-310.

Garcia, C.B., and W.I. Zangwill, A Flex Simplicial Algorithm, **Numerical Solution of Highly Nonlinear Problems,** ed. W. Forster, Amsterdam: North Holland, (1980).

Garcia, C.B., and W.I. Zangwill, **Pathways to Solutions, Fixed Points, and Equilibria,** Prentice-Hall Series in Computational Mathematics, Cleve Moler, Advisor, Prentice-Hall, Inc., New Jersey, (1981).

Gould, F.J., and J.W. Tolle, A Unified Approach to Complementarity in Optimization, **Discrete Mathematics 7** (1974) 225-271.

Grünbaum, B., **Convex Polytopes,** Wiley, New York (1967).

Hansen, T., On the Approximation of a Competitive Equilibrium, Ph.D. Thesis, Department of Economics, Yale University (1968).

van der Heyden, L., Restricted Primitive Sets in a Regularly Distributed List of Vectors and Simplicial Subdivisions with Arbitrary Refinement Factors, Disc. Paper Ser. No. 79D, John Fitzgerald Kennedy School of Government, Harvard University (January 1980).

Hirsch, M.W., A Proof of the Nonretractability of a Cell Onto Its Boundary, **Proceedings of the American Mathematical Society 14** (1963) 364-365.

Hirsch, M.W., On Algorithms for Solving f(x) = 0, **Communication on Pure and Applied Mathematics 32** (1979) 281-312.

Hirsch, M.W., and S. Smale, **Differential Equations, Dynamical Systems, and Linear Algebra.** Academic Press, New York (1974).

Kellogg, R.B., T.Y. Li, and J. Yorke, A Constructive Proof of the Brouwer Fixed-Point Theorem and Computational Results, **SIAM Journal of Numerical Analysis 13** (1976) 473-483.

Kojima, M., On the Homotopic Approach to Systems of Equations with Separable Mappings, **Mathematical Programming Study 7,** eds. M.L. Balinski and R.W. Cottle, North-Holland, Amsterdam (1978) 170-184.

Kojima, M., A Modification of Todd's Triangulation J3, **Mathematical Programming 15** (1978) 223-227.

Kojima, M., An Introduction to Variable Dimension Algorithms for Solving Systems of Equations, Research Reports on Information Sciences B-88, Department of Information Sciences, Tokyo Institute of Technology, Tokyo, (October 1980).

Kojima, M., and Y. Yamamoto, Variable Dimension Algorithms, Part I: Basic Theory, Report No. B-77, Tokyo Institute of Technology, Tokyo (1979).

Kojima, M., and Y. Yamamoto, Variable Dimension Algorithms, Part II: Some New Algorithms and Triangulations with Continuous Refinement of Mesh Size, Report No. B-82, Tokyo Institute of Technology, Tokyo (1980).

Kojima, M., and Y. Yamamoto, Variable Dimension Algorithms: Basic Theory, Interpretations and Extensions of Some existing Methods, **Mathematical Programming 24** (1982a) 177-215.

Kojima, M., and Y. Yamamoto, A Unified Approach to Several Restart Fixed Point Algorithms for Their Implementation and a New Variable Dimension Algorithm, Disc. Paper Ser. No. 151(82-18), Institute of Socio-Economic Planning, University of Tsukuba, Sakura, Ibaraki 305, Japan (April 1982b).

Kuhn, H.W., Simplicial Approximation of Fixed Points, **Proceedings of National Academy of Science, U.S.A. 61** (1968) 1238-1242.

Kuhn, H.W., and J.G. MacKinnon, The Sandwich Method for Finding Fixed Points, **J. Optimization Theory and Applications 17** (1975).

van der Laan, G., and A.J.J. Talman, A Restart Algorithm for Computing Fixed Points Without an Extra Dimension, **Mathematical Programming 17** (1979) 74-84.

van der Laan, G., and A.J.J. Talman, An Improvement of Fixed Point Algorithms by Using a Good Triangulation, **Mathematical Programming 18** (1980a) 274-285.

van der Laan, G., and A.J.J. Talman, A New Subdivision for Computing Fixed Points with a Homotopy Algorithm, **Mathematical Programming 19** (1980b) 78-91.

van der Laan, G., and A.J.J. Talman, **Simplicial Fixed Point Algorithms,** Mathematical Centre Tracts 129, Mathematisch Centrum, Amsterdam (1980c).

van der Laan, G., and A.J.J. Talman, Interpretation of a Variable Dimension Fixed Point Algorithm with an Artificial Level, Department of Actuarial Sciences and Econometrics, Free University, Amsterdam, (October 1979, revised January 1981).

van der Laan, G., and A.J.J. Talman, A Class of Simplicial Restart Fixed Point Algorithms Without an Extra Dimension, **Mathematical Programming 20** (1981) 33-48.

Lemke, C.E., and J.T. Howson, Equilibrium Points of Bimatrix Games, **SIAM Review 12** (1964) 413-423.

Lemke, C.E., Bimatrix Equilibrium Points and Mathematical Programming, **Management Science 11** (1965) 681-689.

Luthi, H-J, Komplementatitats- und Fixpunktalgorithmen in der mathematischen Programmierung, Spieltheorie und Okonomie, **Lecture Notes in Economics and Mathematical Systems, No. 129,** Springer-Verlag, Berlin-Heidelberg-New York, (1976).

Merrill, O.H., Applications and Extensions of an Algorithm that Computes Fixed Points of Certain Upper Semi-Continuous Point to Set Mappings, Ph.D. Thesis, Department of Industrial Engineering, University of Michigan, (1972).

Poincaré, Sur les Courbes définies par une équation differéntielle, **IV, J. Math. Pures Appl. 85** (1886) 151-217.

Peitgen, H.O., **Approximation of Fixed Points and Functional Differential Equations,** Springer-Verlag, New York (1979).

Rourke, C.P., and B.J. Sanderson, **Introduction to Piecewise-Linear Topology,** Springer-Verlag, New York-Heidelberg-Berlin (1972).

Saigal, R., Investigations into the Efficiency of Fixed Point Algorithms, **Fixed Points: Algorithms and Applications,** Eds: S. Karamardian, Academic Press (1977a) 203-223.

Saigal, R., On the Convergence Rate of Algorithms for Solving Equations that are Based on Methods of complementarity Pivoting, **Mathematics of Operations Research 2** (1977b) 108-124.

Saigal, R., and M.J. Todd, Efficient Acceleration Techniques for Fixed Point Algorithms, **SIAM J. Numerical Analysis 15** (1978) 997-1007.

Saigal, R., On Piecewise Linear Approximations to Smooth Mappings, **Mathematics of Operations Research 2** (1979) 153-161.

Saigal, R. An Efficient Procedure for Traversing Large Pieces in Fixed Point Algorithms, in **Homotopy Methods and Global Convergene,** Eds: B.C. Eaves, F.J. Gould, H.O. Peitgen, and M.J. Todd), Plenum Press (1983) 239-248.

Saigal, R., A Homotpoy for Solving Large, Sparse and Structured Fixed Point Problems, **Mathematics of Operations Research** (to appear).

Scarf, H., The Approximation of Fixed Points of a Continuous Mapping, **SIAM J. Applied Mathematics 15** 5 (1967a) 1328-1343.

Scarf, H., The Core of an N Person Game, **Econometrica 35** 1 (1967b) 50-69.

Scarf, H., **The Computation of Economic Equilibria,** Yale University Press, New Haven and London (1973).

Shamir, S., Fixed-Point Computational Methods -- Some New High Performance Triangulations and Algorithms, Ph.D. Thesis, Engineering-Economic Systems, Stanford University (October 1979).

Shapley, L.S., On Balanced Games Without Side Payments, **Mathematical Programming,** Ed: T.C. Hu and S.M. Robinson, Academic Press, New York-Londong (1973), 261-290.

Smale, S., A Convergent Process of Price Adjustment and Global Newton Methods, **Journal of Mathematical Economics 3** (1976) 107-120.

Solow, D., Homeomorphisms of Triangulations with Applications to Computing Fixed Points, **Mathematical Programming 20** (1981) 213-224.

Sperner, E., Neur Beweis fur die Invarianz der Dimensionszahl und des Gebietes, **Abh. a.d. Math. Sem. d. Univ. Hamburg 6** (1928) 265-272.

Talman, A.J.J., and G. van der Laan, **Variable Dimension Fixed Point Algorithms and Triangulations,** Mathematical Centre Tracts 128, Mathematisch Centrum, Amsterdam (1980).

Todd, M.J., Union Jack Triangulations, **Conference on Computing Fixed Points with Applications,** Clemson University, (June 1974).

Todd, M.J., The Computation of Fixed Points and Applications, **Lecture Notes in Economics and Mathematical Systems, No. 124,** Springer-Verlag, Berlin- Heidelberg-New York, (1976a).

Todd, M.J., On Triangulations for Computing Fixed Points, **Mathematical Programming 10** (1976b) 322-346.

Todd, M.J., Improving the Convergence of Fixed-Point Algorithms, **Mathematical Programming Study 7** (1978a), Ed: M.L. Balinski and R.W. Cottle, 151-169.

Todd, M.J., Fixed-Point Algorithms that Allow Restarting Without an Extra Dimension, TR No. 379, School of Operations Research and Industrial Engineering, Cornell University, (May 1978b).

Todd, M.J., Traversing Large Pieces of Linearity in Algorithms that Solve Equations by Following Piecewise-linear Paths, TR No. 390, School of Operations Research and Industrial Engineering, Cornell University (September 1978c).

Todd, M.J., Exploiting Structure in Piecewise-Linear Homotopy Algorithms for Solving Equations, **Mathematical Programming 18** (1980) 233-247.

Todd, M.J., and A.H. Wright, A Variable-dimension Simplicial Algorithm for Antipodal Fixed-point Theorems, Tech. Rep. No. 417, School of Operations Research and Industrial Engineering, Cornell University, Ithaca, N.Y., (April 1979).

Veinott, A.F., and G.B. Dantzig, Integral Extreme Points, **SIAM Review 10** No. 3 (1968) 371-372.

Wright, A.H., The Octahedral Algorithm, a New Simplicial Fixed Point Algorithm, **Mathematical Programming 21** (1981) 47-69.

Yamamoto, Y., A New Variable Dimension Algorithm for the Fixed Point Problem, Discussion Paper No. 111(81-12), Institute of Socio-Economic Planning, University of Tsukuba, Sukura, Ibaraki 305, Japan, (March 1981).

Zangwill, W.I., An Eccentric Berycentric Fixed Point Algorithm, **Mathematics of Operations Research 2** 4 (November 1977) 343-359.

Vol. 157: Optimization and Operations Research. Proceedings 1977. Edited by R. Henn, B. Korte, and W. Oettli. VI, 270 pages. 1978.

Vol. 158: L. J. Cherene, Set Valued Dynamical Systems and Economic Flow. VIII, 83 pages. 1978.

Vol. 159: Some Aspects of the Foundations of General Equilibrium Theory: The Posthumous Papers of Peter J. Kalman. Edited by J. Green. VI, 167 pages. 1978.

Vol. 160: Integer Programming and Related Areas. A Classified Bibliography. Edited by D. Hausmann. XIV, 314 pages. 1978.

Vol. 161: M. J. Beckmann, Rank in Organizations. VIII, 164 pages. 1978.

Vol. 162: Recent Developments in Variable Structure Systems, Economics and Biology. Proceedings 1977. Edited by R. R. Mohler and A. Ruberti. VI, 326 pages. 1978.

Vol. 163: G. Fandel, Optimale Entscheidungen in Organisationen. VI, 143 Seiten. 1979.

Vol. 164: C. L. Hwang and A. S. M. Masud, Multiple Objective Decision Making – Methods and Applications. A State-of-the-Art Survey. XII, 351 pages. 1979.

Vol. 165: A. Maravall, Identification in Dynamic Shock-Error Models. VIII, 158 pages. 1979.

Vol. 166: R. Cuninghame-Green, Minimax Algebra. XI, 258 pages. 1979.

Vol. 167: M. Faber, Introduction to Modern Austrian Capital Theory. X, 196 pages. 1979.

Vol. 168: Convex Analysis and Mathematical Economics. Proceedings 1978. Edited by J. Kriens. V, 136 pages. 1979.

Vol. 169: A. Rapoport et al., Coalition Formation by Sophisticated Players. VII, 170 pages. 1979.

Vol. 170: A. E. Roth, Axiomatic Models of Bargaining. V, 121 pages. 1979.

Vol. 171: G. F. Newell, Approximate Behavior of Tandem Queues. XI, 410 pages. 1979.

Vol. 172: K. Neumann and U. Steinhardt, GERT Networks and the Time-Oriented Evaluation of Projects. 268 pages. 1979.

Vol. 173: S. Erlander, Optimal Spatial Interaction and the Gravity Model. VII, 107 pages. 1980.

Vol. 174: Extremal Methods and Systems Analysis. Edited by A. V. Fiacco and K. O. Kortanek. XI, 545 pages. 1980.

Vol. 175: S. K. Srinivasan and R. Subramanian, Probabilistic Analysis of Redundant Systems. VII, 356 pages. 1980.

Vol. 176: R. Färe, Laws of Diminishing Returns. VIII, 97 pages. 1980.

Vol. 177: Multiple Criteria Decision Making-Theory and Application. Proceedings, 1979. Edited by G. Fandel and T. Gal. XVI, 570 pages. 1980.

Vol. 178: M. N. Bhattacharyya, Comparison of Box-Jenkins and Bonn Monetary Model Prediction Performance. VII, 146 pages. 1980.

Vol. 179: Recent Results in Stochastic Programming. Proceedings, 1979. Edited by P. Kall and A. Prékopa. IX, 237 pages. 1980.

Vol. 180: J. F. Brotchie, J. W. Dickey and R. Sharpe, TOPAZ – General Planning Technique and its Applications at the Regional, Urban, and Facility Planning Levels. VII, 356 pages. 1980.

Vol. 181: H. D. Sherali and C. M. Shetty, Optimization with Disjunctive Constraints. VIII, 156 pages. 1980.

Vol. 182: J. Wolters, Stochastic Dynamic Properties of Linear Econometric Models. VIII, 154 pages. 1980.

Vol. 183: K. Schittkowski, Nonlinear Programming Codes. VIII, 242 pages. 1980.

Vol. 184: R. E. Burkard and U. Derigs, Assignment and Matching Problems: Solution Methods with FORTRAN-Programs. VIII, 148 pages. 1980.

Vol. 185: C. C. von Weizsäcker, Barriers to Entry. VI, 220 pages. 1980.

Vol. 186: Ch.-L. Hwang and K. Yoon, Multiple Attribute Decision Making – Methods and Applications. A State-of-the-Art-Survey. XI, 259 pages. 1981.

Vol. 187: W. Hock, K. Schittkowski, Test Examples for Nonlinear Programming Codes. V. 178 pages. 1981.

Vol. 188: D. Bös, Economic Theory of Public Enterprise. VII, 142 pages. 1981.

Vol. 189: A. P. Lüthi, Messung wirtschaftlicher Ungleichheit. IX, 287 pages. 1981.

Vol. 190: J. N. Morse, Organizations: Multiple Agents with Multiple Criteria. Proceedings, 1980. VI, 509 pages. 1981.

Vol. 191: H. R. Sneessens, Theory and Estimation of Macroeconomic Rationing Models. VII, 138 pages. 1981.

Vol. 192: H. J. Bierens: Robust Methods and Asymptotic Theory in Nonlinear Econometrics. IX, 198 pages. 1981.

Vol. 193: J. K. Sengupta, Optimal Decisions under Uncertainty. VII, 156 pages. 1981.

Vol. 194: R. W. Shephard, Cost and Production Functions. XI, 104 pages. 1981.

Vol. 195: H. W. Ursprung, Die elementare Katastrophentheorie. Eine Darstellung aus der Sicht der Ökonomie. VII, 332 pages. 1982.

Vol. 196: M. Nermuth, Information Structures in Economics. VIII, 236 pages. 1982.

Vol. 197: Integer Programming and Related Areas. A Classified Bibliography. 1978 – 1981. Edited by R. von Randow. XIV, 338 pages. 1982.

Vol. 198: P. Zweifel, Ein ökonomisches Modell des Arztverhaltens. XIX, 392 Seiten. 1982.

Vol. 199: Evaluating Mathematical Programming Techniques. Proceedings, 1981. Edited by J.M. Mulvey. XI, 379 pages. 1982.

Vol. 200: The Resource Sector in an Open Economy. Edited by H. Siebert. IX, 161 pages. 1984.

Vol. 201: P. M. C. de Boer, Price Effects in Input-Output-Relations: A Theoretical and Empirical Study for the Netherlands 1949–1967. X, 140 pages. 1982.

Vol. 202: U. Witt, J. Perske, SMS – A Program Package for Simulation and Gaming of Stochastic Market Processes and Learning Behavior. VII, 266 pages. 1982.

Vol. 203: Compilation of Input-Output Tables. Proceedings, 1981. Edited by J. V. Skolka. VII, 307 pages. 1982.

Vol. 204: K.C. Mosler, Entscheidungsregeln bei Risiko: Multivariate stochastische Dominanz. VII, 172 Seiten. 1982.

Vol. 205: R. Ramanathan, Introduction to the Theory of Economic Growth. IX, 347 pages. 1982.

Vol. 206: M.H. Karwan, V. Lotfi, J. Telgen, and S. Zionts, Redundancy in Mathematical Programming. VII, 286 pages. 1983.

Vol. 207: Y. Fujimori, Modern Analysis of Value Theory. X, 165 pages. 1982.

Vol. 208: Econometric Decision Models. Proceedings, 1981. Edited by J. Gruber. VI, 364 pages. 1983.

Vol. 209: Essays and Surveys on Multiple Criteria Decision Making. Proceedings, 1982. Edited by P. Hansen. VII, 441 pages. 1983.

Vol. 210: Technology, Organization and Economic Structure. Edited by R. Sato and M.J. Beckmann. VIII, 195 pages. 1983.

Vol. 211: P. van den Heuvel, The Stability of a Macroeconomic System with Quantity Constraints. VII, 169 pages. 1983.

Vol. 212: R. Sato and T. Nôno, Invariance Principles and the Structure of Technology. V, 94 pages. 1983.

Vol. 213: Aspiration Levels in Bargaining and Economic Decision Making. Proceedings, 1982. Edited by R. Tietz. VIII, 406 pages. 1983.

Vol. 214: M. Faber, H. Niemes und G. Stephan, Entropie, Umweltschutz und Rohstoffverbrauch. IX, 181 Seiten. 1983.

Vol. 215: Semi-Infinite Programming and Applications. Proceedings, 1981. Edited by A.V. Fiacco and K.O. Kortanek. XI, 322 pages. 1983.

Vol. 216: H.H. Müller, Fiscal Policies in a General Equilibrium Model with Persistent Unemployment. VI, 92 pages. 1983.

Vol. 217: Ch. Grootaert, The Relation Between Final Demand and Income Distribution. XIV, 105 pages. 1983.

Vol. 218: P. van Loon, A Dynamic Theory of the Firm: Production, Finance and Investment. VII, 191 pages. 1983.

Vol. 219: E. van Damme, Refinements of the Nash Equilibrium Concept. VI, 151 pages. 1983.

Vol. 220: M. Aoki, Notes on Economic Time Series Analysis: System Theoretic Perspectives. IX, 249 pages. 1983.

Vol. 221: S. Nakamura, An Inter-Industry Translog Model of Prices and Technical Change for the West German Economy. XIV, 290 pages. 1984.

Vol. 222: P. Meier, Energy Systems Analysis for Developing Countries. VI, 344 pages. 1984.

Vol. 223: W. Trockel, Market Demand. VIII, 205 pages. 1984.

Vol. 224: M. Kiy, Ein disaggregiertes Prognosesystem für die Bundesrepublik Deutschland. XVIII, 276 Seiten. 1984.

Vol. 225: T.R. von Ungern-Sternberg, Zur Analyse von Märkten mit unvollständiger Nachfragerinformation. IX, 125 Seiten. 1984

Vol. 226: Selected Topics in Operations Research and Mathematical Economics. Proceedings, 1983. Edited by G. Hammer and D. Pallaschke. IX, 478 pages. 1984.

Vol. 227: Risk and Capital. Proceedings, 1983. Edited by G. Bamberg and K. Spremann. VII, 306 pages. 1984.

Vol. 228: Nonlinear Models of Fluctuating Growth. Proceedings, 1983. Edited by R.M. Goodwin, M. Krüger and A. Vercelli. XVII, 277 pages. 1984.

Vol. 229: Interactive Decision Analysis. Proceedings, 1983. Edited by M. Grauer and A.P. Wierzbicki. VIII, 269 pages. 1984.

Vol. 230: Macro-Economic Planning with Conflicting Goals. Proceedings, 1982. Edited by M. Despontin, P. Nijkamp and J. Spronk. VI, 297 pages. 1984.

Vol. 231: G.F. Newell, The $M/M/\infty$ Service System with Ranked Servers in Heavy Traffic. XI, 126 pages. 1984.

Vol. 232: L. Bauwens, Bayesian Full Information Analysis of Simultaneous Equation Models Using Integration by Monte Carlo. VI, 114 pages. 1984.

Vol. 233: G. Wagenhals, The World Copper Market. XI, 190 pages. 1984.

Vol. 234: B.C. Eaves, A Course in Triangulations for Solving Equations with Deformations. III, 302 pages. 1984.